Technological Aspects in Fatigue Design of Metallic Structures

Technological Aspects in Fatigue Design of Metallic Structures

Editor

Martin Leitner

MDPI • Basel • Beijing • Wuhan • Barcelona • Belgrade • Manchester • Tokyo • Cluj • Tianjin

Editor
Martin Leitner
Graz University of
Technology
Graz
Austria

Editorial Office
MDPI
St. Alban-Anlage 66
4052 Basel, Switzerland

This is a reprint of articles from the Special Issue published online in the open access journal *Metals* (ISSN 2075-4701) (available at: hhttps://www.mdpi.com/journal/metals/special_issues/technological_fatigue_design_metal).

For citation purposes, cite each article independently as indicated on the article page online and as indicated below:

LastName, A.A.; LastName, B.B.; LastName, C.C. Article Title. *Journal Name* **Year**, *Volume Number*, Page Range.

ISBN 978-3-0365-7378-6 (Hbk)
ISBN 978-3-0365-7379-3 (PDF)

Contents

About the Editor

Martin Leitner

Martin Leitner is a full professor and head of the Institute of Structural Durability and Railway Technology at Graz University of Technology in Austria. His research focusses on the fatigue design of mechanical engineering components and lightweight structures, lifetime assessment considering technological influences and load aspects, and the development of elaborate methods to analyze structural durability. He has published more than 100 papers and received several research awards.

Editorial

Technological Aspects in Fatigue Design of Metallic Structures

Martin Leitner

Institute of Structural Durability and Railway Technology, Graz University of Technology, 8010 Graz, Austria; martin.leitner@tugraz.at

1. Introduction and Scope

Traditional manufacturing processes, such as welding and casting, and modern techniques, such as additive manufacturing, can significantly affect the local material properties of metallic materials. To ensure the safe and reliable operation of engineering components and structures, fundamental knowledge of manufacturing effects on fatigue performance is of the utmost importance. Hence, this Special Issue focuses on the fatigue design of metallic structures, considering the influence of technological aspects. Approaches based on local stress/strain, as well as fracture-mechanics-based concepts are applied, considering local manufacturing-process-dependent characteristics, such as microstructure, hardness, porosity/defects, surface topography, and/or residual stress state. Moreover, probabilistic methods are utilized which enable a link between statistically distributed local characteristics and failure/survival probability, enabling the advanced fatigue design of metallic structures.

2. Contributions

Ten papers have been published dealing with different manufacturing processes, such as welding, casting, and additive manufacturing as well as post-treatments, applying various fatigue design concepts ranging from a stress-based fatigue assessment to the fracture mechanical crack growth. Subsequently, an overview of the contributions is given:

In [1], the local characteristics of welded joints are characterized by digital scanning. Particular focus is laid on fillet welds of S700 high-strength steel T-joints, and corresponding fatigue test results are given. In the course of the scanning methodology, the plate angle, weld toe radius, and angle, as significant local weld geometry parameters, are measured and implemented in numerical analyses, which act as inputs for an elaborated probabilistic fatigue model. This model was then utilized to assess an adequate scanning sampling resolution to accurately predict the failure probability of welded joints.

The study presented in [2] focuses on an assessment of the fatigue life of notched specimens with geometries and microstructure representative of welded steel joints. Thereby, welded and non-welded specimens exhibiting different artificially notched geometries, which are located in varying material zones, are investigated. Applying fracture mechanics to assess the crack growth, it is shown that the total fatigue life can be realistically predicted, whereas the estimation of the lifetime until macroscopic crack initiation is more uncertain in the case of sharply notched specimens in comparison to mild notches. Further evaluations regarding the effect of the applied notch stress range on the short and long fatigue crack initiation to fracture ratio as well as correlations between the slope of the S-N curve and the notch acuity are given, therefore contributing to a holistic investigation of the fatigue behavior of welded steel joints as a basis for further design.

The study presented in [3] deals with an analysis of the fatigue strength of repaired welded joints considering both low- and high-strength steels. Specimens were cyclically pre-cracked and the cracks were detected by non-destructive testing (NDT) by applying penetrant and magnetic testing. Afterwards, the fatigue cracks were repaired by removing the material around the crack and further re-welding with a gas metal arc weld process.

Citation: Leitner, M. Technological Aspects in Fatigue Design of Metallic Structures. *Metals* **2023**, *13*, 610. https://doi.org/10.3390/met13030610

Received: 13 March 2023
Accepted: 16 March 2023
Published: 18 March 2023

Investigations of the microstructure and hardness revealed no major difference between the state before and after repairing. Moreover, the stress concentration factor was significantly lower in the case of the repaired specimens due to smaller flank angles. The experiments finally show that almost all repaired specimens reached at least the fatigue life span of the original condition, and for all cases, the corresponding fatigue class of the IIW recommendations was achieved.

Another study [4] demonstrated a damage-based assessment of the fatigue crack initiation site in high-strength steel welded joints which were post-treated by the High Frequency Mechanical Impact (HFMI) treatment technique. Comprehensive experimental investigations involving fatigue tests and X-ray diffraction residual stress measurements in order to analyze the effect of peak stresses in variable amplitude loads on the resulting residual stress state and crack location are performed. The results show that high peak stresses lead to a significant reduction of the beneficial compressive stresses and varying crack initiation sites depending on the level of peak stress occurring. For a detailed assessment, a numerical simulation considering the measured residual stress field, fatigue loading, HFMI weld geometry, and non-linear material behavior was set up, and the local stress/strain was used to evaluate the damage parameter using Smith, Watson, and Topper's method (SWT), which was utilized to identify the critical failure location. It was concluded that the numerical results confirmed the experiments.

A final analysis in regard to welding as a manufacturing process is presented in [5], showing a holistic view on design implications and opportunities of considering fatigue strength, manufacturing variations, and predictive life-cycle costs (PLCC) in welded structures. Based on different design cases and scenarios, the study concludes that welding and production costs are negligible in relation to re-occurring repair costs for all the considered design cases and fatigue scenarios. Further, repair and maintenance costs are higher compared to operational costs for the more severe fatigue scenarios considered. Moreover, it was shown that for welded box structures, increased flange plate thicknesses are most effective to reduce life-cycle costs, as it can increase the intervals between repairs significantly. The paper therefore highlights the importance of incorporating manufacturing variations in an early design stage to ensure an overall minimized life-cycle cost.

The study in [6] emphasizes the effects of residual stress state, surface roughness, and peak load on the micro-cracking of the base material, with a particular focus on sandblasted S690 high-strength steel plates. A two-dimensional finite-element model is developed to cover the measured surface topography and residual stress state. Furthermore, a ductile fracture criterion considering stress triaxiality is applied in order to assess local damage. The results show that under peak load conditions, surface roughness exhibits a dominant influence on micro-crack formation compared to the effects of the residual stress state. It is further concluded that the effect of peak load range on damage formation and crack size is significantly higher than the influence of the residual stress condition. Hence, this paper scientifically contributes to the elaborated investigation of local manufacturing-process-induced properties, such as surface roughness and residual stress, on the damage behavior of high-strength steel plates, which are widely used in engineering metallic structures.

A probabilistic fatigue assessment of surface layers in EN AC-46200 sand castings is presented in [7]. Comprehensive investigations involving areal surface topography scans, microstructural analyses, fatigue tests covering as-cast and hot isostatically pressed (HIP) specimens, as well as detailed fracture surface inspections to determine the local fatigue crack origin, were performed. An advanced fatigue model based on a neural network covering both surface topography notch effect and its interaction with local defects/pores in the surface layer was applied, and validation with the experiments revealed the sound applicability of the probabilistic fatigue approach.

A further study, [8], investigates the same aluminum cast alloy, but is focused on the statistical size effect due to internal defects in the bulk material. Based on numerous fatigue tests incorporating two specimen types with different highly stressed volumes, the probability of occurrence of critical defect sizes and corresponding parameters to set up a generalized extreme value distribution are defined. Furthermore, a probabilistic fatigue design model applying the fracture-mechanics-based Kitagawa–Takahashi approach is presented, leading to a local Weibull factor depending on the return period of the highly stressed volume and the statistically distributed defect population. A final comparison of the fatigue model and the experiments shows a sound agreement and verifies the applicability of the presented approach to cover the statistical size effect in fatigue design.

The fatigue behavior of wire and arc additively manufactured (WAAM) titanium alloy Ti-6Al-4V is investigated in [9]. Fatigue tests and fracture surface analyses highlight that pores/defects in the bulk material and at the surface are the cause for fatigue crack initiation. Therefore, a fracture-mechanics-based fatigue model covering both failure types by a stress intensity equivalent value is developed which considers the manufacturing-process-dependent defect sizes by means of an extreme value distribution by Gumbel. The results reveal that the modified approach using the introduced stress intensity equivalent parameter leads to an improved probabilistic fatigue assessment of both fatigue life as well as strength, and is therefore well applicable for the fatigue design of WAAM Ti-6Al-4V parts.

In the final study [10] of this Special Issue, the microstructural impact on the fatigue crack growth behavior of the alloy 718, which is commonly used for forged structural components in the aircraft industry, is explored. Dependent on the manufacturing process route, different material series are produced, exhibiting varying microstructures characterized by grain-size parameters and strength. Based on the results of numerous fatigue crack propagation tests, it is concluded that the threshold of stress-intensity factor range depends only on grain size and is mainly governed by roughness-induced crack closure. This outcome once more proves the significant effect of local manufacturing-process-induced characteristics on the fatigue performance of metallic materials.

3. Conclusions and Outlook

The scientific contributions in this Special Issue present comprehensive investigations in regard to technological effects on the fatigue behavior of different metallic materials. Based on the results, it is shown that local manufacturing-process-dependent characteristics, such as microstructure, hardness, porosity/defects, surface topography, and/or residual stress state can significantly impact the fatigue performance and need to be considered in the fatigue design. To statistically cover these influences, it is shown that probabilistic approaches are suitable and can lead to an accurate fatigue assessment if the model parameters are well defined. As a future research direction, further work may focus on the interaction of several fatigue-influencing parameters to holistically consider technological aspects in the fatigue design of metallic structures and set up proper fatigue assessment approaches.

Conflicts of Interest: The author declares no conflict of interest.

References

1. Hultgren, G.; Myrén, L.; Barsoum, Z.; Mansour, R. Digital Scanning of Welds and Influence of Sampling Resolution on the Predicted Fatigue Performance: Modelling, Experiment and Simulation. *Metals* **2021**, *11*, 822. [CrossRef]
2. Braun, M.; Fischer, C.; Baumgartner, J.; Hecht, M.; Varfolomeev, I. Fatigue Crack Initiation and Propagation Relation of Notched Specimens with Welded Joint Characteristics. *Metals* **2022**, *12*, 615. [CrossRef]
3. Schubnell, J.; Ladendorf, P.; Sarmast, A.; Farajian, M.; Knödel, P. Fatigue Performance of High- and Low-Strength Repaired Welded Steel Joints. *Metals* **2021**, *11*, 293. [CrossRef]
4. Ono, Y.; Yıldırım, H.C.; Kinoshita, K.; Nussbaumer, A. Damage-Based Assessment of the Fatigue Crack Initiation Site in High-Strength Steel Welded Joints Treated by HFMI. *Metals* **2022**, *12*, 145. [CrossRef]
5. Hagnell, M.K.; Khurshid, M.; Åkermo, M.; Barsoum, Z. Design Implications and Opportunities of Considering Fatigue Strength, Manufacturing Variations and Predictive LCC in Welds. *Metals* **2021**, *11*, 1527. [CrossRef]
6. Nafar Dastgerdi, J.; Sheibanian, F.; Remes, H.; Hosseini Toudeshky, H. Influences of Residual Stress, Surface Roughness and Peak-Load on Micro-Cracking: Sensitivity Analysis. *Metals* **2021**, *11*, 320. [CrossRef]
7. Pomberger, S.; Oberreiter, M.; Leitner, M.; Stoschka, M.; Thuswaldner, J. Probabilistic Surface Layer Fatigue Strength Assessment of EN AC-46200 Sand Castings. *Metals* **2020**, *10*, 616. [CrossRef]
8. Oberreiter, M.; Pomberger, S.; Leitner, M.; Stoschka, M. Validation Study on the Statistical Size Effect in Cast Aluminium. *Metals* **2020**, *10*, 710. [CrossRef]

9. Springer, S.; Leitner, M.; Gruber, T.; Oberwinkler, B.; Lasnik, M.; Grün, F. Fatigue Assessment of Wire and Arc Additively Manufactured Ti-6Al-4V. *Metals* **2022**, *12*, 795. [CrossRef]
10. Gruber, C.; Raninger, P.; Maierhofer, J.; Gänser, H.-P.; Stanojevic, A.; Hohenwarter, A.; Pippan, R. Microstructural Impact on Fatigue Crack Growth Behavior of Alloy 718. *Metals* **2022**, *12*, 710. [CrossRef]

Article

Digital Scanning of Welds and Influence of Sampling Resolution on the Predicted Fatigue Performance: Modelling, Experiment and Simulation

Gustav Hultgren [1], Leo Myrén [1], Zuheir Barsoum [1] and Rami Mansour [2,*]

[1] Lightweight Structures, Department of Engineering Mechanics, KTH Royal Institute of Technology, SE-100 44 Stockholm, Sweden; gustavhu@kth.se (G.H.); lmyren@kth.se (L.M.); zuheir@kth.se (Z.B.)

[2] Solid Mechanics, Department of Engineering Mechanics, KTH Royal Institute of Technology, SE-100 44 Stockholm, Sweden

* Correspondence: ramimans@kth.se

Abstract: Digital weld quality assurance systems are increasingly used to capture local geometrical variations that can be detrimental for the fatigue strength of welded components. In this study, a method is proposed to determine the required scanning sampling resolution for proper fatigue assessment. Based on FE analysis of laser-scanned welded joints, fatigue failure probabilities are computed using a Weakest-link fatigue model with experimentally determined parameters. By down-sampling of the scanning data in the FE simulations, it is shown that the uncertainty and error in the fatigue failure probability prediction increases with decreased sampling resolution. The required sampling resolution is thereafter determined by setting an allowable error in the predicted failure probability. A sampling resolution of 200 to 250 μm has been shown to be adequate for the fatigue-loaded welded joints investigated in the current study. The resolution requirements can be directly incorporated in production for continuous quality assurance of welded structures. The proposed probabilistic model used to derive the resolution requirement accurately captures the experimental fatigue strength distribution, with a correlation coefficient of 0.9 between model and experimental failure probabilities. This work therefore brings novelty by deriving sampling resolution requirements based on the influence of stochastic topographical variations on the fatigue strength distribution.

Keywords: probabilistic fatigue model; topographical variations; weld quality; quality assurance

Citation: Hultgren, G.; Myrén, L.; Barsoum, Z.; Mansour, R. Digital Scanning of Welds and Influence of Sampling Resolution on the Predicted Fatigue Performance: Modelling, Experiment and Simulation. *Metals* **2021**, *11*, 822. https://doi.org/10.3390/met11050822

Academic Editor: Tilmann Beck

Received: 12 April 2021
Accepted: 12 May 2021
Published: 18 May 2021

Publisher's Note: MDPI stays neutral with regard to jurisdictional claims in published maps and institutional affiliations.

1. Introduction

A wide range of high-strength steel (HSS) engineering structures, such as loader cranes, transport vehicles and construction machinery, rely on welded joints for proper structural integrity. These joints are often the limiting factor for the fatigue strength of such load carrying structures. In order to assure durability and structural integrity in these applications, proper quality assurance methods of the welds are of the outmost importance.

Due to the stochastic nature of the welding process [1,2], local topographical variations in weld geometry are inevitable. This may result in local stress raising effects, such as sharp transitions and adverse undercuts. These quantities need to be quality assessed in production according to a weld quality system, to ensure the durability of the structure [3]. Traditionally, manual audits in which the actual weld geometry is measured, have been used for weld quality assurance. However, such manual systems, often hand-held simple tools, have shown to have limitations in accurately capturing local variations in the weld [4]. These limitations may be overcome by implementing digitalised weld quality assurance systems in production. This allows for a continuous quality control of welded joints with higher efficiency and accuracy than manual audits [4–8].

Given the scanned weld geometry obtained from the digitalized measurement system, the influence of local topographical variability can be assessed [9,10]. For this purpose, simulation approaches based on finite element (FE) analysis of the weld are predominantly used. Lang R. et al. [11–13], as well as Lener et al. [14], presented a framework for scanning and simulation of the weld geometry using a statistical approach and a Weakest-link fatigue model. It is noted that the influence of local surface defects is phenomenologically included in the Weakest-link model [15]. Niederwanger et al. [16] presented a comprehensive study where different fatigue modelling concepts were compared with regard to both idealised and measured weld geometry, captured using laser scanning. Kaffenberger and Vormwald [17,18] proposed modelling recommendations for idealized weld geometry simulations, based on the notch stress concept and 3D scanning of overlap and tee joints. Hou [19,20] and Chaudhuri et al. [21] modelled the measured 3D weld geometry to predict experimentally observed beachmark locations and crack measurements, respectively, based on weld toe stress concentration factors (SCF). Aldén et al. [22] performed detailed 3D simulations on cruciform joints of different weld classes showing that, while the maximum von Mises stress gives a good indication on the fatigue failure initiation location, it gives ambiguous predictions of the fatigue strength. Liinalampi et al. [23] investigated the fatigue strength of laser-hybrid welded fully penetrated butt joints based on 2D simulations of the measured weld geometry, captured using both stereo camera measurements as well as 3D laser microscopy. Ladinek et al. [24,25] studied the influence of measured weld geometry on fatigue life using a strain-life concept and Lang E. et al. [26,27] found that numerical simulations of the measured 3D geometry reproduce the experimental strain data for butt joints in the low cycle fatigue regime.

Although the aforementioned studies include the measured geometry of welded joints in detailed FE simulations, they lack the ability to identify the level of detail that is required for a proper fatigue assessment. In the current study, the sampling resolution is investigated for digitalized weld quality assurance systems which is required to accurately predict the fatigue performance. Based on the scanned topographies, a probabilistic weakest-link approach [28–30] is proposed. This work therefore provides the following scientific contribution.

1. A framework for determining the scanning resolution needed for digital quality assurance of welded joints that is assessed on non-load carrying tee joints. The resolution requirements can then be directly incorporated in production for continuous quality assurance of welded structures.
2. A modelling approach to predict the fatigue strength probability distribution based on measured weld geometry variation.

An overview of the workflow is presented in Figure 1. Laser scanning of welded tee joints is first performed to accurately capture the measured weld topography, which is then evaluated using both an industrially commercial quality assurance system and detailed FE analysis. Based on the FE analysis of the scanned weld geometry, the fatigue failure probability is computed using a Weakest-link area model. The parameters in the model are determined by minimizing the error between experimental and simulated failure probabilities. Finally, the sensitivity of the computed failure probability for different sampling resolutions is determined based on FE simulations with different levels of topographical details.

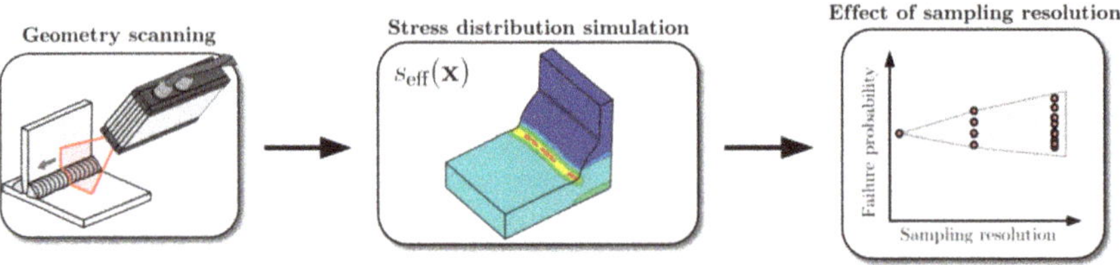

Figure 1. Workflow of the present study.

2. Experimental Investigation

Single pass, non-load carrying, tee joint test specimens produced out of S700 steel were used as the basis for the probabilistic fatigue investigation in the present study. The geometry of the specimen is presented in Figure 2, together with the desired weld throat thickness and weld penetration. Post-treatment using High Frequency Mechanical Impact (HFMI) [31] was carried out on one side of the tee joint specimens to create an uneven distribution in the fatigue strength between the two welded sides. Only the un-treated weld will be further considered as the fatigue strength of this weld will be detrimental for the fatigue strength of the complete specimen.

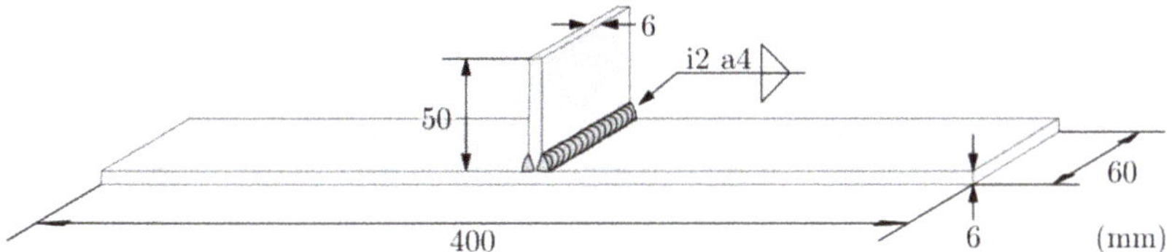

Figure 2. Fatigue test specimen, single-pass tee joint weld specimen.

The specimens were produced using a gas metal arc welding (GMAW) process in the PA position using a pushing travelling angle and a heat input of 0.62 kJ/mm. Sandblasting was used after the welding to remove slag and oxides, as this step helps ensure that the geometry captured in the scanning procedure is the actual geometry of the weld and not the oxide layer.

2.1. Topographic Scanning

The complete geometry of the un-treated weld seams is captured using a scanCON-TROL 2950-50-line laser (Micro-Epsilon, Ortenburg, Germany). This unit has a pixel density of 1280 px/profile and the nominal length of the laser line that is 50 mm. The laser is operated using a robot arm which moves the projected laser line along the surface of the welded joint at a constant velocity while the laser unit continuously captures linear, 2D profiles at fixed intervals. This gives a point cloud database for each specimen containing the complete geometry of the scanned profiles with location data for each data point captured by the sensor matrix. The position of the measured points in each profile is given with reference to the measurement unit.

There are two resolutions that need to be considered when scanning welded joints— the reference profile resolution and the sampling resolution. The reference profile resolution is determined in the direction of the laser line by the pixel density of the sensor matrix and the nominal measurement length of the laser line. For this specific laser, the reference profile resolution is 39 µm. The sampling resolution is given as the distance between two sample profiles and is calculated using the scanning velocity and the scanning frequency, as presented in Figure 3. The highest resolution investigated in this study is 50 µm, which is also the resolution used when fitting the model parameters.

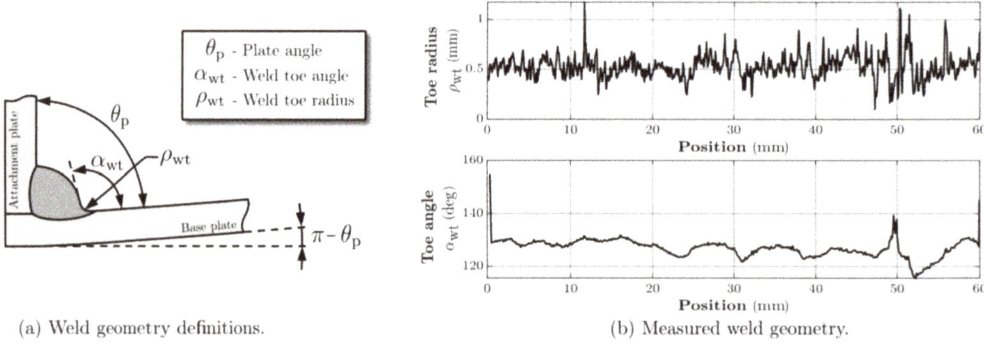

	Scanning speed (mm/s)					
Scanning frequency (Hz)	5	10	20	25	50	100
100	0.05	0.1	0.2	0.25	0.5	1
50	0.1	0.2	0.4	0.5	1	2
25	0.2	0.4	0.8	1	2	4
20	0.25	0.5	1	1.25	2.5	5
10	0.5	1	2	2.5	5	10
5	1	2	4	5	10	20

(mm)

Figure 3. Sampling resolution (mm) for different combinations of scanning speeds (mm/s) and sampling frequencies (Hz).

2.2. Weld Geometry Evaluation

The captured weld geometry is analysed using the commercial quality assurance system Winteria ® qWeld (Winteria, Hudiksvall, Sweden) [32]. This is done to investigate and present the scatter in weld geometry for the investigated group of specimens independently from the simulations that is later carried out. This quality assurance system quantifies the weld geometry into geometric definitions as throat thickness, weld leg length, undercut, weld toe radius and weld toe angle using internal algorithms developed by Stenberg et al. [5,6]. The algorithm goes through each weld section with an internal algorithm which locates the weld toes and determines the weld toe radii and angle by a fitting routine. The accuracy of the method has been proven to be comparable to other commercial programs when the same reference block has been studied [5]. The accuracy has also been studied in [6]. The latter two geometrical parameters are presented in Figure 4 along the weld surface of specimen 24.

(a) Weld geometry definitions.   (b) Measured weld geometry.

Figure 4. (**a**) Weld geometry definitions and (**b**) variation measured by the Winteria® qWeld [32] system for specimen 24.

The variation in radius and weld toe angle between the tested specimens is presented in Figure 5. The scatter of the mean weld toe radius between the specimens is less than 0.5 mm and the corresponding scatter for the weld toe angle is approximately 20 degrees. The mean value of the weld toe radius for the investigated specimens is 0.56 mm with a standard deviation of 0.21 mm for the group. There are some specimens where outliers with smaller radii are visible, which potentially indicates the presence of local regions where high stress concentrations can be expected. One of these is the previously mentioned specimen 24 where a sharp corner can be seen at the 50 mm position. This specimen is of interest as its mean values for both presented parameters are representative for the group of specimens, while it has a large number of outliers below the mean for both parameters.

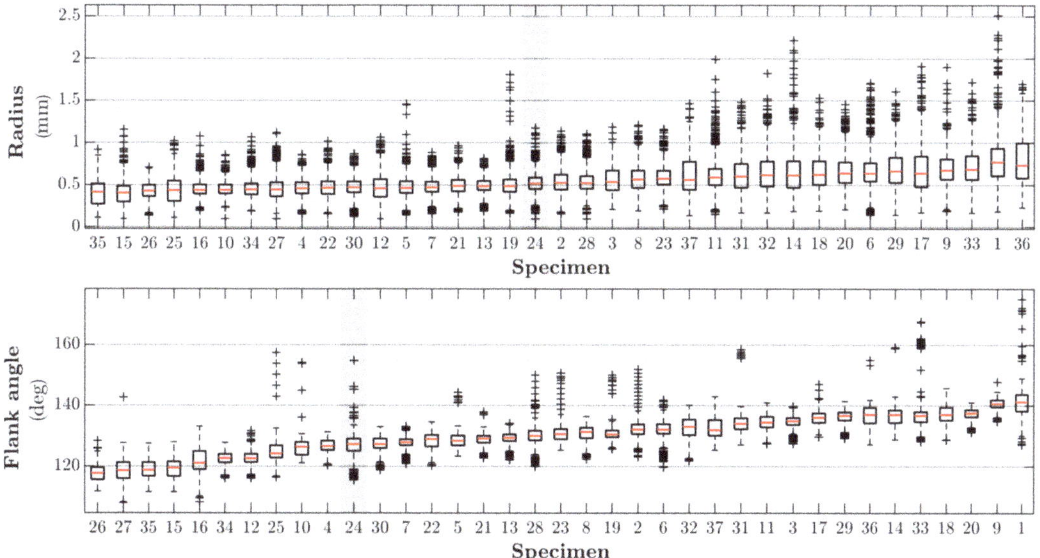

Figure 5. Variation of geometry within the tested group of specimens.

2.3. Uniaxial Fatigue Testing

Constant amplitude fatigue testing was carried out using an MTS High-Force Servo hydraulic tensile tester (MTS, Eden Prairie, MN, USA) at a load ratio of 0.1. All specimens were tested at a load level of 180 MPa. A total of 51 specimens were tested, and 14 of those were stopped at 5 million cycles as no failure had occurred. The test results for the failed specimens are presented both in Figure 6 and Table 1. Only the specimens that failed are processed further in this study.

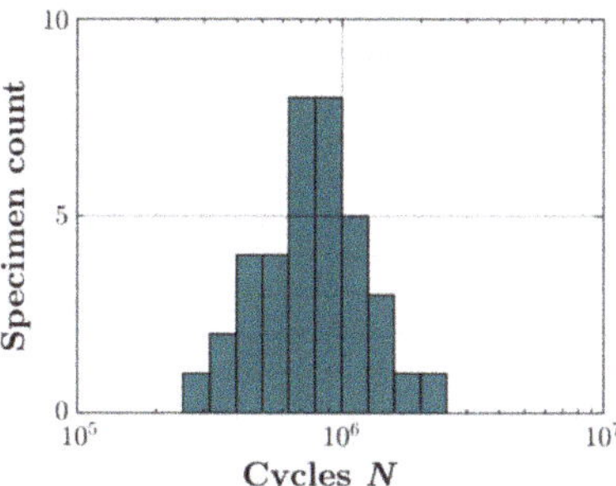

Figure 6. Scatter of the uniaxial fatigue test data.

Table 1. Fatigue test data for single-pass tee joint test specimens and median rank estimates. All specimens are fatigue-loaded at $R = 0.1$ and amplitude 180 MPa.

Failure Order i	Cycles [-]	Median Rank, $p_{f,i}^{exp}$	Failure Order i	Cycles	Median Rank, $p_{f,i}^{exp}$
1	296319	0.0187	20	796070	0.5267
2	319754	0.0455	21	844067	0.5535
3	358569	0.0722	22	874843	0.5802
4	450880	0.0989	23	875982	0.6070
5	479592	0.1257	24	911631	0.6337
6	491902	0.1524	25	942420	0.6604
7	499422	0.1791	26	978798	0.6872
8	554425	0.2059	27	998173	0.7139
9	577447	0.2326	28	1017389	0.7406
10	601596	0.2594	29	1067794	0.7674
11	621224	0.2861	30	1102915	0.7941
12	637281	0.3128	31	1159417	0.8209
13	644313	0.3396	32	1199808	0.8476
14	697257	0.3663	33	1259817	0.8743
15	707011	0.3930	34	1478800	0.9010
16	712686	0.4198	35	1563887	0.9278
17	739705	0.4465	36	1960249	0.9545
18	761384	0.4733	37	2418827	0.9813
19	781303	0.5			

The experimental failure probability was determined for the 37 specimens that failed using an approximation for median ranks proposed by A. Benard [33],

$$p_{f,i}^{exp} \approx \frac{i - 0.3}{n_{spec} + 0.4} \tag{1}$$

This gives an approximative failure probability for the i:th ranked failure out of a population of $n_{spec} = 37$ specimens at a 50% confidence level. The cumulative distribution function for the given group of specimens using Benard's approximation is presented in Figure 7.

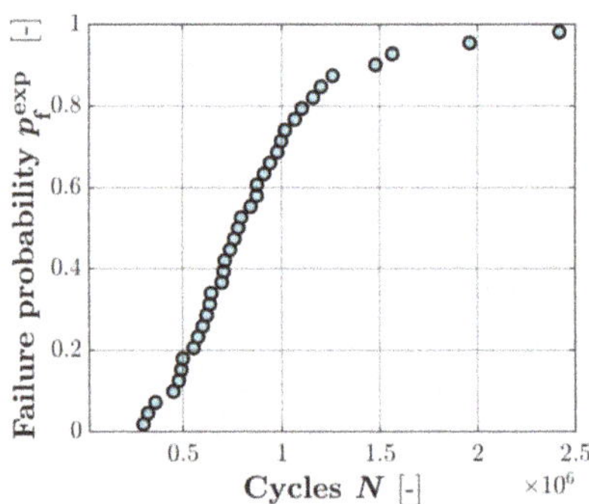

Figure 7. Experimental failure probability using the median rank method.

3. Probabilistic Fatigue Model

In the following, the probability distribution for the fatigue strength is derived based on the Weakest-link model [28,29].

3.1. Weakest-Link Area Model

The Weakest-link area model is derived from the assumption that fatigue failure is caused by cracks initiated at critical surface defects and that the distance between the defects is large enough not to cause any interaction between them. Assume that the number of such defects on a reference surface area A_{ref} is q. Assume further that the area A_{ref} is divided into k small sub-areas and that all sub-areas are subjected to the same stress amplitude S. The probability to find a critical defect in the sub-area is then q/k. The probability of not finding a critical defect in a total number of k sub-areas is therefore $(1 - q/k)^k$. The probability of failure, i.e., the probability of finding a critical defect in A_{ref}, can therefore be written as

$$p_f^{WL} = 1 - \lim_{k \to \infty} \left(1 - \frac{q}{k}\right)^k = 1 - \exp(-q). \tag{2}$$

The number of critical defects is assumed to increase with increased stress level s. A function proposed by Weibull that shows good agreement with experimental data is $q = (s/\lambda)^\beta$ which results in

$$p_f^{WL} = F_S(s) = 1 - \exp\left[-\left(\frac{s}{\lambda}\right)^\beta\right], \tag{3}$$

where $F_S(s)$ is the Weibull probability distribution of the fatigue strength with scale and shape parameters λ and β, respectively. It is noted that Equation (3) gives the fatigue failure probability for a uniaxial stress amplitude S at a given number of cycles to failure n. The dependence on the number of cycles is given by the scale parameter $\lambda(n)$. An analytical expression of the form

$$\lambda(n) = \lambda_0 \left(\frac{n_0}{n}\right)^{\frac{1}{m}} \exp\left[\frac{c_{EM}}{\beta}\right] \tag{4}$$

has been proposed by the authors [30], where m is the inverse Basquin exponent, $c_{EM} \approx 0.5772$ is the Euler–Mascheroni constant, $n_0 = 2 \times 10^6$ is a reference number of cycles to failure and λ_0 is a fitting parameter. A value of $m = 3$ is recommended for as-welded components with normal weld quality [34]. Using Equation (4), it is noted that the expression for the failure probability according to Equation (3) has two unknown parameters, λ_0 and β, which will be determined in Section 5.

3.2. Multiaxial Considerations

The expression in Equation (3) can be generalized to be applicable for non-uniform stress distributions and multiaxial stress states acting on an area of arbitrary size. Assume that the measured weld area A, see Figure 8a, is subjected to a multiaxial stress distribution. Assume further that $s_{eff}(x)$ is an effective local stress amplitude, where x is a location vector. The probability of failure according to Equation (3) can be written as

$$p_f^{WL} = 1 - \exp\left[-\left(\frac{s_{equ}}{\lambda}\right)^\beta\right], \tag{5}$$

where an equivalent stress amplitude is introduced according to

$$s_{equ} = \left[\frac{1}{A_{ref}} \int_A s_{eff}(x)^\beta dA\right]^{\frac{1}{\beta}}. \tag{6}$$

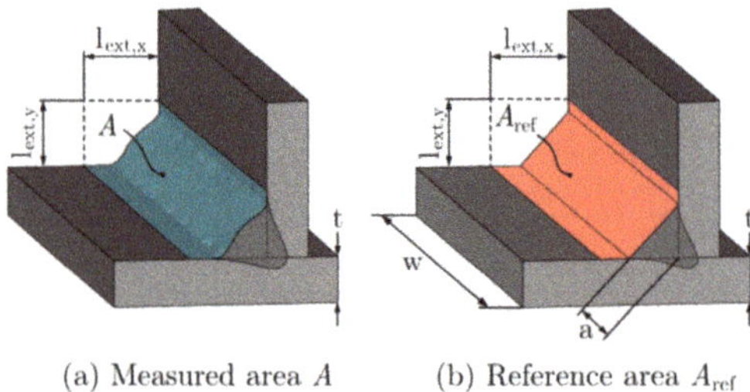

Figure 8. (a) Measured area and (b) reference area of the welded joint used in the Weakest-link area model.

It is noted that the integration is performed over the surface area A. From Equations (5) and (6) it is seen that if $A = A_{ref}$ and $s_{eff}(x) = s$, then $s_{equ} = s$ and Equation (3) is recovered. It should be emphasized that Equation (5) is a conditional probability given a deterministic applied load [35] and a known measured topography of the weld area. Including random uncertainties in load and stress field is possible using methods from structural reliability [36–40] but is out of the scope of the present work.

The reference area, A_{ref}, is taken as the local area of the idealised weld surface. This is idealised as the throat area of the weld bead with two extension regions at both weld toes with lengths $l_{ext,x}$ and $l_{ext,y}$, as seen in Figure 8 and Table 2. It is possible to use either the measured mean value of the throat thickness for the group of specimens or the desired throat thickness, as the fitting procedure will account for any deviation between the two. The latter one is used in the present study. The distances of the extension regions are chosen so that all stress-raising effects near the weld toes are included in A_{ref}. The idealised weld surface is not influenced by the measured geometry of individual specimens. The reference area must be large enough to accurately capture the stress distribution in the weld and the highly stressed region near the load-carrying weld toe for all investigated specimens. This needs to be fulfilled even for the most extreme variations in the local weld geometry seen in the group of specimens. The reference area is also chosen so it is smaller than the total area measured by the laser scanner, as this ensures that no extrapolation in the dataset is needed. The reference area is calculated as

$$A_{ref} = w\left[\left(1 - \sqrt{2}\right)2a + l_{ext,x} + l_{ext,y}\right]. \tag{7}$$

Table 2. Reference area parameters.

Parameter	Implemented Value
a	4 mm
w	60 mm
$l_{ext,x}$	10 mm
$l_{ext,y}$	8 mm
t	6 mm
A_{ref}	881.2 mm^2

A common choice of effective stress is based on the largest principal stress $\sigma_1(x,t)$ which varies with time t. During the complete load cycle, the principal stress attains a maximum $\sigma_1^{max}(x)$ and a minimum $\sigma_1^{min}(x)$. The effective stress can be evaluated as [41]

$$s_{eff}(x) = \frac{1}{2}\left[\max(0, \sigma_1^{max}(x)) - \min\left(0, \sigma_1^{min}(x)\right)\right], \tag{8}$$

which is applicable when the direction of $\sigma_1(x,t)$ is little affected by t. It is noted that a pulsating stress will be more decisive for the fatigue life than an alternating stress and that a point with only compressive stresses throughout the complete load cycle will not contribute to the failure probability.

It should be emphasized that the influence of surface defects is phenomenologically included in the Weakest-link model. Consider a hypothetical case where the surface topography is perfectly smooth without any variability, which corresponds to constant local stresses along the surface. The Weakest-link (WL) model according to Equation (5) still predicts a variability in the fatigue strength. The origin of this variability stems from the assumption that fatigue failure is caused by cracks initiated at randomly distributed critical surface defects that do not interact with each other. The size of this variability in fatigue strength is determined by the Weibull parameters, which are directly affected by surface defects, such as micropores and inclusions.

4. Numerical Implementation of True Weld Geometry

The detailed sectional geometry measurements were stitched together in a pre-processing algorithm to form the complete weld surface. Any unwanted skewness in the surface resulting from misalignments in the trajectory of the laser unit when scanning is corrected by affine transformations based on the relative angles of the plate and the transverse stiffener. Erroneous or missing data points are replaced with linearly interpolated points, no extrapolation is carried out as the final size of the reconstructed surface is truncated at the ends. Short wavelength noise in the recreated data set is removed using a Gaussian filter with a sampled Gaussian kernel. This is essential, as the un-filtered surface includes discontinuations not seen in the true weld surface, which can reduce the accuracy.

The pre-processed surface is then sent as database sampling points to ANSYS (2020 R2, ANSYS Inc, Canonsburg, PA, USA) where the local model of the welded joint is created, one weld profile at a time, using non-uniform rational basis spline (NURBS) surfaces. Once the complete local weld is created, it is merged with the nominal geometry to produce the complete specimen.

The solid model is discretised using tetrahedral and hexahedral brick elements with quadratic shape functions (SOLID 186 in Ansys). A structured mesh is prescribed on the weld surface so that the corner nodes of each element at the surface coincides with the measured weld cross-section profiles, as seen in Figure 9. The number of element facets that are tangential to the weld surface between two weld section profiles will therefore be the same throughout the topographic surface. The side length Δ_ξ of the element surface facets in the ξ direction does thus vary to a small extent. The mean value of Δ_ξ in the region close to the fatigue loaded weld toe is prescribed as 15 µm, whereas the remaining element surface facets have a mean value of 150 µm. The side length Δ_η of the element surface facets in the η direction is set to 50 µm.

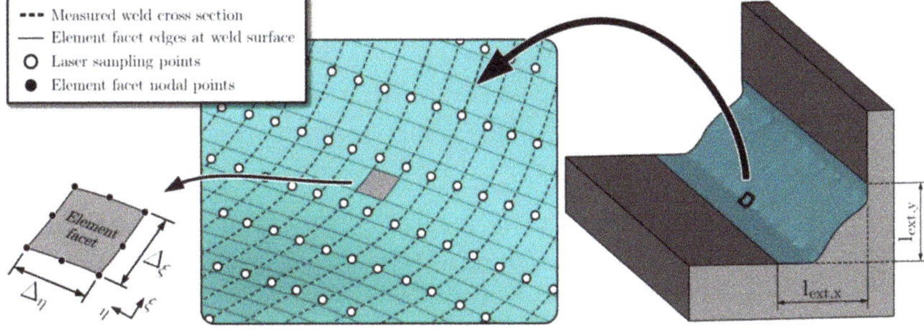

Figure 9. Structured discretisation of the weld geometry surface.

Half of the specimen, including the non-treated weld, is modelled as a representative unit of fatigue strength for the complete specimen. This is motivated by the fact that no fatigue failure was initiated at the HFMI-treated weld. A symmetry boundary condition is applied on the symmetry plane going through the middle of the stiffener. The nodes located at the far end of the specimen, where the fatigue clamps grip, are slave nodes connected rigidly to remote master nodes that lie in corresponding z-planes, as schematically presented in Figure 10. These nodes are constraints from translation in all directions except for in the x-direction and they are only free to rotate around z.

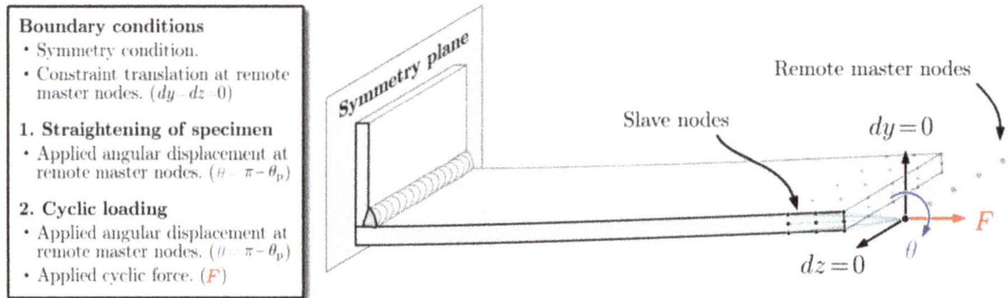

Figure 10. Boundary condition applied in the numerical simulations of the stress distribution.

Each specimen is simulated in two analysis steps—the first step represents the straightening of the specimen in the clamping phase (prescribed rotation $\theta = \pi - \theta_p$) and the second corresponds to the cyclic loading of the specimen (prescribed rotation $\theta = \pi - \theta_p$ and applied nominal force F). Linear elastic material behaviour is implemented in the simulation model with a Young's modulus of 200 GPa and a Poisson's ratio of 0.3. The stress distribution for specimen 24 under cyclic loading is presented in Figure 11. The implemented simulation process is completely automated from start to finish to ensure that all specimens are analysed using the same conditions. The only manual input needed during the process is the location of the fatigue critical weld toe in the scanning data. The CPU (Intel(R) Core(TM) i9-10940X (14Core, 3.30GHz)/64GB RAM) simulation time for each specimen is around 1 h per specimen.

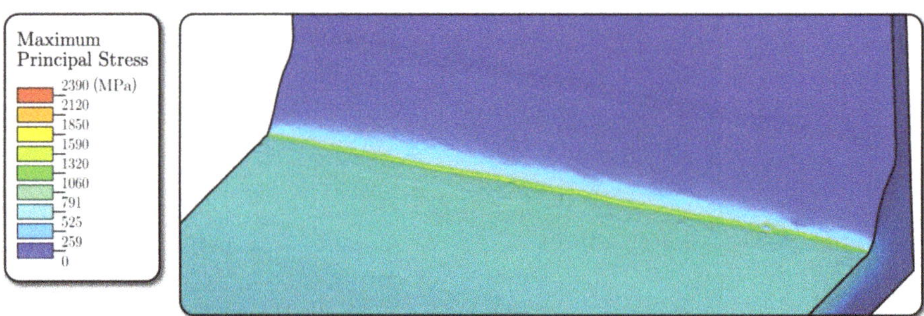

Figure 11. Stress distribution at the weld surface of the specimen 24 (failure order 17 in Table 1) for a nominal stress of 180 MPa.

5. Evaluation of Failure Probability and Determination of Model Parameters

For each specimen, the simulated stress distribution for the clamping phase and the cyclic loading are used to determine the distribution of the maximum principal stress amplitude at the weld surface. The failure probability for the specimen is calculated by writing the area integral in Equation (6) into a double integral

$$s_{\text{equ}} = \left[\frac{1}{A_{\text{ref}}} \int \int s_{\text{eff}}(\eta, \xi)^{\beta} d\xi d\eta \right]^{\frac{1}{\beta}}, \tag{9}$$

where the first integration direction, ξ, is along the weld profiles and the second integration direction, η, is in the scanning direction. This is presented schematically in Figure 9. The integral is evaluated numerically as

$$
s_{equ} = \left[\frac{1}{A_{\text{ref}}} \sum_i \sum_j s_{\text{eff}}(\eta_i, \xi_j) \Delta \xi_j \Delta \eta_j \right]^{\frac{1}{\beta}},
\tag{10}
$$

where η_i and ξ_j are the nodal positions and $s_{\text{eff}}(\eta_i, \xi_j)$ is the effective stress value evaluated at the corner nodes. The latter is computed by extrapolation of the stress values from the Gauss points, followed by an averaging at each node.

The two model parameters, β and λ_0 in Equations (4) and (5), are fitted by minimising the mean square error (MSE) given by

$$
MSE = \frac{1}{n_{\text{spec}}} \sum_{i=1}^{n_{\text{spec}}} \left(p_{\text{f},i}^{\text{exp}} - p_{\text{f},i}^{\text{WL}} \right)^2.
\tag{11}
$$

The parameters that minimised the MSE are presented in Table 2. One specimen was excluded from the fitting process as it singly influenced the fitted parameters and increased the MSE more than any other specimen. The same specimen after the fitting was verified as an outlier as it was more than 1.5 times the interquartile range [42] away from the upper quartile or below the lower quartile of the relative error in estimated failure probability

$$
e_i = \left| p_{\text{f},i}^{\text{exp}} - p_{\text{f},i}^{\text{WL}} \right|.
\tag{12}
$$

The parameter e_i is the fitting error of the Weakest-link model. The fitting accuracy is quantified using Pearson's correlation coefficient

$$
r = \frac{n_{\text{spec}} \sum p_{\text{f},i}^{\text{exp}} p_{\text{f},i}^{\text{WL}} - \sum p_{\text{f},i}^{\text{exp}} \sum p_{\text{f},i}^{\text{WL}}}{\sqrt{n_{\text{spec}} \sum \left(p_{\text{f},i}^{\text{exp}} \right)^2 - \left(\sum p_{\text{f},i}^{\text{exp}} \right)^2} \sqrt{n_{\text{spec}} \sum \left(p_{\text{f},i}^{\text{WL}} \right)^2 - \left(\sum p_{\text{f},i}^{\text{WL}} \right)^2}}
\tag{13}
$$

and is presented together with the root mean square error $RMSE = \sqrt{MSE}$ in Table 3. The fitting accuracy of each specimen is presented in Figure 12 where the model failure probabilities are compared with the experimental values derived from the median rank approximation. A high correlation of 0.9 is computed between experimental and model failure probability. Therefore, the proposed modelling approach captures the influence of weld topography on the specimen fatigue failure probability well.

Table 3. Fitting parameters and accuracy metrics.

Weakest-Link Fitting Parameters		Accuracy Metrics	
β (-)	λ_0 (MPa)	$RMSE$	r
8.22	314	0.121	0.903

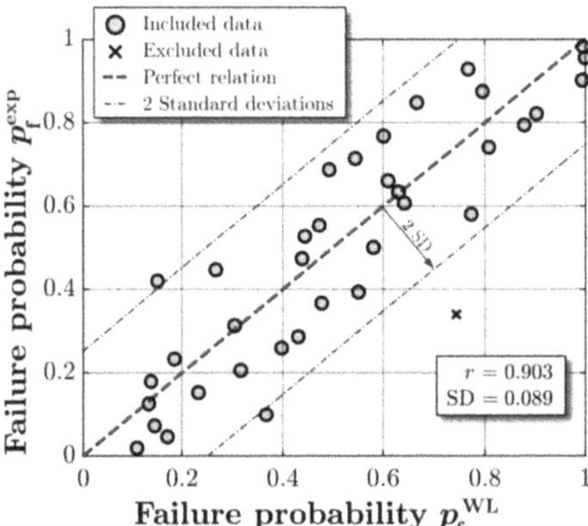

Figure 12. Numerical fatigue strength estimations at the corresponding fatigue life determined from experimental testing.

6. Influence of Sampling Resolution on Predicted Fatigue Failure Probability

In the following, the Weakest-link model is implemented using different sampling resolutions and sampling sequences, in order to quantify the sampling-induced uncertainty. Based on the quantified uncertainty, an optimal sampling resolution is thereafter recommended.

6.1. Sampling-Induced Uncertainty in Computed Fatigue Failure Probability

Consider the weld geometry of the specimen shown in Figure 12, which is scanned using a high resolution of 50 µm. The geometry of the specimen is resampled with lower sampling resolutions and different sampling sequences, as schematically shown in Figure 13. As can be seen, each sampling sequence corresponds to a different scanning start position. By taking every other weld profile into consideration, the sampling resolution is decreased to 100 µm with two possible starting positions. Decreasing the sampling resolution further to every fourth weld profile decreases the resolution to 200 µm with four possible starting positions. It is noted that that the highest sampling resolution of 50 µm corresponds to 1200 weld profiles for the studied 60 mm long weld.

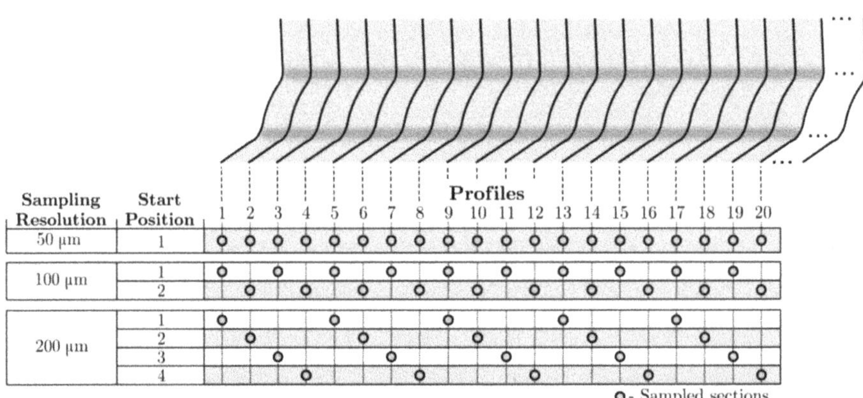

Figure 13. Resampling of scanning data by excluding weld profiles. Each row represents a sampled sequence including only the profiles with filled cells.

The weld geometry of the specimen is scanned from the different scanning starting positions using a sampling resolution of 50, 100, 200, 250, 500, 1000, 2000 and 5000 µm.

This corresponds to more than 100 scanned weld surfaces with different combinations of sampling resolutions and scanning start positions. An FE simulation is thereafter performed for each of the scanned surfaces and the failure probability is computed using Equation (9). The computed failure probabilities for different sampling starting positions are presented in Figure 14a–e for each sampling resolution. In Figure 14a, the failure probabilities for four different scanning start positions are shown. These four start positions are illustrated in Figure 13 for a resolution of 200 µm (start positions 1, 2, 3 and 4). If the resolution is reduced by a factor of 10, see Figure 14c, a total of 40 different scanning start positions results in the shown variation in the computed failure probability. As can be seen, the computed failure probability appears to have a sinusoidal relation to the scanning start position. It is noted that all probabilities are computed at $n = 761,384$ cycles, which is the experimental number of cycles to failure for the considered specimen.

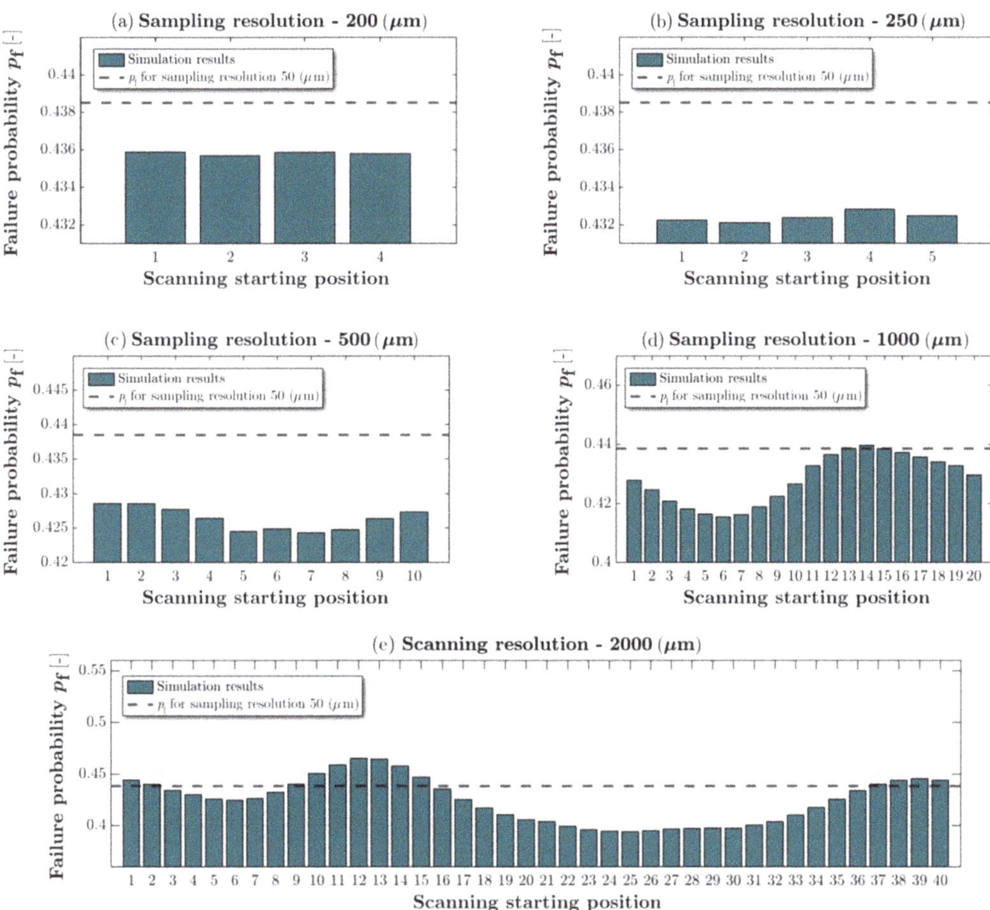

Figure 14. Influence of starting position on the model failure probability at different sampling resolutions. (**a**) 200 µm sampling resolution, (**b**) 250 µm sampling resolution, (**c**) 500 µm sampling resolution, (**d**) 1000 µm sampling resolution, (**e**) 2000 µm sampling resolution.

6.2. Required Sampling Resolution

To determine the required scanning accuracy, all computed probabilities in Figure 14a–f are plotted as a function of the sampling resolution, see Figure 15a. For each resolution, the marked crosses correspond to different scanning start positions as described in Figure 13. The mean and variance of the failure probability are determined for each sampling resolution as

$$\overline{p}_{\mathrm{f}}^{\mathrm{WL}} = \frac{1}{n_{\mathrm{seq}}} \sum_{j=1}^{n_{\mathrm{seq}}} p_{\mathrm{f},j}^{\mathrm{WL}} \tag{14}$$

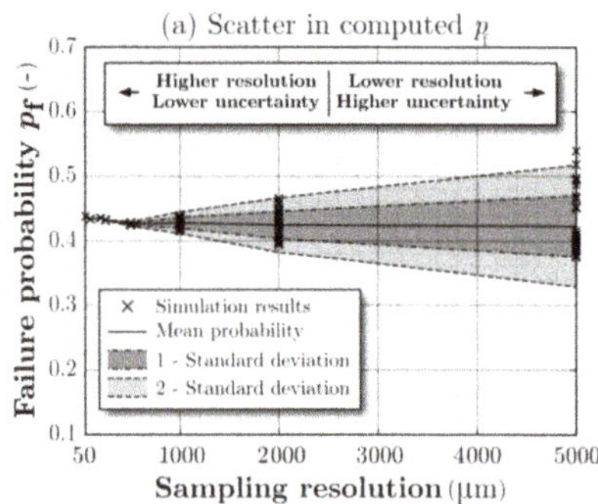
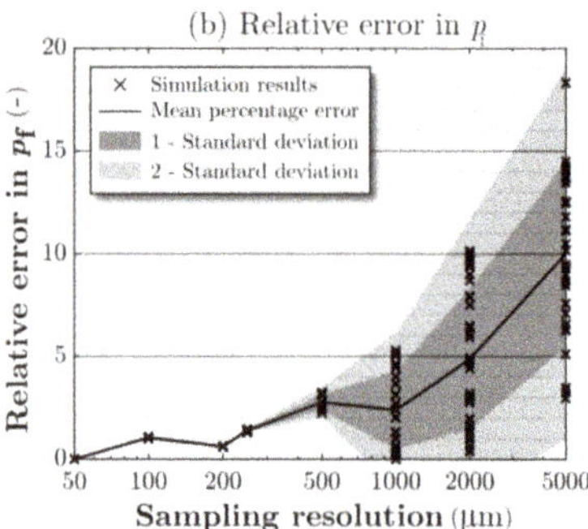

Figure 15. Influence of sampling resolution on (**a**) the computed Weakest-link failure probability of 1 digitally scanned specimen and (**b**) the relative absolute error in the computed probability with respect to the highest resolution (50 μm). At each sampling resolution, different sampling sequences are marked by crosses.

And

$$\mathrm{Var}\left[p_{\mathrm{f}}^{\mathrm{WL}}\right] = \sqrt{\frac{1}{n_{\mathrm{seq}}} \sum_{j=1}^{n_{\mathrm{seq}}} \left(p_{\mathrm{f},j}^{\mathrm{WL}} - \overline{p}_{\mathrm{f}}^{\mathrm{WL}}\right)} \tag{15}$$

respectively, where $p_{\mathrm{f},j}^{\mathrm{WL}}$ is the Weakest-link probability computed using a sampling sequence j and n_{seq} is the number of sampling sequences for the considered resolution (see Figure 14). As can be seen, the standard deviation $\sqrt{\mathrm{Var}\left[p_{\mathrm{f}}^{\mathrm{WL}}\right]}$ decreases with decreasing sampling resolution. It is further noted that, although the mean probability slightly decreases with decreased resolution, this change is relatively small.

In Figure 15b, the relative error in the failure probabilities, with respect to the probability computed at the highest resolution of 50 μm, is plotted as a function of sampling resolution based on

$$\mathrm{error}_i = \frac{\left|p_{\mathrm{f}}^{\mathrm{WL}}\big|_{50\mu\mathrm{m}} - p_{\mathrm{f},i}^{\mathrm{WL}}\right|}{p_{\mathrm{f}}^{\mathrm{WL}}\big|_{50\mu\mathrm{m}}} \times 100\% \tag{16}$$

It is from Figure 15b possible to determine the required sampling resolution by setting an allowable relative error. It should be noted that the tolerated uncertainty in predicted failure probability is highly dependent on the application. A higher degree of certainty is generally needed for structurally critical components in the high cycle fatigue regime. For the studied specimen, a sampling resolution of 200 to 250 μm results in an acceptable mean relative error in p_{f} of less than 2%.

7. Concluding Remarks

In this study, the influence of sampling resolution in digital scanning of welds for fatigue quality control was investigated. The method for determining the optimal resolution consisted of five steps.

1. Digital scanning—the local weld geometries of more than 50 welded tee joints were measured with a high resolution of 50 μm.
2. Fatigue testing—all measured specimens were fatigue tested at the same load level. The experimental fatigue failure probability was computed using median rank for each specimen that failed within 5 million cycles.
3. Finite-element analysis—for each of the failed specimens during fatigue testing, the local stresses on the weld surface were computed from FE analysis using the Digital scanning data of the weld topography.
4. Weakest-link failure probability—a two-parameter weakest-link area model was applied to model the fatigue failure probability based on the local stresses computed from the finite element analysis. The weakest-link parameters were determined by fitting the model probabilities to the experimental probabilities determined from the fatigue testing.
5. Sensitivity analysis—the sensitivity of the computed Weakest-link failure probability to a reduction in sampling resolution was studied based on an arbitrarily chosen specimen. The digital scanning data was down-sampled to sampling resolutions in the range of 100 μm to 5 mm with different scanning start positions. A finite-element analysis was performed for each of the down-sampled scanned geometries and the corresponding failure probability was computed. The error and uncertainty in the computed probabilities due to the down-sampling was quantified and the required sampling resolution was determined by setting an allowable mean error.

Based on the above steps, a sampling resolution of 200 to 250 μm has been shown to be adequate for fatigue quality control of the studied welded joints. This sampling resolution results in a mean relative error of less than 2% in the computed fatigue failure probability compared to highest studied resolution in this work (50 μm). The present study therefore indicates that the local weld geometry needs to be captured with a relatively high resolution to accurately predict the failure probability.

The adequate sampling resolution presented in this study would, in a production environment, be translated to scanning speeds of around 20–25 mm weld per second scanning using a high sampling frequency of 100 Hz (illustrated in Figure 3). This means that the time it would take to scan 1 m of weld is 40–50 s.

This study also suggests that a two-parameter Weakest-link area model using the maximum principal stress as a multiaxial effective stress measure, is appropriate to model the failure probability of welds. This is demonstrated by a correlation coefficient of 0.9 between model and experimental probabilities for the studied tee joints.

Finally, this work paves the way to quantifying the influence of local geometrical variations on the fatigue failure probability by combining advanced statistical spatial field analysis [43–45] of the scanned data and Weakest-link fatigue models. A comparison between Weakest-link area [46] and volume models [47] as well as the choice of multiaxial effective stress measure [48] need to be investigated further.

Author Contributions: Conceptualization, G.H. and R.M.; methodology, G.H. and R.M.; software, G.H. and L.M.; validation, G.H. and R.M.; formal analysis, G.H. and R.M.; investigation, G.H. and R.M.; resources, Z.B.; data curation, G.H.; writing—original draft preparation, G.H. and R.M.; writing—review and editing, G.H., R.M. and Z.B.; visualization, G.H.; supervision, R.M. and Z.B.; project administration, Z.B.; funding acquisition, Z.B. All authors have read and agreed to the published version of the manuscript.

Funding: This research was financially supported by SSAB and Sweden's Innovation Agency (Vinnova) programme for Strategic vehicle research and innovation (FFI) through the Q-IN-MAN project (contract number: 2017-05533). The support is grate-fully acknowledged.

Data Availability Statement: The data that support the findings of this study are available from Swerim AB—Centre for Joining and Structures but restrictions apply to the availability of these data, which were used under license for the current study, and so are not publicly available. Data are however available from the authors upon reasonable request and with permission of FATscat project members.

Acknowledgments: The authors would like to acknowledge the FATscat project members that shared the data. The Q-IN-MAN project group are also acknowledged for valuable discussions.

Conflicts of Interest: The authors declare no conflict of interest.

Nomenclature

α_{wt}	Weld toe angle	k, q	Number of sub-areas and number of defects
β, λ	Weibull shape and scale parameter		
Δ_ξ, Δ_η	Element facet dimension	$l_{ext,x}, l_{ext,y}$	Extension of reference area
θ	Rotational degree of freedom	m	Basquin slope exponent
θ_p	Plate angle	MSE	Mean square error
λ_0	Fitting parameter	n	Cycles to failure
ξ, η	Local reference system parameters	n_{seq}	Number of scanning sequences
ρ_{wt}	Weld toe radius	n_{spec}	Number of tested specimens
σ_1	Largest principal stress	p_f^{exp}	Experimental failure probability
a	Weld throat thickness	p_f^{WL}	Weakest-link failure probability
A_{ref}	Reference surface area	R	Fatigue load ratio
c_{EM}	Euler-Mascheroni constant	r	Pearson correlation coefficient
dy, dz	Translational degree of freedom	s_{eff}	Effective stress amplitude
e_i	Fitting error	s_{equ}	Equivalent stress amplitude
F	Applied nodal force	w	Specimen width
F_S	Weibull probability distribution	x	Spatial position

References

1. Mansour, R.; Zhu, J.; Edgren, M.; Barsoum, Z. A probabilistic model of weld penetration depth based on process parameters. *Int. J. Adv. Manuf. Technol.* **2019**, *105*, 499–514. [CrossRef]
2. Tomaz, I.D.V.; Colaço, F.H.G.; Sarfraz, S.; Pimenov, D.Y.; Gupta, M.K.; Pintaude, G. Investigations on quality characteristics in gas tungsten arc welding process using artificial neural network integrated with genetic algorithm. *Int. J. Adv. Manuf. Technol.* **2021**, *113*, 3569–3583. [CrossRef]
3. Jonsson, B.; Samuelsson, J.; Marquis, G.B. Development of weld quality criteria based on fatigue performance. *Weld. World* **2011**, *55*, 79–88. [CrossRef]
4. Hammersberg, P.; Technology, M.; Olsson, H. Statistical evaluation of welding quality in production. In Proceedings of the Swedish Conference on Light Weight Optimized Welded Structures, Borlänge, Sweden, 24–25 March 2010; pp. 148–162.
5. Stenberg, T.; Lindgren, E.; Barsoum, Z. Development of an algorithm for quality inspection of welded structures. *Proc. Inst. Mech. Eng. Part B J. Eng. Manuf.* **2012**, *226*, 1033–1041. [CrossRef]
6. Stenberg, T.; Barsoum, Z.; Åstrand, E.; Ericson Öberg, A.; Schneider, C.; Hedegård, J. Quality control and assurance in fabrication of welded structures subjected to fatigue loading. *Weld. World* **2017**, *61*, 1003–1015. [CrossRef]
7. Barsoum, Z.; Stenberg, T.; Lindgren, E. Fatigue properties of cut and welded high strength steels-Quality aspects in design and production. *Procedia Eng.* **2018**, *213*, 470–476. [CrossRef]
8. Hultgren, G.; Barsoum, Z. Fatigue assessment in welded joints based on geometrical variations measured by laser scanning. *Weld. World* **2020**, *64*, 1825–1831. [CrossRef]
9. Stasiuk, P.; Karolczuk, A.; Kuczko, W. Analysis of correlation between stresses and fatigue lives of welded steel specimens based on real three-dimensional weld geometry. *Acta Mech. Autom.* **2016**, *10*, 12–16. [CrossRef]
10. Alam, M.M.; Barsoum, Z.; Jonsén, P.; Kaplan, A.F.H.; Häggblad, H.Å. The influence of surface geometry and topography on the fatigue cracking behaviour of laser hybrid welded eccentric fillet joints. *Appl. Surf. Sci.* **2010**, *256*, 1936–1945. [CrossRef]
11. Lang, R.; Lener, G.; Schmid, J.; Ladinek, M. Welded seam evaluation based on 3D laser scanning—Practical application of mobile laser scanning systems for surface analysis of welds—Part 1. *Stahlbau* **2016**, *85*, 336–343. [CrossRef]
12. Lang, R.; Lener, G. Assessment of welds based on 3D laser scanning. Practical application of a mobile laser scan system for the surface assessment of welds—Part 2. *Stahlbau* **2016**, *85*, 395–408. [CrossRef]
13. Lang, R.; Lener, G. Application and comparison of deterministic and stochastic methods for the evaluation of welded components' fatigue lifetime based on real notch stresses. *Int. J. Fatigue* **2016**, *93*, 184–193. [CrossRef]
14. Lener, G.; Lang, R.; Ladinek, M.; Timmers, R. A numerical method for determining the fatigue strength of welded joints with a significant improvement in accuracy. *Procedia Eng.* **2018**, *213*, 359–373. [CrossRef]
15. Vuherer, T.; Maruschak, P.; Samardžić, I. Behaviour of coarse grain heat affected zone (HAZ) during cycle loading. *Metalurgija* **2012**, *51*, 301–304. Available online: https://hrcak.srce.hr/8 (accessed on 30 April 2021).

16. Niederwanger, A.; Warner, D.H.; Lener, G. The utility of laser scanning welds for improving fatigue assessment. *Int. J. Fatigue* **2020**, *140*, 105810. [CrossRef]
17. Kaffenberger, M.; Vormwald, M. Fatigue resistance of weld ends—Analysis of the notch stress using real geometry. *Materwiss. Werksttech.* **2011**, *42*, 874–880. [CrossRef]
18. Kaffenberger, M.; Vormwald, M. Application ofthe notch stress concept to the real geometry ofweld end points. *Materwiss. Werksttech.* **2011**, *42*, 289–297. [CrossRef]
19. Hou, C.Y. Fatigue analysis of welded joints with the aid of real three-dimensional weld toe geometry. *Int. J. Fatigue* **2007**, *29*, 772–785. [CrossRef]
20. Hou, C.Y. Computer simulation of weld toe stress concentration factor sequence for fatigue analysis. *Int. J. Struct. Integr.* **2019**, *10*, 792–808. [CrossRef]
21. Chaudhuri, S.; Crump, J.; Reed, P.A.S.; Mellor, B.G. High-resolution 3D weld toe stress analysis and ACPD method for weld toe fatigue crack initiation. *Weld. World* **2019**, *63*, 1787–1800. [CrossRef]
22. Aldén, R.; Barsoum, Z.; Vouristo, T.; Al-Emrani, M. Robustness of the HFMI techniques and the effect of weld quality on the fatigue life improvement of welded joints. *Weld. World* **2020**, *64*, 1947–1956. [CrossRef]
23. Liinalampi, S.; Remes, H.; Lehto, P.; Lillemäe, I.; Romanoff, J.; Porter, D. Fatigue strength analysis of laser-hybrid welds in thin plate considering weld geometry in microscale. *Int. J. Fatigue* **2016**, *87*, 143–152. [CrossRef]
24. Ladinek, M.; Niederwanger, A.; Lang, R.; Schmid, J.; Timmers, R.; Lener, G. The strain-life approach applied to welded joints: Considering the real weld geometry. *J. Constr. Steel Res.* **2018**, *148*, 180–188. [CrossRef]
25. Ladinek, M.; Niederwanger, A.; Lang, R. An individual fatigue assessment approach considering real notch strains and local hardness applied to welded joints. *J. Constr. Steel Res.* **2018**, *148*, 314–325. [CrossRef]
26. Lang, E.; Rudolph, J.; Beier, T.; Vormwald, M. Low Cycle Fatigue Behavior of Welded Components: A New Approach—Experiments and Numerical Simulation. In Proceedings of the Pressure Vessels and Piping Conference, Toronto, ON, Canada, 15–19 July 2012; pp. 289–298.
27. Lang, E.; Rudolph, J.; Beier, H.T.; Vormwald, M. Geometrical influence of a butt weld in the low cycle fatigue regime. *Procedia Eng.* **2013**, *66*, 73–78. [CrossRef]
28. Wormsen, A.; Sjödin, B.; Härkegård, G.; Fjeldstad, A. Non-local stress approach for fatigue assessment based on weakest-link theory and statistics of extremes. *Fatigue Fract. Eng. Mater. Struct.* **2007**, *30*, 1214–1227. [CrossRef]
29. Sandberg, D.; Mansour, R.; Olsson, M. Fatigue probability assessment including aleatory and epistemic uncertainty with application to gas turbine compressor blades. *Int. J. Fatigue* **2017**, *95*, 132–142. [CrossRef]
30. Hultgren, G.; Mansour, R.; Barsoum, Z.; Olsson, M. Fatigue probability model for AWJ-cut steel including surface roughness and residual stress. *J. Constr. Steel Res.* **2021**, *179*, 106537. [CrossRef]
31. Marquis, G.B.; Barsoum, Z. *IIW Recommendations for the HFMI Treatment*; IIW Collection; Springer: Singapore, 2016; ISBN 9789811025037.
32. Winteria | Laser Scanning Systems for Quality Assurance. Available online: https://winteria.se (accessed on 7 April 2021).
33. Benard, A.; Bos-Levenbach, E.C. Het uitzetten van waarnemingen op waarschijnlijkheids-papier. *Stat. Neerl.* **1953**, *7*, 163–173. [CrossRef]
34. Hobbacher, A.F. *Recommendations for Fatigue Design of Welded Joints and Components*; IIW Collection; Springer International Publishing: Cham, Switzerland, 2016; ISBN 9783319237565.
35. Mansour, R.; Olsson, M. Efficient Reliability Assessment with the Conditional Probability Method. *J. Mech. Des. Trans. ASME* **2018**, *140*. [CrossRef]
36. Hasofer, A.M.; Lind, N.C. Exact and Invariant Second-Moment Code Format. *ASCE J. Eng. Mech. Div.* **1974**, *100*, 111–121. [CrossRef]
37. Mansour, R.; Olsson, M. A closed-form second-order reliability method using noncentral chi-squared distributions. *J. Mech. Des. Trans. ASME* **2014**, *136*. [CrossRef]
38. Mansour, R.; Olsson, M. Response surface single loop reliability-based design optimization with higher-order reliability assessment. *Struct. Multidiscip. Optim.* **2016**, *54*, 63–79. [CrossRef]
39. Hu, Z.; Du, X. Saddlepoint approximation reliability method for quadratic functions in normal variables. *Struct. Saf.* **2018**, *71*, 24–32. [CrossRef]
40. Park, J.W.; Lee, I. A Study on Computational Efficiency Improvement of Novel SORM Using the Convolution Integration. *J. Mech. Des. Trans. ASME* **2018**, *140*. [CrossRef]
41. Olsson, E.; Olander, A.; Öberg, M. Fatigue of gears in the finite life regime—Experiments and probabilistic modelling. *Eng. Fail. Anal.* **2016**, *62*, 276–286. [CrossRef]
42. Tukey, J.W. *Exploratory Data Analysis*; Addison-Wesley: Boston, MA, USA, 1977; ISBN 978-0-201-07616-5.
43. Genton, M.G.; Kleiber, W. Cross-Covariance Functions for Multivariate Geostatistics. *Stat. Sci.* **2015**, *30*, 147–163. [CrossRef]
44. Mansour, R.; Kulachenko, A.; Chen, W.; Olsson, M. Stochastic Constitutive Model of Isotropic Thin Fiber Networks Based on Stochastic Volume Elements. *Materials* **2019**, *12*, 538. [CrossRef]
45. Alzweighi, M.; Mansour, R.; Lahti, J.; Hirn, U.; Kulachenko, A. The influence of structural variations on the constitutive response and strain variations in thin fibrous materials. *Acta Mater.* **2021**, *203*, 116460. [CrossRef]
46. Lanning, D.B.; Nicholas, T.; Palazotto, A. HCF notch predictions based on weakest-link failure models. *Int. J. Fatigue* **2003**, *25*, 835–841. [CrossRef]

47. Tomaszewski, T.; Strzelecki, P.; Mazurkiewicz, A.; Musiał, J. Probabilistic Estimation of Fatigue Strength for Axial and Bending Loading in High-Cycle Fatigue. *Materials* **2020**, *13*, 1148. [CrossRef] [PubMed]
48. Norberg, S.; Olsson, M. The effect of loaded volume and stress gradient on the fatigue limit. *Int. J. Fatigue* **2007**, *29*, 2259–2272. [CrossRef]

metals

Article

Fatigue Crack Initiation and Propagation Relation of Notched Specimens with Welded Joint Characteristics

Moritz Braun [1,*], Claas Fischer [2], Jörg Baumgartner [3,4], Matthias Hecht [3] and Igor Varfolomeev [5]

[1] Institute of Ship Structural Design and Analysis, Hamburg University of Technology, 21073 Hamburg, Germany

[2] TÜV NORD EnSys GmbH & Co. KG, 22525 Hamburg, Germany; cfischer@tuev-nord.de

[3] Research Group System Reliability, Adaptive Structures Machine Acoustics SAM, Mechanical Engineering Department, Technical University of Darmstadt, 64287 Darmstadt, Germany; joerg.baumgartner@sam.tu-darmstadt.de (J.B.); matthias.hecht@sam.tu-darmstadt.de (M.H.)

[4] Fraunhofer Institute for Structural Durability and System Reliability, 64289 Darmstadt, Germany

[5] Fraunhofer Institute for Mechanics of Materials, 79108 Freiburg, Germany; igor.varfolomeev@iwm.fraunhofer.de

* Correspondence: moritz.br@tuhh.de; Tel.: +49-40-42878-6091

Abstract: This study focuses on predicting the fatigue life of notched specimens with geometries and microstructure representative of welded joints. It employs 26 series of fatigue tests on welded and non-welded specimens containing notches located in different material zones, including the parent material, weld metal, and heat-affected zone. Overall, 351 test samples made of six structural steels are included in the present evaluation. For each individual specimen, the stress concentration factor, as well as the stress distribution in the notched section, was determined for subsequent fracture mechanics calculation. The latter is employed to estimate the fraction of fatigue life associated with crack propagation, starting from a small surface crack until fracture. It was shown that the total fatigue life can be realistically predicted by means of fracture mechanics calculations, whereas estimates of the fatigue life until macroscopic crack initiation are subject to numerous uncertainties. Furthermore, methods of statistical data analyses are applied to explore correlations between the S–N curves and the notch acuity characterized by the notch radius, opening angle, and the stress concentration factor. In particular, a strong correlation is observed between the notch acuity and the slope of the S–N curves.

Keywords: notch fatigue analysis; finite element analysis; fracture mechanics; stress gradient; notch acuity; S–N curves; statistical methods; artificial notches

Citation: Braun, M.; Fischer, C.; Baumgartner, J.; Hecht, M.; Varfolomeev, I. Fatigue Crack Initiation and Propagation Relation of Notched Specimens with Welded Joint Characteristics. *Metals* **2022**, *12*, 615. https://doi.org/10.3390/met12040615

Academic Editor: Francesco Iacoviello

Received: 9 February 2022
Accepted: 28 March 2022
Published: 2 April 2022

Publisher's Note: MDPI stays neutral with regard to jurisdictional claims in published maps and institutional affiliations.

1. Introduction

The fatigue life of engineering structures is a complex process that can be divided into several phases. Simplifying this process, three superordinate phases—crack initiation, fatigue crack propagation, and unstable crack growth—are typically considered for the design of engineering structures. In fact, these phases are usually assessed individually. The first phase ends when a small crack has nucleated which is of the size of characteristic microstructure dimensions, e.g., grain size [1]. The second phase is defined by steady fatigue crack propagation. Both phases are primarily governed by the magnitude of cyclic loads. Finally, the third phase is described by unstable crack propagation when approaching the load-carrying capacity of a structure or component. While the first two phases govern the majority of service life, the last phase can be neglected in the fatigue life analysis.

Typically, different fatigue assessment methods are applied to estimate the duration of the fatigue crack initiation and propagation phase, e.g., stress-based fatigue concepts for crack initiation or linear elastic fracture mechanics for crack propagation. Even advanced methods that enable a joint estimation of fatigue crack initiation and propagation behavior—such as the IBESS procedure [2]—rely on a large number of assumptions or require additional tests to define relevant input parameters.

Different methods have been developed and are still being developed for a holistic fatigue life assessment; however, distinguishing the different phases for welded components is a complex procedure and often relies on assumptions, such as transition crack sizes [3,4]. This is exacerbated by the fact that the ratio between the number of cycles to crack initiation and propagation is typically influenced by the magnitude of external loads: the lower the applied loading, the longer the crack initiation portion [1,5]. Hence, this study aims at shedding light on the relationship between fatigue crack initiation and propagation in welded joints using artificially notched specimens with welded joint characteristics. For this purpose, an extensive experimental database is evaluated including 351 welded and non-welded specimens made of constructional steels of various strength levels, containing notches with varying notch tip radius, opening angles, and notch depths, and positioned in different material zones. The effect and significance of different influencing factors on fatigue crack initiation and propagation, as well as on the shape of stress-life (S–N) curves, is assessed both experimentally and numerically. Compared to other investigations in the literature, e.g., [2,3,6–14], this study employs experimental fatigue data obtained on specimens with both well-defined notch geometry and distinct notch location. This allows for precise stress calculations and for distinguishing between crack propagation behavior in different microstructural zones. Furthermore, for a significant number of the reference fatigue tests, fatigue lives associated with short and long crack initiation are reported, and this information is then used for validating the analysis approach. In addition, a large number of the reference tests include specimens in stress-relieved conditions, which allows for separating the effect of welding residual stresses from that of the notch acuity and microstructure. A particular intention of this paper is to introduce a reproducible analysis approach for predicting fatigue lives of welded joints using, where possible, conventional fatigue crack growth tests, correlations between the hardness and strength properties of particular weld zones, and a unique definition of an initial "technical" crack size. Such an approach, described in Section 4.2, is shown to yield a good accuracy with a rather moderate computational effort and a limited amount of material data required.

This paper, which is an extension of an earlier study [15], is organized as follows. In Section 2, basic relations are introduced which are used further for the analysis of fatigue crack initiation and propagation with a focus on welded joints. Section 3 describes the test series and specimens involved in the present analysis, test procedures, and details of the crack detection in some of the test specimens. Analysis methods applied for evaluating the test results, including finite element and fatigue crack growth calculations, are described in Section 4. A comparison of fatigue lives obtained experimentally and predicted by the analysis, as well as their statistical evaluation, are given in Section 5. Finally, the main findings of this study are discussed in Section 6.

2. Remarks on Fatigue Life Assessment of Welded Joints

Welded joints generally contain fabrication-induced notches that reduce fatigue strength regardless of the presence of other (macro-geometrical) notch effects, as well as imperfections. Typically, the S–N curves for welded joints are steeper than those of base metal, which is related to the notch effect. The respective slope is often fixed to $k = 3$ [16–19], being argued by the fact that the crack initiation phase is short, and the majority of the fatigue life is spent in crack propagation. It can be shown that under simplified conditions, integration of the Paris–Erdogan law ($da/dN = C \cdot \Delta K^m$) leads to an S–N curve with a slope k equal to the exponent m [20].

As mentioned above, a fixed slope $k = 3$ is considered in most rules and recommendations; however, in some documents, shallower slopes of $k = 3.5$ or $k = 4.0$ are recommended for welded details with a low-stress concentration, such as high-quality welds or welds with weld reinforcement removed [21]. Another difference between the major guidelines is the position of the knee point. In some documents, it is set to $N_k = 5 \cdot 10^6$

cycles [18,19] or to $N_k = 10^7$ [16,21], whereas other documents assume a variable position of the knee point depending on the weld details [22,23].

Typically, cracks initiate from one or more locations in weld transitions, such as weld toes, starting from an imperfection or from the point of the highest stress concentration, and retain an approximately semi-elliptical shape during their growth. Since the weld geometry often varies considerably along the weld seam [24], a simplified weld shape is usually considered in assessment procedures. Even though a considerable part of fatigue life relates to crack propagation, welded joints are typically assessed using stress-based concepts [16–18]. The respective design curves already include a variable amount of the crack propagation portion that depends on the specimen shape, in particular, on the notch acuity. In a more general case, the total fatigue life of a welded component until failure, N_f, can be represented by a sequence of the crack initiation and fatigue crack growth (FCG) events, as given by

$$N_f = N_i + N_{FCG} \tag{1}$$

The first term in Equation (1), N_i, is usually estimated using a material S–N curve, an appropriate multi-axial fatigue criterion together with the information on the local stress gradient in the notch root, the highly loaded area and/or material volume, surface roughness, etc. [1,19,25–27]. In the second term, N_{FCG}, can be calculated based on the fracture mechanics methodology [28], while starting from an initial macroscopic surface crack with a size selected according to either the crack detection limit or some simplified rules, see e.g., [29]. Such an initial crack is often considered to have a semi-elliptical shape with a depth of a_0 and the length of $2c_0$. For example, Fiedler et al. [27] define the initial fatigue crack with a surface length ranging between $2c_0 = 0.25$ and 3 mm.

Within the framework of the linear-elastic fracture mechanics, the number of cycles associated with the crack propagation stage can be calculated by numerically integrating an FCG equation of

$$da/dN = f(\Delta K, R_K, C_i) \tag{2}$$

with $\Delta K = K_{max} - K_{min}$ being the stress intensity factor range, $R_K = K_{min}/K_{max}$ the stress intensity ratio, and C_i denoting material parameters derived by fitting to experimental data. The number of such parameters depends on the type of equation selected for the particular application. In particular, for the analysis of the test data presented in Section 3.1, a description of the FCG rates is required, including the stress ratio effect and a proper estimation of the threshold value ΔK_{th}. This is achieved, e.g., by using an FCG equation of the NASGRO type [30].

As the notch stress may exceed the yield strength, the initial surface crack $a_0 \times 2c_0$ can be completely located within the plastically deformed material. Consequently, the applicability of the linear-elastic fracture mechanics and ΔK as the crack driving force parameter becomes questionable. In such a case, a properly defined cyclic J-integral, ΔJ, [31–33] can be employed instead of ΔK to correlate with the fatigue crack growth rates. Engineering estimates of ΔJ [34,35] can be derived based on the reference stress method incorporated in the failure assessment diagram approach [28,36].

3. Experimental Data

3.1. Specimens

For this study, welded joints were manufactured with artificial notches of different notch acuity (different radii and opening angle) to separate the effect of varying notch shape along weld seams from other important influencing factors (loading, geometry, and material characteristics) and to determine the effect of those on fatigue crack initiation and propagation, as well as on the shape of the S–N curves. In addition, well-documented fatigue data on similar specimens from the literature (Fischer et al. [37], Baumgartner [38]) were included in the evaluation, Tables 1 and 2. This is the reason for the range of different specimen geometries.

Table 1. Overview of artificially notched welded specimens.

Steel Type		Opening Angle ω [°]	Notch Radius r [mm]	Notch Depth d [mm]	Stress Concentration Factor K_t [-]	Width in Crack Growth Direction w [mm]	Thickness t [mm]	Number of Notches	Nominal Stress Ratio R	Series Number
S235JR + N	HAZ	0	0.15	3	8.33	9.5	20	1	−1	1
	HAZ	135	0.15	3	5.64	9.5	20	1	−1	2
S355MC	HAZ	15	0.05	5	12.3	8	8	2	0	23
	HAZ	135	0.05	5	6.55	8	8	2	0	22
	HAZ	180	∞^1	5	~1	8	8	–	0	24
S355J2 + N	HAZ	0	0.15	3	8.33	9.5	20	1	−1	5
	HAZ	135	0.15	3	5.64	9.5	20	1	−1	6
	HAZ	135	0.5	3	3.82	9.5	20	1	−1	7
	WM	135	0.15	3	5.64	9.5	20	1	−1	8
S690QL1	HAZ	0	0.15	3	8.33	9.5	20	1	−1	10
	HAZ	135	0.15	3	5.64	9.5	20	1	−1	11
	WM	160	10	10	1.44	10	15	2	−1	16
	WM	0	0.15	3	8.33	9.5	15	1	−1	17
S960QL	HAZ	15	0.05	5	12.3	8	8	2	0	26
	HAZ	15	0.5	5	4.08	8	8	2	0	25
	HAZ	135	0.05	5	6.55	8	8	2	0	29
	HAZ	135	0.5	5	3.10	8	8	2	0	28
	HAZ	180	∞^1	5	~1	8	8	–	0	27

[1] unnotched specimens with a constant cross-section.

Table 2. Overview of base material-type specimens.

Steel Type	Open-ing Angle ω [°]	Notch Radius r [mm]	Notch Depth d [mm]	Stress Concentration Factor K_t [-]	Width in Crack Growth Direction w [mm]	Thickness t [mm]	Number of Notches	Nominal Stress Ratio R	Series Name
S355J2 + N	135	0.15	3	5.64	9.5	20	1	−1	4
	0	0.15	3	8.33	9.5	20	1	−1	13
QStE380TM	15	0.05	5	12.3	8	8	2	0	18
	15	0.5	5	4.08	8	8	2	0	20
	135	0.05	5	6.55	8	8	2	0	19
	135	0.5	5	3.10	8	8	2	0	21
S690QL1	0	0.15	3	8.33	9.5	15	1	−1	15
	160	10	10	1.44	10	15	2	−1	14

The specimens show typical weld characteristics: firstly, they have different sharp notches with a notch opening angle of $0° \leq \omega \leq 15°$ (weld root failure at partial penetration welds), $\omega = 135°$ (weld toe failure at fillet welds), $\omega \geq 160°$ (mild weld toe at the butt joint) and $\omega = 180°$ (butt joints ground flush). Secondly, the notch radius is typically very small but varies strongly depending on the welding process. This fact is considered by using radii down to $r = 0.05$ mm. The majority of specimens were machined from MAG-welded butt joints with the notch placed in the heat-affected zone (HAZ) or the weld metal (WM), Table 1. These specimens were stress relieved by annealing after welding. Additionally, some specimens were investigated from non-welded plates, Table 2, referred to as base metal (BM) specimens. Overall, six different steel grades were considered, ranging from S235 to S960.

Figure 1 shows schematically the specimen geometries employed in this study. These can be divided into two groups: some of the specimens have two symmetrical notches, whereas the others are notched asymmetrically, which assures crack initiation at the specimen side with a sharper notch.

3.2. Specimen Characterisation

To characterize the specimens, measurements of the actual geometry after specimen preparation and of hardness values were performed. The results of the hardness measurements are presented in Figure 2. For the welded joints, measurements were performed either for all relevant zones (BM, HAZ, and WM) and for top, middle and bottom layers (S235JR + N, S355J2 + N, S690QL1) or only for the zones where the crack initiation occurred (S355MC, QStE380, S960QL).

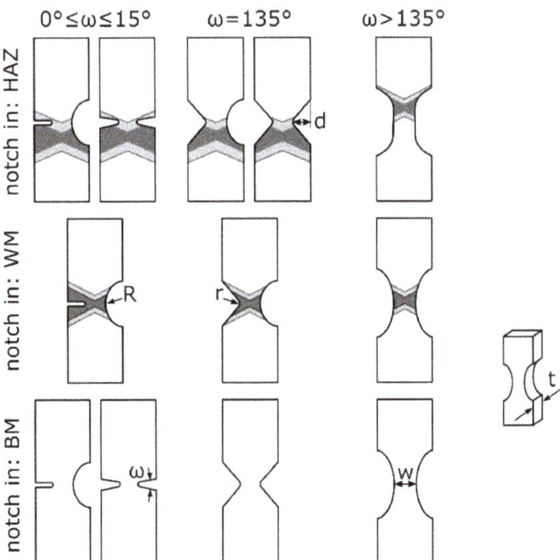

Figure 1. Overview of the specimens used for the evaluation (schematic, not to scale).

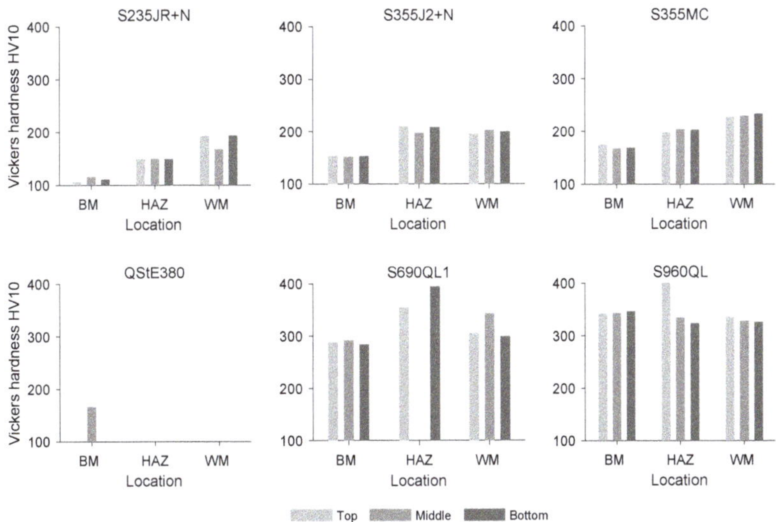

Figure 2. Hardness measurements for the six steel grades with results for the different zones and layers of the welded plates according to ISO 6507-1:2005, data from [39].

3.3. Specimen Preparation

The specimens for test series 1–17 (asymmetric notch geometry) were taken from welded plates with reinforcements removed by grinding. For this, two plate strips were joined by manual metal active gas welding using an X-butt joint with corresponding weld preparation and several alternating weld passes (eight layers for S235 and S355; 14 layers for S690). Most of the plate strips were fixed in a relatively stiff frame during welding in order to reduce the axial misalignment and the induced angular distortion. Afterward, stress annealing was applied to the welded plates.

In the next step, the individual specimens of each series were saw-cut from the butt-welded plates. The specimens were 400 mm long, 20 mm wide, and 15 mm thick.

Then the side surfaces in the direction of the welding were polished and acid-treated to make the HAZ visible with the aim to place the notch tip in a controlled way. Finally, the intended surface notches were produced by wire electro-erosion, resulting in a net thickness $w = 9.5$ mm.

In case of relatively large misalignments on a specimen, its ends were additionally milled over 60 mm in length. Thus, high additional stresses caused by clamping the specimens in the testing machine were avoided.

Furthermore, each individual specimen's geometry and the existing axial and angular misalignments were measured. Because of misaligned specimens, the notch depth d of the rounding and the notch vary amongst specimens, but w is fixed. The individual notch depths were also measured.

The specimens for test series 18–29 were taken from plates with a thickness of $t = 10$ mm. Whereas the plate made from QStE380TM was non-welded, the plates made from S355MC and S960QL were MAG-welded as a butt joint in an X-configuration. Both sides of all plates were ground to reach a final thickness of $t = 8$ mm. The final specimen geometry was extracted from the plates by wire electro-erosion, Figure 3. For test series with notches in the heat-affected zone, the notches were positioned after etching the plate edges.

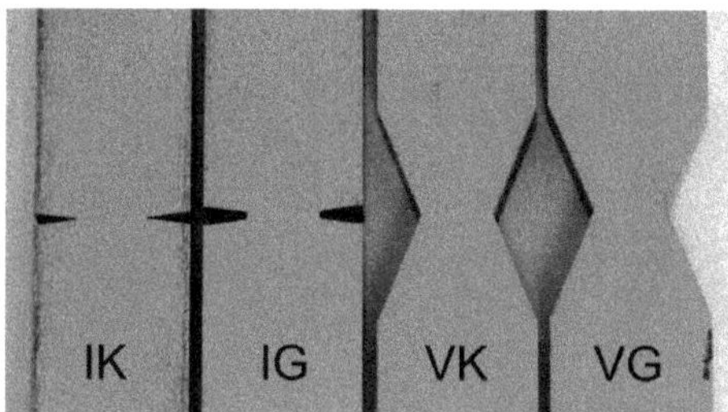

Figure 3. Specimen of test series 18–21.

3.4. Test Procedure

The existing misalignments on specimens of series 1–17 cause additional bending stresses after clamping which interfere with the applied nominal tensile stress. Both stresses are included in the structural stress, which was measured by strain gauge 2 (denoted as "DMS 2" in Figure 4) during a static pre-test for each specimen. The gauge was bonded on the specimen surface at a distance of 160 mm from the end so that the related signal was not affected by the notch. The opposite strain gauge "DMS 3" was applied to a few specimens only to verify the stress distribution over the thickness. Moreover, the strain gauge "DMS 1" was used to detect crack initiation by measuring an altered strain distribution due to the presence of a crack.

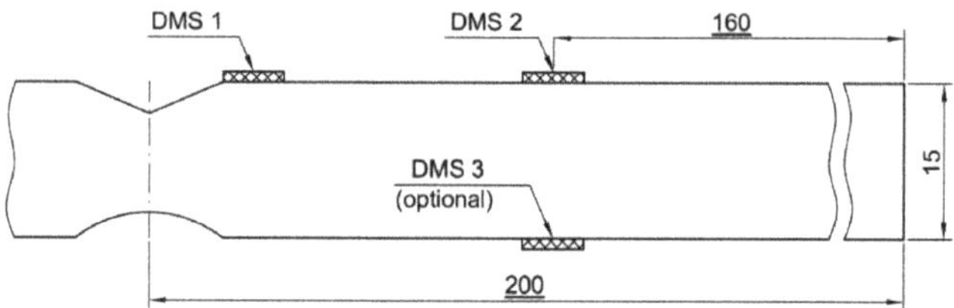

Figure 4. Schematic presentation of the specimen geometry for series 1–17 including the strain gauge positions to detect crack initiation and misalignment-related secondary bending stresses, adapted from [37].

The specimens at the ends that are not milled had different local stress ratio's R_{notch} at the notch, which was caused by the angular misalignments. For the consistency of their evaluation in terms of S–N curves, all test results were adjusted to $R_{notch} = 0$. For this purpose, a mean stress correction $f(R_{notch})$ was applied. For notched specimens with $\omega = 135°$ from Fischer et al. [37], $f(R_{notch})$ was taken from the recommendations of the International Institute of Welding [16] for stress-relieved welded joints. The correction factor used for the sharp notch was experimentally determined by evaluating test results gained under two nominal stress ratios $R = -1$ and $R = 0$. The difference with respect to the mean fatigue strength (survival probability of $P = 50\%$) at $N = 2 \cdot 10^6$ cycles is the mean stress effect, being a factor of 1.67, which is experimentally determined. For the base material specimens, the values of $f(R_{notch})$ were estimated according to the FKM guidelines [19]. Finally, the corrected structural stress ranges were used as the loading condition in the finite element (FE) analyses, aiming to compute the local fatigue parameter at the notch tip (the stress gradient, for instance).

Fatigue testing machines with hydraulic clamps were used in all tests. The tests were performed under a constant force range, at room temperature, at a frequency of about $f = 32$ Hz, on average. The nominal stress ratio was $R = 0$ for test series 1–17 and $R = -1$ for series 18–29. Each test was conducted until the specimen's failure with the related number of cycles denoted by N_f; however, a test was stopped between $N_G = 2 \cdot 10^6$ and $N_G = 1 \cdot 10^7$ cycles when either an initial crack was not visually detected and a decrease in the strain gauge signal, if available, was not noted. Then, the specimen was marked as a run-out and was re-tested to fracture under a larger nominal stress range. In case a crack initiation was detected before N_G cycles, the test was continued until failure.

3.5. Crack Detection

Two methods were applied for crack detection. Crack length foils were bonded to the side surface of specimens of the series 13, 15, and 17. For series 13, the first resistor strand was located approx. 0.5 mm below the notch root. The measures were aimed at originally catching the crack propagation quantitatively. For series 15 and 17, a similar method was used with 0.1 mm spacing between the strands and 0.1 mm distance to the notch root, see Figure 5a. For series 18–29, a digital camera was used for the optical detection of crack initiation and propagation on some of the specimens, see Figure 5b, with a resolution of approx. 150 pixels per mm. Pictures of the notches were taken at isochronous timing. For every test specimen, a minimum of 50 pictures were recorded. In order to improve the resolution of the crack initiation detection, a very small layer of a mixture of zinc oxide and glycerin was applied on the notch surface. With this approach, crack length on the surface down to 0.2 mm could be made visible. For the evaluation of cycles to crack initiation, a crack length of 0.5 mm on the surface was used as the failure criteria.

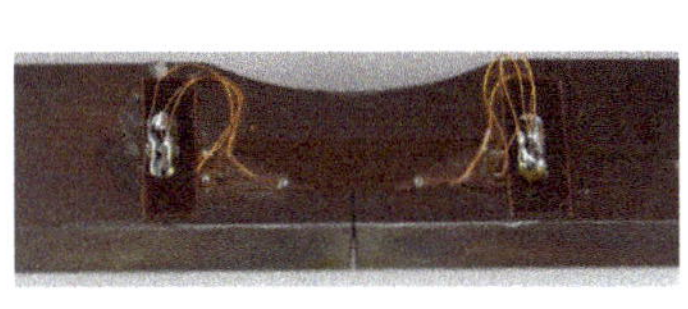

(a)

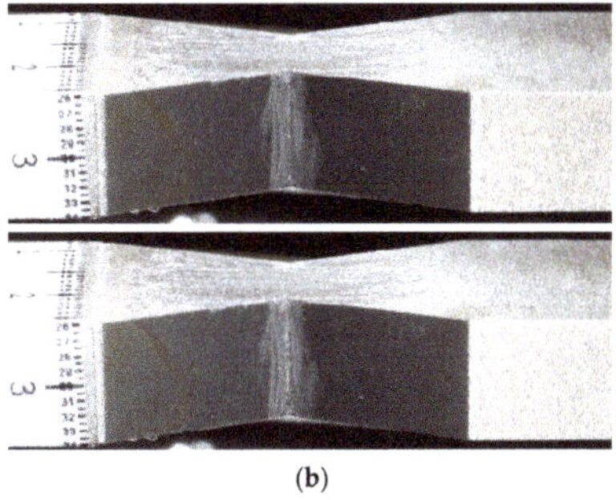

(b)

Figure 5. (**a**) Specimen with attached crack gauge after removal of weld overfill and wire erosion of notch into the middle of the weld metal and (**b**) two consecutive pictures of crack detection with photographs (specimen from test series 22).

The pictures were evaluated visually. As the first crack initiation criterion, a crack length on the surface of 1.0 mm was chosen as a *small crack initiation*. As soon as the first cracks appeared on the side surface of the specimen, the failure criterion of the *long crack initiation* was assigned. Similarly, crack initiation and propagation were monitored and assigned to the different stages by assessing the changing strain signals of the crack detection gauges, see Figure 6.

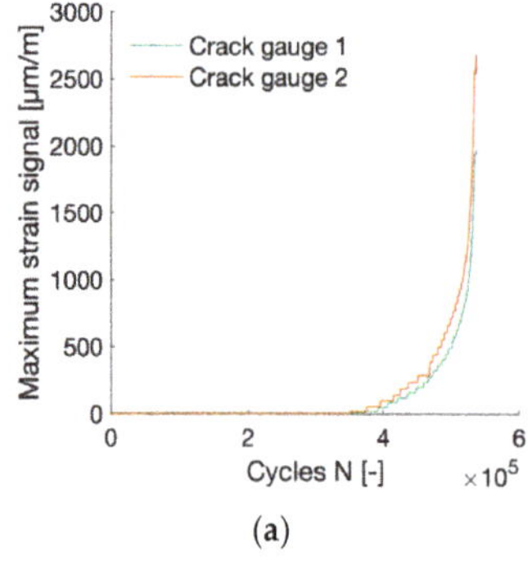

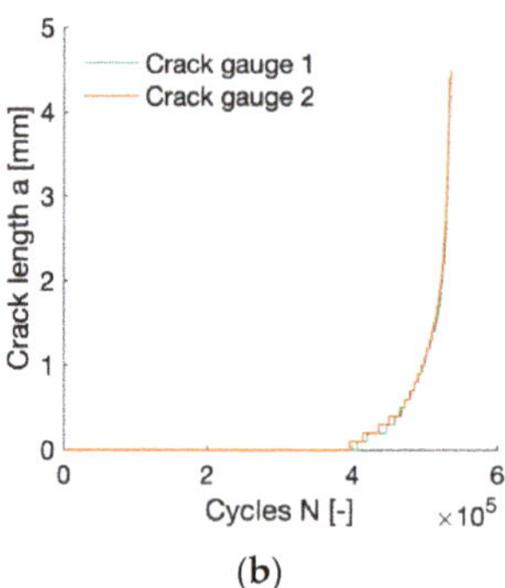

Figure 6. (a) Strain signal measurement of two crack gauges at top and bottom side over a number of cycles for a specimen of the series 15; (b) Crack lengths derived from strain signals.

3.6. Statistical Evaluation of Fatigue Test Results

In the statistical evaluation, the following parameters of the S–N curves were identified:

- endurable nominal stress at the knee point $\Delta\sigma_{n,k}$,
- number of cycles at the knee point N_k,
- slope of the S–N curve in the high cycle fatigue regime ($N \leq N_k$),
- slope of the S–N curve in the very high cycle regime ($N > N_k$).

Since k^* is quite difficult to identify [40], it was set to $k^* = 45$ according to Sonsino [41].

The statistical evaluation of the S–N curve was performed using the maximum likelihood method [40]. A mathematical best fit S–N curve was calculated, Figure 7 (left), by varying the knee point, Figure 7 (right), and identifying the one with the highest probability, i.e., with the smallest value of the support function [42] that is the natural logarithm of the probability. It must be mentioned that this location of the knee point does not necessarily align with the one leading to the smallest standard deviation or the scatter defined as $T_S = \Delta\sigma(P_S = 90\%)/\Delta\sigma(P_S = 10\%)$. The results of the statistical evaluation are the S–N curve parameters for all test series, Table A1.

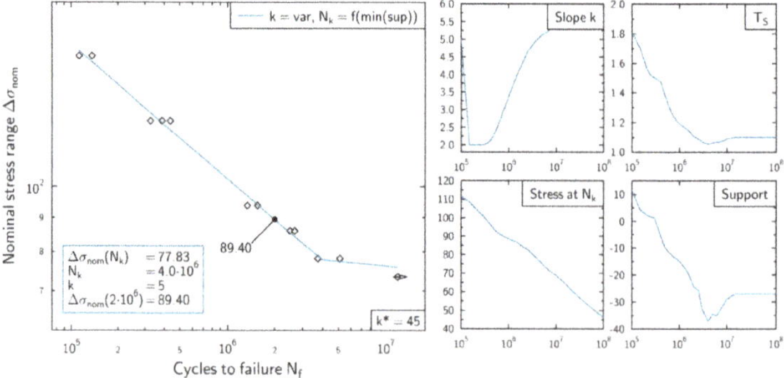

Figure 7. Example of a statistical evaluation of an S–N curve with maximum likelihood (test series 18). Left: Evaluated S-N curve, right: S-N curve parameters in relation to the position of the knee point N_k

4. Calculations

4.1. Stress Calculations

In order to perform the crack initiation or crack propagation assessment, FE models were set up based on the recommendations for mesh refinement for different effective stress methods by Braun et al. [43]. Accordingly, 32 elements per 360° with quadratic shape function are sufficient to accurately determine effective stresses at the notch, see Figure 8. As mentioned in Section 3.4, for the specimens affected by misalignment, the structural stress at strain gauge "DMS 2" consists of membrane stress and bending stresses that are caused either by clamping or tensioning the axially misaligned specimen with milled ends (secondary bending stress), see Fischer et al. [37]; therefore, the structural stress range applied in the FE analyses is equal to the membrane stress because the constant clamping stress vanishes and the secondary bending stress is negligibly small. In the case of milled specimen ends, the clamping stress does not exist, hence, the applied structural stress range results from a superposition of membrane and bending stress.

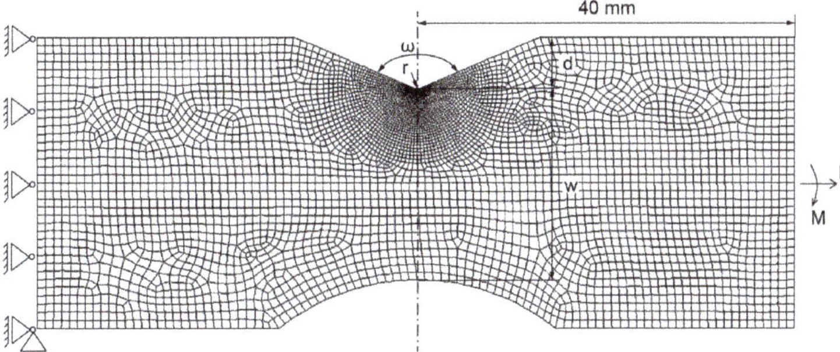

Figure 8. Parametric finite element model of the investigated specimens presented exemplarily for one specimen of series 7.

For each specimen, the stress profiles are then obtained from the location of the first principal stress (at the notch root) and in the crack growth direction. Examples of stress gradients obtained for all specimens of series 6 (S355, $r = 0.15$ mm, $\omega = 135°$, WM) and 1 (S235, $r = 0.15$ mm, $\omega = 0°$, HAZ) are presented in Figure 9. Differences in stress gradients are clearly visible from the comparison, which are due to varying applied nominal stresses as well as different ratios between the membrane and bending stress; however, the normalized gradients are almost identical. Hence, only one line is visible on the right-hand side in Figure 9.

4.2. Fatigue Crack Growth Calculations

To support fracture mechanics calculations, fatigue crack growth data for construction steels were collected describing the FCG rates both in the threshold and in the Paris regime at various stress ratios. Among the results available in the literature, Zerbst [44] provides comprehensive data for the materials S355NL and S960NL, shown as symbols in Figure 10. Note that a considerable number of welded specimens described in Section 3 exhibited misalignment and thus were subjected to superimposed tension and bending loading during fatigue testing (Section 4.1). Consequently, the nominal stress ratio, R, varied from specimen to specimen, while the stress intensity ratio, R_K, additionally varied in the course of crack propagation as a function of the crack depth. To consider this effect, the experimental data [44] were smooth curve fitted by the NASGRO-type function [30], as shown by the curves in Figure 10. These were employed for all specimens of the same material designation, however, without distinguishing between the base material, weld metal, and heat-affected zone. Moreover, since no sufficient data were found for other materials specified in Tables 1 and 2, the following assumptions were made:

- FCG curves for S355NL, Figure 10a, were adopted for S235JR + N, S355MC, S355J2 + N and QStE380TM,
- FCG curves for S960NL, Figure 10b, were adopted for S690QL.

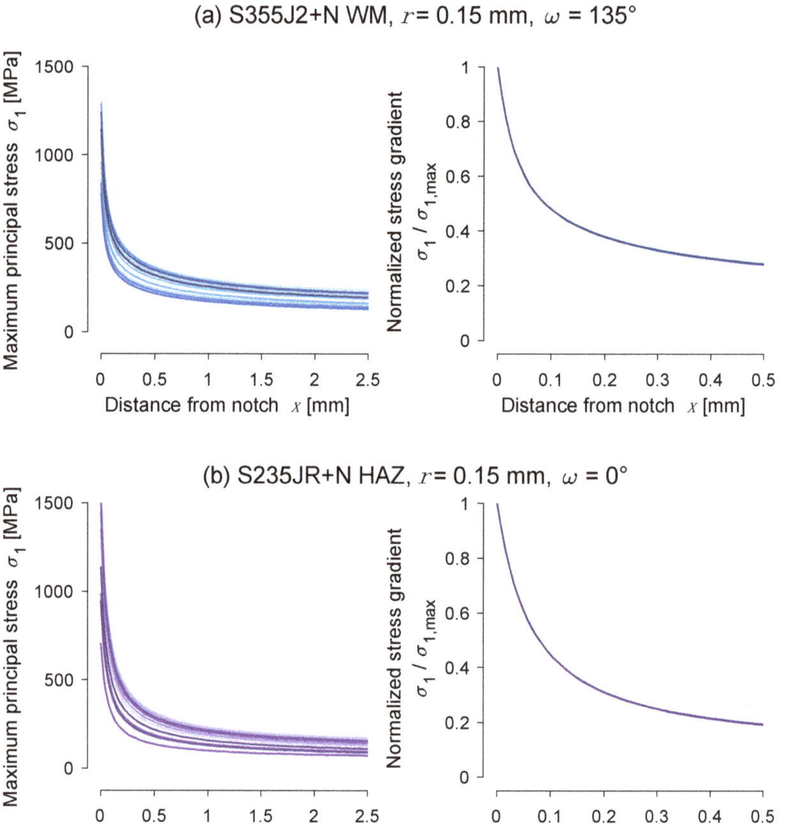

Figure 9. Exemplary stress gradients in crack growth direction of the specimens of series 6 (**a**) and 1 (**b**), as well as corresponding stress gradients normalized with the maximum principal stress at the notch tip.

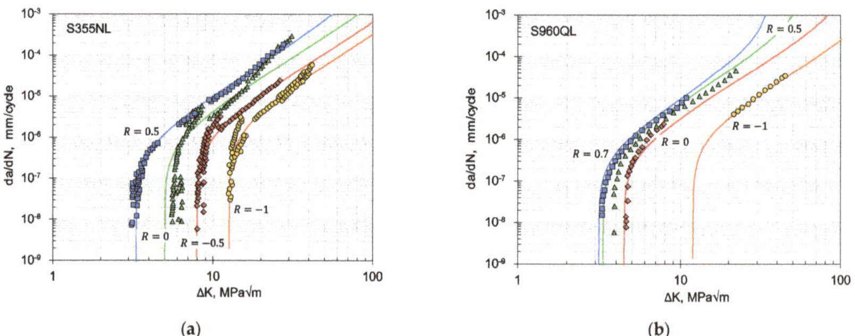

Figure 10. FCG curves for S355NL (**a**) and S960QL (**b**). Symbols: experimental data from [44], curves: analytical fit data from [30]. Symbols and curves of the same color correspond to the same R value.

In the fatigue crack growth calculations, an initial semi-circular surface crack was considered to be located in the notch root at the center of the specimen thickness, t. Such a crack, referred to as a short crack (index "sc"), was assumed to have unique dimensions for all test specimens: the crack depth of $a_{sci} = 0.5$ mm, and the crack length of $2c_{sci} = 1$ mm similar to the recommendation by Radaj et al. [1]. Thereafter, two stages of crack propagation were analyzed. The first one corresponds to the growth of the initial short crack until complete penetration through the specimen thickness, ending up at $2c = t$ and $a = a_{lc}$. Here, a_{lc} denotes the depth of a resulting through-thickness, or long crack. Subsequent crack growth, starting at $a = a_{lc}$ until final fracture, corresponds to the second stage of crack propagation. The total fatigue life associated with fatigue crack propagation, N_{FCG} in Equation (1), is then determined by

$$N_{FCG} = N_{sc} + N_{lc} \tag{3}$$

where N_{sc} and N_{lc} are the load cycles associated with the first and the second stages, respectively. Given the total fatigue life, N_f, and provided the quantities on the right-hand side of Equation (3) can be estimated by fracture mechanics calculations, the number of load cycles until short crack and long crack initiation can be deduced from

$$N_{sci} = N_f - N_{sc} - N_{lc} \tag{4}$$

and

$$N_{lci} = N_f - N_{lc} \tag{5}$$

respectively.

High-stress concentration factors (maximum elastic stress at the notch root divided by the nominal stress, see Section 4.1), reported in Tables 1 and 2, suggest that a significant amount of the tested specimens revealed plastic deformations in the notch root at a potential crack initiation location. As mentioned in Section 2, the linear-elastic fracture mechanics methods need to be modified in such a case to account for material yielding. For this purpose, an analytical approach is adopted in the present study, as described below.

Given the upper and the lower stress intensity factors in a load cycle, K_{max} and K_{min}, their plasticity corrected values are calculated based on the FAD methodology:

$$K_{J,max} = \frac{K_{max}}{f(L_{r,max})}, \; K_{J,min} = \frac{K_{min}}{f(L_{r,min})} . \tag{6}$$

Here, $f(L_r)$ is the failure line according to [36], and L_r the plasticity parameter determined from the reference stress or, alternatively, from the plastic limit load for a particular crack configuration. Thereby, the plasticity correction applies only to an open crack: in the case of $K_{min} \leq 0$ or $K_{min} \leq K_{max} \leq 0$, the denominator in Equation (6) is set to $f(L_r) = 1$. Accordingly, the following two parameters

$$\Delta K_J = K_{J,max} - K_{J,min}, \; R_{K_J} = \frac{K_{J,min}}{K_{J,max}} . \tag{7}$$

substitute the ΔK and R_K values in the fatigue crack growth equation.

Note that this definition of an "effective" stress intensity factor range, ΔK_J, differs from those suggested in [34] or [35]. First, it does not involve a crack closure term for which the estimate is rather ambiguous. Second, it seems that using the whole range ΔL_r instead of L_r in Equation (6) may lead to an overestimation of the crack driving force at $R_K < 0$ and vice versa at $R_K > 0$. Note also that no residual stresses were considered in the analyses since most of the welded specimens were stress annealed after their manufacturing.

When calculating the first FCG stage, N_{sc} or short crack growth, the stress intensity factor and the parameter L_r were evaluated using the corresponding solutions for a plate with a semi-elliptical surface crack according to [45,46]. In the second FCG stage, N_{lc} or long crack growth was analyzed based on the model of an extended surface crack of a constant depth, and respective K and L_r solutions from [47,48].

The calculations of the function $f(L_r)$ require the knowledge of the material yield strength, $R_{p0.2}$, and the ultimate strength, R_m. Those properties were not available for all the material zones and steel grades considered. Therefore, and for the sake of consistency, the strength properties were estimated in all cases based on correlations with the hardness HV [49]:

$$R_{p0.2} = -90.7 + 2.876HV, \quad R_m = -99.8 + 3.734HV \tag{8}$$

The results of the hardness measurements are summarized in Figure 2.

5. Results

5.1. Assessment of Fatigue Test Results

To be able to assess the accuracy of the fatigue crack growth calculations, the effects on the measured crack initiation and propagation behavior were first assessed. Typically, the ratio between the number of cycles until crack initiation and fracture is influenced by the magnitude of the applied loading, see Radaj et al. [1] and Murakami [5]. As specimens with a wide range of stress concentration factors K_t were tested, the ratio between experimental cycles to short $N_{sci,exp}$ and long crack initiation $N_{lci,exp}$ to the cycles to fracture $N_{f,exp}$ are plotted over the applied notch stress range $\Delta\sigma_{notch}$ in Figure 11. The latter is defined as the product of nominal stress $\Delta\sigma_{nom}$ and stress concentration factor K_t. All results are corrected to $R_{notch} = 0$ with the aforementioned mean stress correction functions $f(R_n)$.

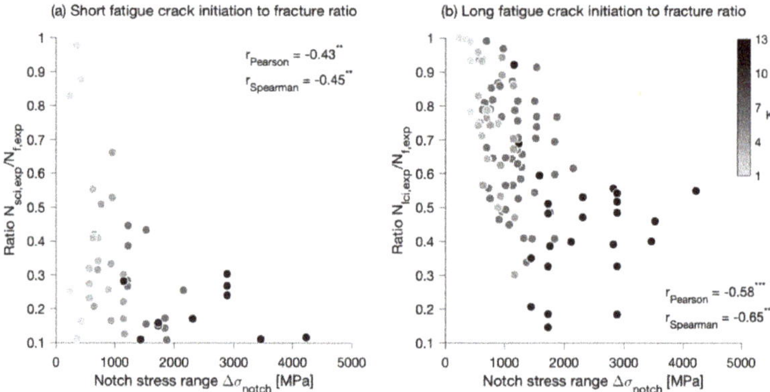

Figure 11. Comparison of ratios between (**a**) cycles until short or (**b**) long crack initiation with cycles to fracture (notch stress range mean stress corrected to $R_{notch} = 0$).

Two different types of correlation coefficients are determined to assess the statistical interference between the applied notch stress range $\Delta\sigma_{notch}$ and the ratios $N_{sci,exp}/N_{f,exp}$ and $N_{lci,exp}/N_{f,exp}$. The first one is the Pearson correlation coefficient, which determines a linear correlation between two variables based on their covariance, divided by the product of their standard deviations. The second one is Spearman's rank correlation, which assesses whether there is a monotonic relation between both input parameters. Hence, it is also capable of assessing non-linear relations. Pearson's correlation coefficient is not suitable for such relations. This is an important aspect, as the relation between the number of cycles to failure and the applied loading follows a power law (Basquin's equation). This effect is expected to be accounted for more effectively by Spearman's rank correlation than by Pearson's correlation.

In addition, Spearman's rank correlation is relatively robust against outliers [50]. Definitions of moderate to strong correlations vary in the literature; however, values above $|r_{xy}| = 0.7$ are often associated with strong correlation and values below $|r_{xy}| = 0.3$ are typically considered weak.

From Figure 11a, moderate Pearson's and Spearman's type correlations between the ratio $N_{sci,exp}/N_{f,exp}$ are observed. In contrast, an almost strong correlation is determined between the ratio $N_{lci,exp}/N_{f,exp}$ in Figure 11b. This effect is thought to be related to two aspects. First, there is more data available for long crack initiation $N_{lci,exp}$, as the crack detection gauges cannot detect a semi-elliptical surface crack on the surface of the notch, and second, it is more difficult to determine a small semi-elliptical crack reliably. This can also be seen by the larger scatter of results for small to moderate notch stress ranges $\Delta\sigma_{notch} < 500$ MPa in Figure 11a—ranging from almost immediate crack initiation to crack initiation just before final fracture.

In summary, the results agree with the general understanding that the crack initiation portion dominates in the high-cycle fatigue regime whereas, for crack propagation in the medium- and low-cycle fatigue regime, see Radaj et al. [1].

5.2. Comparison of Fatigue Test Results and Fatigue Life Estimates Based on FCG Calculations

Figure 12a is a plot of fatigue lives for the notched specimens measured experimentally, $N_{f,exp}$, and those predicted by fracture mechanics calculations, $N_{FCG,pre}$. The latter does not include the load cycles N_{sci} required for short crack initiation and, thus, tend to underestimate the total fatigue life. Indeed, for 77% of all data, the calculated points are allocated below the 1:1 line in the diagram. At the same time, almost all data are located within a threefold scatter band bound by the 1:3 and 3:1 lines. An exception is mainly for the tests classified as runouts (no crack initiation), while an initial crack was inevitably assumed in the fracture mechanics calculations. Another exceptional point is located above the 1:3 line, in the area of non-conservative prediction, and relates to an S690QL HAZ sample with the notch characteristics $\omega = 135°$, $R = 0.15$ mm, and $K_t = 5.64$ (test series 11). This outlier can be explained by (i) uncertainties in the material data assumed for S690QL and its HAZ, and (ii) the fact that the respective sample was tested at a stress level considerably exceeding the yield strength. As a consequence, that sample revealed the shortest lifetime, $N_f = 10,056$ cycles, of all samples in test series 11.

Figure 12b compares predicted fatigue lives until short crack initiation, Equation (4), with the corresponding experimental results. On average, most data points in the diagram are allocated around the 1:1 line, thus suggesting that the fracture mechanics model reasonably predicts short crack initiation. A large scatter of the results can be explained by numerous uncertainties in the input data and underlying assumptions, e.g., material properties and inaccuracies of their approximation (Figure 10), initial crack size and crack initiation site selected to be fixed for all specimens, deterministic approach adopted in this study. Additional uncertainties may arise due to the limited capabilities of the detection of short cracks during the tests. For the sharply notched specimens, no uncertainties should result from defects or material flaws due to the manufacturing process, since geometrically well-defined artificial notches were included. Only for the unnotched specimens could flaws in the material have had an impact. Some of the uncertainties mentioned above are eliminated after a long crack has been initiated. This is demonstrated in Figure 12c, which shows a good agreement between fatigue lives predicted until long crack initiation, Equation (5), and the related experimental observations.

A more detailed view of the results can be gained by plotting them in an S–N diagram for each individual test series, Figure 13. For test series 18, where specimens have a sharp notch ($r = 0.05$ mm, $\omega = 15°$), a good correlation between the number of cycles calculated by fracture mechanics (solid black line) and the total fatigue life (red symbols) can be observed. Accordingly, the lifetime is slightly underestimated by a factor of 2. The ratio between the cycles until long crack initiation and fracture is comparable to the fatigue tests and calculations. The respective ratio corresponds to the difference between the orange and red symbols, on the one hand, and between the grey and black lines, on the other hand. However, the number of cycles to crack initiation $N_{sci,pre}$, estimated by subtracting the total cycles calculated for crack growth from the experimentally determined cycles to fracture, appears to be considerably overestimated as compared to the experimental results. This result for the sharply notched specimens, Figure 12a, is mainly explained by a small fraction of the $N_{sci,exp}$ cycles in the total fatigue life, $N_{f,exp}$, which constitutes 2–12%. Consequently, any inaccuracy relating to the fracture mechanics model may have a large impact on the estimated number of cycles $N_{sci,pre}$. In contrast, a rather good correlation between experimentally determined and numerically calculated crack initiation cycles is obtained for test series 21 involving specimens with a comparatively mild notch ($r = 0.5$ mm, $\omega = 135°$), especially those tested in the HCF regime. A larger deviation between $N_{sci,pre}$ and $N_{sci,exp}$ at higher stress amplitudes is probably caused by overestimating crack growth rates.

5.3. Statistical Assessment of Influencing Factors

The parameters of the S–N curve were derived from a statistical evaluation using the maximum likelihood method; however, from this evaluation, no measure of the accuracy can be deduced. To test the reliability of the results of the statistical evaluation, a bootstrapping approach was used. For each test series, 1000 resamples with replacements were evaluated. Each resample had the same number of cycles as the original test series and was again analyzed statistically by maximum likelihood. To avoid unrealistically steep slopes of the S–N curves, the minimum slope was set to $k_{min} = 1$. As a result, 1000 S–N curves were derived for one test series with identified individual values of $\Delta\sigma_k$, N_k and k. The variation of the S–N curve parameters and subsequently the accuracy of these values can be visualized in distribution plots, Figure 14.

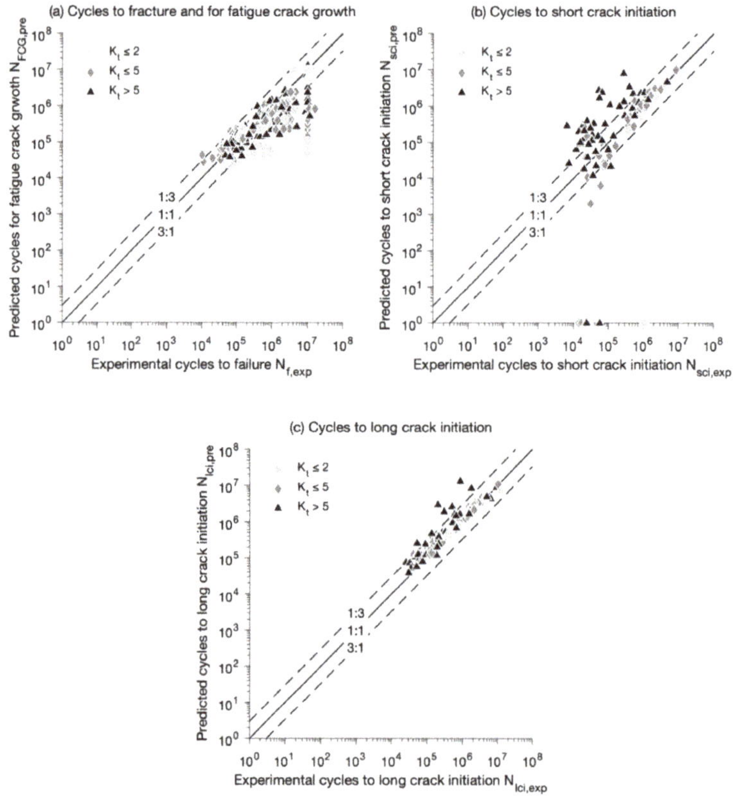

Figure 12. Comparison of (**a**) cycles to fracture and cycles for fatigue crack growth, (**b**) cycles to crack initiation, and (**c**) cycles to long crack initiation.

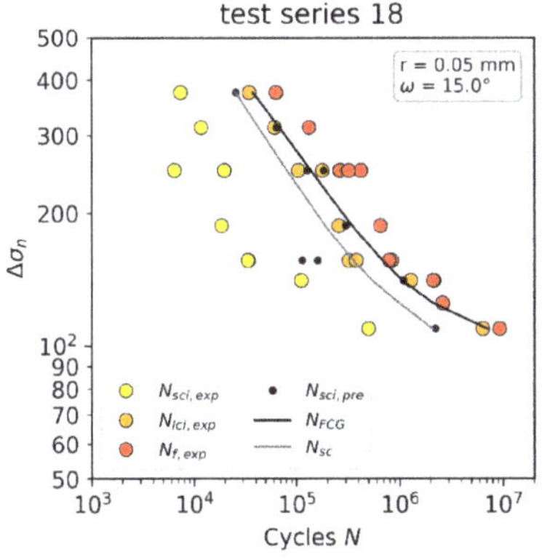

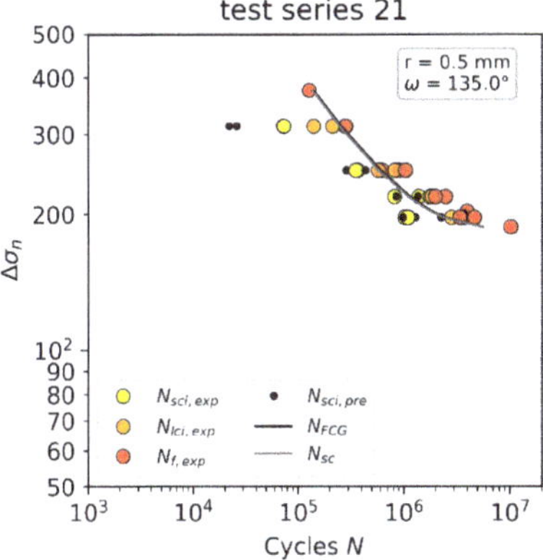

Figure 13. Comparison of S–N data: experimentally derived for three failure criteria, numerically calculated cycles for short crack growth and total crack growth, and derived crack initiation cycles. Material QStE380TM BM.

The width of the filled areas in Figure 14 represents the distribution of the parameters k, N_k and $\Delta\sigma_{n,a,k}$. A small variation of the parameters is indicated if the distribution has a small height, such as that shown for slope k of test series 1. A high variation, as identified, for example, for the slope of test series 2, stands for a high scatter in the S–N data. It indicates that the determined (mean) values might not represent the real parameters. In addition, the derived mean values are plotted with a white dot. The parameters in a range between the 1st and 3rd quartile can be identified by the black line. Next to the distribution of the S–N parameters, their coefficient of variation (CV) is also plotted showing a standardized measure of the dispersion of the probability distribution.

As can be seen, some test series show quite a low scatter in all parameters k, N_k and $\sigma_{n,a,k}$, such as 1, 4, or 20. For some other test series, such as 2, 16, or 26–27, quite a high scatter can be identified. This can directly be attributed to a comparatively high scatter in the original S–N data and a small number of overall tests. Even if the values of the slope and the knee point show high variations, the variation of the endurable stresses at the knee point is comparatively low. It is important to mention that the distributions determined by bootstrapping should be interpreted with care since they are based only on a few data points, the available S–N data; however, the distributions give a good understanding of the reliability of the determined parameters.

In a subsequent step, a correlation analysis between

- shape parameters of the S–N curve k and N_k,

- geometric properties of the specimens r and ω and
- stress concentration factor K_t

was performed. For this, the average values of the slope and the knee point of all 1000 resamples, k_{mean} and N_{mean}, as well as their standard deviation, k_{std} and N_{std}, were calculated.

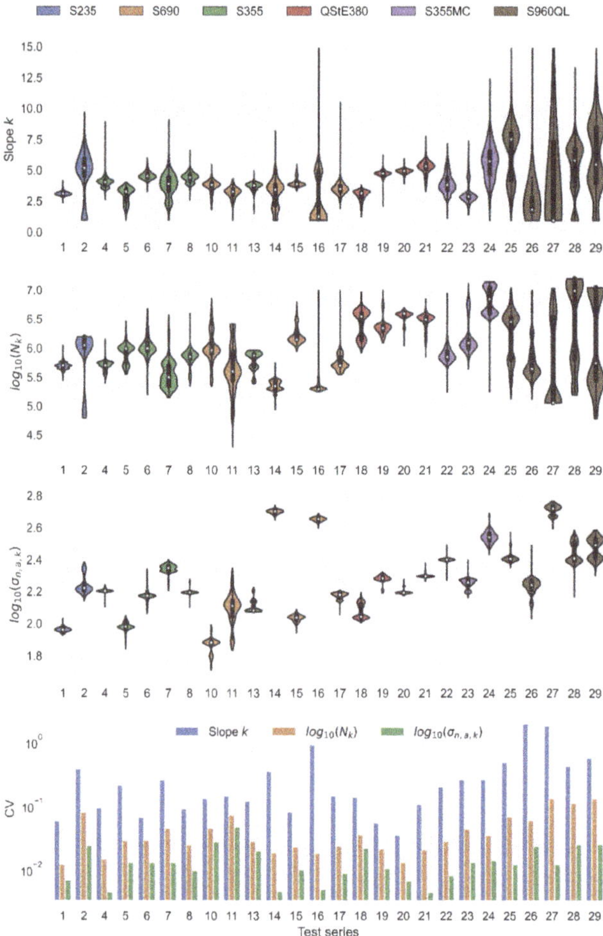

Figure 14. Evaluation of the slope k, location of N_k, $\sigma_{n,k,a}$ and the coefficient of variation (CV) of all test series determined by bootstrapping (minimum slope was set to $k_{min} = 1$).

Again, two different types of correlation coefficients, Pearson's and Spearman's, are determined to assess the statistical interference between the parameters of the S–N curve derived by the maximum likelihood method and other influencing factors (geometrical parameters and hardness). The first one is the Pearson correlation coefficient and the second is Spearman's rank correlation, which is capable of assessing non-linear relations. This is an important aspect, as some parameters are typically expected to have a non-linear impact, e.g., the notch radius on the stress concentration factor. As mentioned before, a correlation above $|r_{xy}| = 0.7$ is often associated with a strong correlation and values below $|r_{xy}| = 0.3$ are typically considered weak.

The results for both types of correlation coefficients are presented in Figure 15 as correlation matrices. Therein, the magnitude of the correlation is displayed by its color intensity. In other words, darker colors on both sides of the contrasting spectrum are associated with stronger correlations.

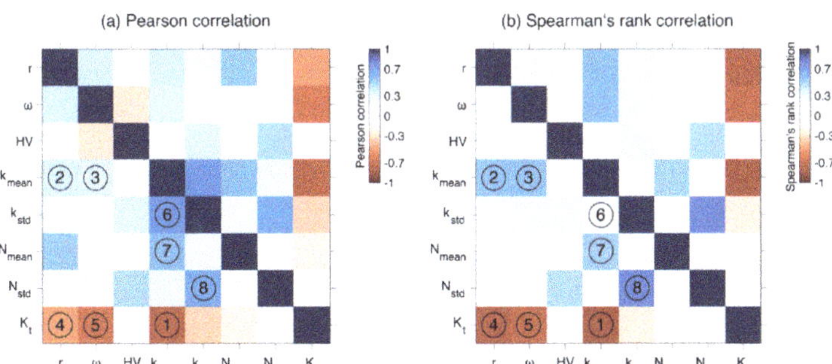

Figure 15. Correlation between maximum likelihood parameters and influencing factors based on (**a**) Pearson (linear) and (**b**) Spearman's (non-linear) rank correlation coefficient.

The main findings from the assessment of correlation coefficients are as follows:

- The highest correlation (marked by "1" in Figure 15) between two variables is observed between the stress concentration factors and the mean slope exponents determined by bootstrapping ($|r_{Pearson}| = 0.72$ and $|r_{Spearman}| = 0.79$). In addition, moderate correlations are determined between the mean slope exponents and the notch radius ($|r_{Pearson}| = 0.37$ and $|r_{Spearman}| = 0.55$), as well as an opening angle ($|r_{Pearson}| = 0.33$ and $|r_{Spearman}| = 0.56$), marked by "2" and "3", respectively.
- Unsurprisingly, moderate to strong correlations are also observed for the relation between stress concentration factors, and either the notch radius ($|r_{Pearson}| = 0.48$ and $|r_{Spearman}| = 0.77$) or the opening angle ($|r_{Pearson}| = 0.65$ and $|r_{Spearman}| = 0.71$), marked by "4" and "5", respectively.
- The reason for the higher Spearman-type correlations for the aforementioned parameters is expected to be related to their non-linear interaction, which is expected to be better assessed using a non-linear type of correlation coefficient.
- While there is a Pearson correlation (marked by "6") between the mean and scatter of the slope exponents ($|r_{Pearson}| = 0.7$), there is no such correlation according to Spearman's rank correlation. Assessing the results, it is found that the strong Pearson correlation is related to one outlier. This supports the assumption that a Pearson correlation coefficient is more prone to outliers.
- Additionally, a moderate correlation is observed between the means of the knee points and the means of the slope exponents ($|r_{Pearson}| = 0.54$ and $|r_{Spearman}| = 0.49$) and a fairly strong correlation between the standard deviations of both variables ($|r_{Pearson}| = 0.65$ and $|r_{Spearman}| = 0.76$), marked by "7" and "8", respectively. The second correlation is thought to be systematically related to the general increase in scattering in the fatigue test results with decreasing notch acuity.

In addition to the correlation matrixes, a symbolic regression was conducted with the same input data using the Python toolkit *sklearn*. As identified in the correlation matrices, the slope k has a strong correlation to K_t and can be expressed by Equation (7). The slope gets steeper with increasing stress concentration.

$$k = 6.6 - \sqrt{K_t} \tag{9}$$

In contrast, no meaningful and interpretable correlation between the position of the knee point and the geometric and material parameters of the specimens could be achieved. This result is also in line with the correlation matrices, in Figure 15.

6. Discussion

A large number of artificially notched specimens differing with respect to the notch acuity, material strength, and microstructural properties, were tested until fracture, leading to different fatigue characteristics. Overall, 26 test series, including 351 specimens, were considered in the analysis. For each test series, the material was characterized (geometry, misalignment, and hardness measured) and the influence of misalignments caused by welding was quantified by means of the strain gauges. As a basis for the fatigue crack growth calculations, a parametric FE model was created and used to determine the actual stress profiles along the expected crack path for each specimen. Thereby, it was possible to model the specimens in the most accurate way, including misalignment-induced secondary bending stresses.

For the majority of specimens, crack initiation was detected visually or per gauges showing a dependency of crack initiation duration on the notch acuity. Principally, fatigue cracks initiated early in the case of crack-like notches, whereas the fraction of load cycles before crack initiation increased with decreasing notch stress range, as indicated in Figure 11b.

The statistical evaluation of the S–N curves with the maximum likelihood approach proved to be a good method to provide information on the S–N curve, i.e., its parameters knee point N_k, slope k and the endurable stress at the knee point $\sigma_{n,a,k}$; however, the determined position of N_k at the location of the smallest value of the support function is accompanied by uncertainties, especially in cases where the support function does not have a pronounced minimum, see Figure 7, bottom right.

A measure of the accuracy of the values can be derived by bootstrapping the S–N data. With this approach, a standard deviation as a measure of the accuracy can be assigned to each value. It must be mentioned that the bootstrapping might lead, in some cases, to unrealistic results that increase in number with the scatter of the S–N curve and the number of resamples. Therefore, extreme values, as can be identified in Figure 14 by a small distribution width, should not be over-interpreted.

The coefficients of variation of the three parameters k, N_k, and $\sigma_{a,n,k}$ from bootstrapping show clearly that the values of the slope k have the highest variation, followed by the ones of the knee point N_k; however, it was observed that the value ranges are not directly comparable. A change in slope from, e.g., 5 to 6 is, in terms of the endurable number of cycles, small compared to a change of the knee point from $log10(N_k) = 5$ to 6, i.e., from 10^5 to 10^6.

The results in Section 5.1 suggest that the total fatigue life of notched specimens can rather accurately be predicted by a fracture mechanics approach; however, this requires comprehensive material data, including FCG curves and tensile properties for different weld zones, including the base metal and HAZ. As such data are rarely available, they can be estimated. In this study, only two experimental datasets on FCG curves were involved— for S355NL and S960QL steel grades, both related to the base metal. Subsequently, those FCG curves were used for other materials with similar strength properties and for other microstructural zones of the same steel grade. Another problem faced in the strength assessment of welded components is missing information on the local strength properties of the weld metal and HAZ. As a pragmatic solution, both the yield strength and the tensile strength were estimated in all cases, including the base metal, from the hardness measurements. Finally, a unique definition of an initial surface crack, referred to as a short crack, was employed in all fracture mechanics calculations. Accordingly, an assumption of a semi-elliptical crack with the depth of $a_{sci} = 0.5$ mm and the length of $2c_{sci} = 1$ mm was found to yield an optimal accuracy in predicting fatigue lives of all test series. An essential feature of sharply notched specimens is a high-stress level at the notch root and, consequently, plastic deformations achieved for a considerable number of specimens analyzed. These effects were accounted for by means of a proper plasticity correction applied to the stress intensity factors.

The predicted fatigue lives associated with crack propagation, $N_{FCG,pre}$, are in good agreement with the total fatigue lives, $N_{f,exp}$, determined in the tests. For 77% of all specimens, the fracture mechanics approach yields $N_{FCG,pre} < N_{f,exp}$ (see Figure 12) and, thus, allow for estimating the number of cycles until crack initiation. In the other cases, the fracture mechanics model results in a non-conservative lifetime prediction and, thus, in negative estimates of the lifetime until crack initiation, $N_{sci,pre}$. The latter conclusion holds for some of the specimens of the following six test series:

- Series 13: material S355J2+N, base metal, notch opening angle $\omega = 0°$
- Series 5 and 6: material S355J2+N, HAZ, notch opening angle $\omega = 0°$ and $\omega = 135°$, respectively.
- Series 10 and 11: material S690QL, HAZ, notch opening angle $\omega = 0°$ and $\omega = 135°$, respectively.
- Series 24: material S355MC, HAZ, unnotched specimens, $\omega = 180°$

For test series 13, $N_{FCG,pre}$ and $N_{f,exp}$ are in good agreement: all points are bounded by the 1:3 and 3:1 lines, and an approximately equal number of points are located above and below the 1:1 line. Hence, non-conservative lifetime prediction in this test series is mainly attributed to the experimental data scatter and the fact that the sharp notch leads to very early crack initiation. For the other five test series, all with the crack position in the HAZ, the main reason for the non-conservative prediction is likely due to an inaccurate estimation of the respective FCG curves. For better confidence in a fracture mechanics model, additional FCG tests are required on specimens extracted from the HAZ with different microstructure and strength properties.

In addition, multiple crack initiation was observed during fatigue testing for some specimens. These cracks coalesced in the following cycles and formed a longer crack. Similarly, some cracks did not initiate in the middle of the notch but closer to one of the edges. For such scenarios, deviations in prediction accuracy are inevitable. This also influences the crack propagation behavior and therefore leads to lower prediction accuracy.

Based on the statistical evaluation, it was possible to determine correlations between influencing factors and the shape of the S–N curves. Two different correlation coefficient types, linear Pearson and non-linear Spearman rank, were used, as it is known that some factors follow a non-linear relation. The highest correlation between two variables is observed between the stress concentration factor and the mean slope exponents determined by bootstrapping. In addition, moderate Spearman rank correlations are also determined between the mean slope exponent and the notch geometry, represented by the notch radius as well as the opening angle. Interestingly, there seems to be no effect on the mean and standard deviation of the bootstrapped knee point of the S–N curves. This is in contrast to former studies, e.g., by Hück et al. [51], who determined a logarithmic relation between the stress concentration factor and both the slope k and the knee point N_k of the S–N curves for base materials. The results presented in the current study do not agree with the concept of normalized S–N scatter bands by Haibach [25]. He argued that differences in the slopes of the S–N curves are related to the fact that for lower notch acuity, a large part of fatigue tests performed at high-stress ranges fall into the transition region from the high-cycle to the low-cycle fatigue regime. For plain specimens ($K_t \approx 1$), this effect is even more pronounced due to gross cyclic plastification at high-stress ratios (corresponding to a small number of cycles to failure, $N_f < 10^5$ cycles). As a considerable amount of test data for this study is based on a test of high strength steels with the number of cycles to failure above 10^5 cycles, there is clear evidence that there is a relation between the slope of S–N curves and the notch acuity. In fact, this agrees with one conclusion presented by Hück et al. [51].

As mentioned in Section 2, the slope of welded joints is set to $k = 3$ in most rules and recommendations. This assumption is in contradiction to the relationship identified in this work. Only BS 7608 [21] recommends a slightly shallower slope for weld details with low-stress concentration factors; however, it should be noted, that the specimens considered here have only three weld characteristics, (i) sharp notches, (ii) typical notch opening angles, and (iii) typical microstructural properties in the area of crack initiation and crack propagation. Other characteristics that may have an influence on the course of the S–N curves, such as residual stresses, a varying weld profile, or inner weld imperfections, are not considered. It can be assumed that the quality of the butt joints that build the experimental basis for the recommendation in rules and guidelines was not high, and the joints contained strong irregularities and, consequently, stress raisers.

In this investigation, no correlation between geometrical or metallurgical features and the knee point was identified. This stands in contrast to some guidelines that correlate the knee point to the weld details but likely explains the circumstance that there is no common agreement between all major guidelines. As for the knee point, it has to be mentioned that the specimens investigated in this work show only some features of welded joints. A further re-analysis of (high-quality) fatigue data for welded joints needs to be performed in which all relevant properties are well documented.

7. Conclusions

This study investigated the relationship between fatigue crack initiation and propagation in welded joints using artificially notched specimens with welded joint characteristics of different notch acuity (different radii and opening angles). From the statistical and numerical assessment of the experiments, the following conclusions are obtained:

- The fracture mechanics approach allows for a reasonable prediction of the total fatigue life. In most cases, 77% of the specimens analyzed, this approach leads to conservative estimates of fatigue lives. For the rest of the 23% of the specimens, the fatigue life is overestimated by a maximum factor of 3. The assumption of an initial semi-elliptical crack with the depth of $a_{sci} = 0.5$ mm and the length of $2c_{sci} = 1$ mm appeared to be a good compromise for all 26 test series. In the fracture mechanics calculations, plasticity deformations at the notch root need to be taken into account. This was achieved in this study by applying a plasticity correction to the stress intensity factors, based on the FAD approach. Non-conservative results and relatively large data scatter observed for some of the test series are believed to be partly attributed to assumptions related to fatigue crack growth curves and inaccuracies resulting from their smooth curve fitting. Additional inaccuracies may result from a simplified analysis approach assuming a single crack initiation site in the middle of the specimen thickness, thus ignoring possibilities of multiple crack initiation or crack initiation at the specimen edge. Both latter scenarios would result in a shorter fatigue life as compared to the model adopted in this study.
- For sharply notched specimens, the initiation phase is negligible and the total fatigue life is dominated by fatigue crack propagation. Therefore, the back-calculation of the fatigue crack initiation phase, N_{sci}, is subject to large errors. In contrast, such estimates of N_{sci} are shown to be rather accurate for mild notches.
- The bootstrapping of the S–N data is a suitable statistical method to identify the accuracy of the evaluated S–N parameters and to determine statistically validated estimates for the slope k and the knee point N_k of the S–N curves.
- Depending on the chosen correlation coefficient type, a moderate to almost strong correlation between applied notch stress range $\Delta\sigma_{notch}$ with the ratio between experimental cycles to short $N_{sci,exp}$ or long crack initiation $N_{lci,exp}$. and cycles to fracture $N_{f,exp}$ were determined, Figure 11. In addition, the assessment supports the general understanding that the crack initiation portion dominates in the high-cycle fatigue regime, whereas the crack propagation stage dominates in the medium- and low-cycle fatigue regime, see Radaj et al. [1] and Murakami [5].
- Using artificially notched specimens and statistical methods, it was shown that the slope k but not the knee point N_k. of the S–N curves correlate to the notch acuity, the latter being defined by the notch radius and the notch opening angle, or the stress concentration factor.

Author Contributions: Conceptualization, methodology, validation, formal analysis, and investigation, M.B., J.B., C.F. and I.V.; software, M.B., J.B. and I.V.; data curation, M.B., C.F., J.B., M.H. and I.V.; visualization, M.B. and J.B.; writing—original draft preparation, M.B., C.F., J.B., M.H. and I.V.;

writing—review and editing, M.B., C.F., J.B., M.H. and I.V.; funding acquisition, J.B. and I.V. All authors have read and agreed to the published version of the manuscript.

Funding: The Federal Ministry for Economic Affairs and Energy BMWi funded the work of J.B. and M.H. by the AiF e.V. (Arbeitsgemeinschaft industrieller Forschungsvereinigungen "Otto von Guericke" e.V.) under grant 20.366 BG.

Data Availability Statement: Not applicable.

Acknowledgments: This paper is an extended version of the reference [15] previously published in the conference proceedings of the DVM (German Association for Materials Research and Testing). M.B. would like to thank Franziska Rolof for helping with the specimen's preparation.

Conflicts of Interest: The authors declare no conflict of interest.

Nomenclature

Symbol	Unit	Description
a_{lc}	mm	Depth of a through-thickness (long) crack
a_{sci}	mm	Depth of a semi-elliptical (short) crack
c_{sci}	mm	Half-length of a semi-elliptical (short) crack
C_i	-	Material parameters in crack growth equation
CV	-	Coefficient of variation
d	mm	Notch depth
f	Hz	Test frequency
$f(L_r)$	-	Plasticity correction factor
$f(R_n)$	-	Nominal stress ratio correction factor
HV	-	Vickers hardness
k	-	Slope of the S–N curve at $N < N_k$
k^*	-	Slope of the S–N curve at $N > N_k$
k_{mean}, k_{std}	-	Mean and standard deviation of the slope of the S–N curve in the high cycle fatigue regime ($N \leq N_k$) obtained from bootstrapping
$K_{J,max}, K_{J,min}, \Delta K_J$	MPa$\sqrt{m}$	Maximum and minimum stress intensity factor, and stress intensity factor range including plasticity correction
$K_{max}, K_{min}, \Delta K$	MPa$\sqrt{m}$	Maximum and minimum stress intensity factor, and stress intensity factor range
K_t	-	Stress concentration factor
$L_r, \Delta L_r$	-	Plasticity parameter and its range
N_{FCG}	Cycle	Cycles spend in crack propagation
N_G	Cycle	Maximum number of cycles for which a test is considered a runout
$N_f, N_{f,exp}$	Cycle	Cycles to fracture and experimental cycles to fracture
$N_{sci}, N_{sci,exp}, N_{sci,pre}$	Cycle	Cycles to short crack initiation, respective experimental and predicted values
$N_{lci}, N_{lci,exp}, N_{lci,pre}$	Cycle	Cycles to long crack initiation, respective experimental and predicted values
N_k	Cycle	Cycles at the knee point
P	-	Probability of survival
r	mm	Radius of notch
R	-	Nominal stress ratio
R_K, R_{K_J}	-	Stress intensity ratio related to ΔK and ΔK_J
$R_m, R_{p0.2}$	MPa	Material ultimate strength and yield strength
R_{notch}	mm	Stress ratio of local stresses
$r_{xy}, r_{Pearson}, r_{Spearman}$	-	Correlation coefficient, Pearson's correlation coefficient, and Spearman's rank correlation coefficient
t	mm	Specimen thickness
T_S	-	Scatter ratio in stress direction
x	mm	Distance from the notch root in crack growth direction
w	mm	Specimen width

ΔK_{th}	MPa$\sqrt{m}$	Threshold of the crack intensity factor range
$\Delta\sigma_{n,k}$	MPa	Nominal stress range at the knee point
$\Delta\sigma_{nom}$	MPa	Nominal stress range
$\Delta\sigma_{notch}$	MPa	Stress range at the notch
σ_1	MPa	Maximum principal stress
ω	°	Notch opening angle

Abbreviations

BM   Base material
HAZ Heat affected zone
WM   Weld metal

Appendix A

Table A1. Results of the statistical evaluation of the S–N curves with maximum likelihood.

Test Series	Slope k	Stress S_k	Knee Point N_k [$\times 10^3$]	Scatter 1:T_S	Slope k	Stress S_k	Knee Point N_k [$\times 10^3$]	Scatter 1:T_S
	Maximum Likelihood Evaluation with Original Data				Maximum Likelihood Evaluation from Bootstrapping (Mean Values)			
1	3.11	91.7	501	1.11	3.12	92.0	504	1.10
2	5.32	165.2	1122	1.36	4.79	175.3	1032	1.30
4	4.03	160.9	501	1.06	4.03	161.0	503	1.05
5	3.29	93.3	1000	1.23	3.02	96.5	875	1.18
6	4.54	149.0	1000	1.14	4.57	147.5	1197	1.12
7	3.91	227.1	316	1.18	3.80	221.4	445	1.16
8	4.54	155.7	794	1.08	4.41	159.4	720	1.06
10	3.79	77.1	891	1.16	3.74	74.2	1267	1.14
11	3.28	130.3	398	1.35	3.22	127.4	716	1.31
13	3.78	121.7	794	1.08	3.67	129.8	664	1.06
14	3.49	502.7	251	1.24	3.15	505.3	233	1.22
15	3.83	110.3	1413	1.10	4.02	109.0	1753	1.09
16	1.14	454.9	200	1.23	2.48	453.5	233	1.20
17	3.43	153.1	501	1.13	3.55	150.3	611	1.11
18	3.24	109.7	3981	1.22	3.02	120.7	2985	1.18
19	4.78	193.2	2239	1.09	4.75	192.0	2429	1.08
20	4.99	155.7	3981	1.05	4.95	158.0	3704	1.05
21	5.33	199.6	3162	1.08	5.28	200.1	3120	1.07
22	3.80	254.3	708	1.10	4.00	251.9	891	1.07
23	2.94	181.5	1259	1.14	3.12	179.1	1909	1.11
24	5.79	339.6	8913	1.34	5.74	352.8	7816	1.27
25	7.98	249.0	3981	1.15	6.71	261.6	2646	1.11
26	1.74	179.7	398	1.21	3.16	173.4	1025	1.16
27	5.00	548.8	116	1.29	8.05	520.7	1883	1.16
28	6.37	251.3	10,000	1.41	5.15	284.3	7385	1.31
29	3.92	332.6	282	1.36	6.18	302.9	4879	1.25

References

1. Radaj, D.; Sonsino, C.M.; Fricke, W. *Fatigue Assessment of Welded Joints by Local Approaches*, 2nd ed.; Woodhead Publishing: Cambridge, UK, 2006.
2. Zerbst, U.; Madia, M.; Schork, B.; Hensel, J.; Kucharczyk, P.; Ngoula, D.; Tchuindjang, D.; Bernhard, J.; Beckmann, C. *Fatigue and Fracture of Weldments*; Springer: Berlin/Heidelberg, Germany, 2019. [CrossRef]
3. Baumgartner, J.; Waterkotte, R. Crack initiation and propagation analysis at welds—Assessing the total fatigue life of complex structures. *Mater. Werkst.* **2015**, *46*, 123–135. [CrossRef]
4. Fischer, C.; Fricke, W. Effect of the stress distribution in simple welded specimens and complex components on the crack propagation life. *Int. J. Fatigue* **2016**, *92*, 488–498. [CrossRef]
5. Murakami, Y. *Metal Fatigue: Effects of Small Defects and Nonmetallic Inclusions*, 2nd ed.; Academic Press: Cambridge, MA, USA, 2019. [CrossRef]
6. Mann, T.; Tveiten, B.W.; Härkegård, G. Fatigue crack growth analysis of welded aluminium RHS T-joints with manipulated residual stress level. *Fatigue Fract. Eng. Mater. Struct.* **2006**, *29*, 113–122. [CrossRef]
7. Remes, H. *Strain-Based Approach to Fatigue Strength Assessment of Laser-Welded Joints*; Helsinki Univercity of Technolgy: Espoo, Finland, 2008.
8. Song, W.; Liu, X.; Xu, J.; Fan, Y.; Shi, D.; He, M.; Wang, X.; Berto, F. Fatigue fracture assessment of 10CrNi3MoV welded load-carrying cruciform joints considering mismatch effect. *Fatigue Fract. Eng. Mater. Struct.* **2021**, *44*, 1739–1759. [CrossRef]
9. Toyosada, M.; Gotoh, K.; Niwa, T. Fatigue life assessment for welded structures without initial defects: An algorithm for predicting fatigue crack growth from a sound site. *Int. J. Fatigue* **2004**, *26*, 993–1002. [CrossRef]
10. Fricke, W.; Gao, L.Y.; Paetzold, H. Fatigue assessment of local stresses at fillet welds around plate corner. *Int. J. Fatigue* **2017**, *101*, 169–176. [CrossRef]
11. Barsoum, Z.; Jonsson, B. Fatigue Assessment and LEFM Analysis of Cruciform Joints Fabricated with Different Welding Processes. *Weld. World* **2008**, *52*, 93–105. [CrossRef]
12. Chapetti, M.D.; Steimbreger, C. A simple fracture mechanics estimation of the fatigue endurance of welded joints. *Int. J. Fatigue* **2019**, *125*, 23–34. [CrossRef]
13. Zong, L.; Shi, G.; Wang, Y.-Q.; Li, Z.-X.; Ding, Y. Experimental and numerical investigation on fatigue performance of non-load-carrying fillet welded joints. *J. Constr. Steel Res.* **2017**, *130*, 193–201. [CrossRef]
14. Liu, Y.; Tsang, K.S.; Hoh, H.J.; Shi, X.; Pang, J.H.L. Structural fatigue investigation of transverse surface crack growth in rail steels and thermite welds subjected to in-plane and out-of-plane loading. *Eng. Struct.* **2020**, *204*, 110076. [CrossRef]
15. Braun, M.; Fischer, C.; Baumgartner, J.; Hecht, M.; Varfolomeev, I. Zum Verhältnis von Rissinitiierung und -ausbreitung an gekerbten Proben mit Schweißnahtcharakteristik. In Proceedings of the 54. Tagung Bruchmechanische Werkstoff und Bauteilbewertung: Beanspruchungsanalyse, Prüfmethoden und Anwendungen, Berlin, Germany, 23 February 2022.
16. Hobbacher, A.F. *Recommendations for Fatigue Design of Welded Joints and Components*, 2nd ed.; Springer International Publishing: Cham, Switzerland, 2016. [CrossRef]
17. DNV GL AS. *DNVGL-RP-C203: Recommended Practice for Fatigue Design of Offshore Steel Structures*; DNV GL AS: Høvik, Norway, 2016.
18. *EN 1993-1-9:2005*; Eurocode 3: Design of Steel Structures—Part 1–9: Fatigue. European Committee for Standardization: Brussels, Belgium, 2005.
19. Forschungskuratorium Maschinenbau (FKM). *Analytical Strength Assessment of Components: FKM Guideline*, 6th ed.; VDMA: Frankfurt, Germany, 2012.
20. Baumgartner, J. Review and considerations on the fatigue assessment of welded joints using reference radii. *Int. J. Fatigue* **2017**, *101*, 459–468. [CrossRef]
21. *BS 7608:2014+A1:2015*; Guide to Fatigue Design and Assessment of Steel Products. British Standards Institution BSI: London, UK, 2014.
22. American Welding Society. *AWS D1.1:2010 Structural Welding Code—Steel*; American Welding Society: Miami, FL, USA, 2010.
23. Japanse Society of Steel Construction. *Fatigue Design Recommendations for Steel Structures*; Japanese Society of Steel Construction: Tokyo, Japan, 1995.
24. Renken, F.; von Bock und Polach, R.U.F.; Schubnell, J.; Jung, M.; Oswald, M.; Rother, K.; Ehlers, S.; Braun, M. An algorithm for statistical evaluation of weld toe geometries using laser triangulation. *Int. J. Fatigue* **2021**, *149*, 106293. [CrossRef]
25. Haibach, E. *Betriebsfestigkeit: Verfahren und Daten zur Bauteilauslegung*, 3rd ed.; Springer: Berlin/Heidelberg, Germany; New York, NY, USA, 2006.
26. Socie, D.F.; Marquis, G.B. *Multiaxial Fatigue*; SAE International: Warrendale, PA, USA, 1999.
27. Fiedler, M.; Wächter, M.; Varfolomeev, I.; Vormwald, M.; Esderts, A. *Rechnerischer Festigkeitsnachweis unter Expliziter Erfassung Nichtlinearen Werkstoffverformungsverhaltens für Bauteile aus Stahl, Stahlguss und Aluminiumknetlegierungen*; VDMA Verlag: Frankfurt, Germany, 2019.
28. *BS 7910:2019*; Guide to Methods for Assessing the Acceptability of Flaws in Metallic Structures. British Standards Institution: London, UK, 2019.
29. Bowness, D.; Lee, M.M.K. *Fracture Mechanics Assessment of Fatigue Cracks in Offshore Tubular Structures*; Offshore Technology Report 2000/077; Health and Safety Executive: Bootle, UK, 2002.
30. Forman, R.; Shivakumar, V.; Mettu, S.; Newman, J. Fatigue crack growth computer program NASGRO version 3.0. In *Reference Manual, NASA JSC-22267B*; NASA: Washington, DC, USA, 2000.

31. Paris, P.C.; Bucci, R.J.; Wessel, E.T.; Clark, W.G.; Mager, T.R. Extensive Study of Low Fatigue Crack Growth Rates in A533 and A508 Steels. In *Stress Analysis and Growth of Cracks. Proceedings of the 1971 National Symposium on Fracture Mechanics: Part 1*; ASTM International: West Conshohocken, PA, USA, 1972; Volume 513, pp. 141–176. [CrossRef]
32. Dowling, N.; Begley, J. Fatigue crack growth during gross plasticity and the J-integral. In *Mechanics of Crack Growth*; ASTM International: West Conshohocken, PA, USA, 1976.
33. Wüthrich, C. The extension of the J-integral concept to fatigue cracks. *Int. J. Fract.* **1982**, *20*, R35–R37. [CrossRef]
34. McClung, R.C.; Chell, G.G.; Lee, Y.D.; Russel, D.A.; Orient, G.E. *Development of a Practical Methodology for Elastic-Plastic and Fully Plastic Fatigue Crack Growth*; ASTM International: West Conshohocken, PA, USA; Marshall Space Flight Center: Huntsville, AL, USA, 1999.
35. Zerbst, U.; Madia, M.; Hellmann, D. An analytical fracture mechanics model for estimation of S–N curves of metallic alloys containing large second phase particles. *Eng. Fract. Mech.* **2012**, *82*, 115–134. [CrossRef]
36. British Energy Generation Limited. *R6, Assessment of the Integrity of Structures Containing Defects, Revision 4*; British Energy Generation Limited: London, UK, 2001.
37. Fischer, C.; Fricke, W.; Rizzo, C.M. Fatigue tests of notched specimens made from butt joints at steel. *Fatigue Fract. Eng. Mater. Struct.* **2016**, *39*, 1526–1541. [CrossRef]
38. Baumgartner, J. *Schwingfestigkeit von Schweißverbindungen unter Berücksichtigung von Schweißeigenspannungen und Größeneinflüssen*; Fraunhofer Verlag: Stuttgart, Germany, 2014.
39. *ISO 6507-1:2005*; Metallic Materials—Vickers Hardness Test—Part 1: Test Method. International Standards Organization: Geneva, Switzerland, 2005.
40. Störzel, K.; Baumgartner, J. Statistical evaluation of fatigue tests using maximum likelihood. *Mater. Test.* **2021**, *63*, 714–720. [CrossRef]
41. Sonsino, C. Course of SN-curves especially in the high-cycle fatigue regime with regard to component design and safety. *Int. J. Fatigue* **2007**, *29*, 2246–2258. [CrossRef]
42. Spindel, J.E.; Haibach, E. The method of maximum likelihood applied to the statistical analysis of fatigue data. *Int. J. Fatigue* **1979**, *1*, 81–88. [CrossRef]
43. Braun, M.; Müller, A.M.; Milaković, A.-S.; Fricke, W.; Ehlers, S. Requirements for stress gradient-based fatigue assessment of notched structures according to theory of critical distance. *Fatigue Fract. Eng. Mater. Struct.* **2020**, *43*, 1541–1554. [CrossRef]
44. Zerbst, U. *Analytische bruchmechanische Ermittlung der Schwingfestigkeit von Schweißverbindungen (IBESS-A3)*; Bundesanstalt für Materialforschung und-Prüfung (BAM): Berlin, Germany, 2016.
45. Wang, X.; Lambert, S.B. Stress intensity factors for low aspect ratio semi-elliptical surface cracks in finite-thickness plates subjected to nonuniform stresses. *Eng. Fract. Mech.* **1995**, *51*, 517–532. [CrossRef]
46. Sattari-Far, I.; Dillström, P. Local limit load solutions for surface cracks in plates and cylinders using finite element analysis. *Int. J. Press. Vessel. Pip.* **2004**, *81*, 57–66. [CrossRef]
47. Fett, T.; Munz, D. *Stress Intensity Factors and Weight Functions for One-Dimensional Cracks*; Kernforschungszentrum Karlsruhe: Karlsruhe, Germany, 1994.
48. Willoughby, A.A.; Davey, T.G. Plastic Collapse in Part-Wall Flaws in Plates. In *Fracture Mechanics: Perspectives and Directions (Twentieth Symposium)*; Wei, R.P., Gangloff, R.P., Eds.; ASTM International: West Conshohocken, PA, USA, 1989; Volume ASTM. [CrossRef]
49. Pavlina, E.J.; Van Tyne, C.J. Correlation of Yield Strength and Tensile Strength with Hardness for Steels. *J. Mater. Eng. Perform.* **2008**, *17*, 888–893. [CrossRef]
50. Caruso, J.C.; Cliff, N. Empirical Size, Coverage, and Power of Confidence Intervals for Spearman's Rho. *Educ. Psychol. Meas.* **1997**, *57*, 637–654. [CrossRef]
51. Hück, M.; Thrainer, L.; Schütz, W. *Berechnung von Wöhlerlinien für Bauteile aus Stahl, Stahlguß und Grauguß: Synthetische Wöhlerlinien*, 3rd ed.; Stahleisen: Düsseldorf, Germany, 1983.

Article

Fatigue Performance of High- and Low-Strength Repaired Welded Steel Joints

Jan Schubnell [1],*, **Phillip Ladendorf [2]**, **Ardeshir Sarmast [1]**, **Majid Farajian [1]** and **Peter Knödel [2]**

[1] Fraunhofer Institute for Mechanics of Materials (IWM), 79108 Freiburg, Germany; ardeshir.sarmast@iwm.fraunhofer.de (A.S.); Farajian@slv-duisburg.de (M.F.)

[2] Steel and Leightweight Structures, Karlsruhe Institute of Technology, 76131 Karlsruhe, Germany; philipp.ladendorf@kit.edu (P.L.); peter.knoedel@kit.edu (P.K.)

* Correspondence: jan.schubnell@iwm.fraunhofer.de; Tel.: +49-761-5142-235

Abstract: Large portions of infrastructure buildings, for example highway- and railway bridges, are steel constructions and reach the end of their service life, as a reason of an increase of traffic volume. As lifetime extension of a commonly used weld detail (transverse stiffener) of these structures, a validated approach for the weld repair was proposed in this study. For this, welded joints made of S355J2+N and S960QL steels were subjected to cyclic loading until a pre-determined crack depth was reached. The cracks were detected by non-destructive testing methods and repaired by removal of the material around the crack and re-welding with the gas metal arc welding (GMAW). Then, the specimens were subjected to cyclic loading again. The hardness, the weld geometry, and the residual stress state was investigated for both the original- and the repaired conditions. It was determined that nearly all repaired specimens reached at least the fatigue life of the original specimen.

Keywords: fatigue; repair welding; residual stress; non-destructive testing

Citation: Schubnell, J.; Ladendorf, P.; Sarmast, A.; Farajian, M.; Knödel, P. Fatigue Performance of High- and Low-Strength Repaired Welded Steel Joints. *Metals* **2021**, *11*, 293. https://doi.org/10.3390/met11020293

Academic Editor: Vincenzo Crupi

Received: 11 January 2021
Accepted: 31 January 2021
Published: 8 February 2021

Publisher's Note: MDPI stays neutral with regard to jurisdictional claims in published maps and institutional affiliations.

1. Introduction

A large portion of infrastructural buildings, such as bridges and offshore structures, are steel constructions. Especially railway or highway bridges are exposed to a strongly increasing rail cargo and traffic volume. This leads to an increasing load of such structures and thus, to an increasing fatigue damage. In many cases, weld details are especially affected as a reason of their comparably low fatigue resistance. In the Federal Republic of Germany, nearly 50% of the steel bridges are built before 1980 [1] and around 55% of the railway bridges are built before 1950 [2]. Similar conditions are reported from the United State of America where around 85% of the bridges in Minnesota were built before 1986 [3]. Originally, these structures are designed for significant lower loads and a higher number of repair cases is expected for the future.

The right choice of the repair strategy is an important factor to extend the fatigue life of cracked steel structures [3]. Usually, fatigue cracks are detected by visual inspection or non-destructive testing methods in periodically time intervals. After the detection, possible repair methods are the usage of bolted splices or by gouging and re-welding the material [4]. However, bolted splices are not always efficient for relatively minor fatigue damage especially if the available working space around the damage is limited [5]. Compared to the conventional bolt splice method, repair welding is a more cost-efficient solution and can be performed with less time effort. However, possible weld-defects like splatters, undercuts or cold laps can be induced by re-welding and might only lead to a comparatively small, extended life-span after the retrofit.

Due to countless different weld-details present in steel-bridges and -constructions, many repair cases exist [4,6–12]. Especially, the IIW-document XIII-2284r1-09 [4] contains a large number of repair-cases from the 1960s. Based on many of these cases, it has to be pointed out, that a retrofit of fatigue-damaged steel structures is often carried out by combining many of the before mentioned measures. Hence, it is not possible to quantify the effect of every single one of them. In order to quantify the sole effect of repair-welding, a literature survey has been carried out. Wylde [13] carried out fatigue experiments on transverse and longitudinal stiffeners with fillet welds in the as-welded state until a pre-defined crack length and depth had been reached. A subsequent repair was prepared by removing the damaged material by disc grinding. The repair-welding was fabricated by electrode welding. He concluded that the original fatigue detail class can be restored in case there are no internal defects present.

The fatigue behavior of repair-welded transverse stiffeners were also investigated in [14–16]. However, the repair welding was performed at unloaded and uncracked specimen. Thus, no direct comparison of the welded joints in original condition and in repaired condition was possible.

In the scope of the German FOSTA research project P864 [17], fatigue experiments were conducted at butt welds and specimens with longitudinal stiffeners and fillet welds. In the case of the butt welds, the repaired weld toes could not be tested up to a fatigue failure due to a failure of a non-repaired weld toe. Therefore, no conclusion for the butt welds can be drawn. Instead of this, the fatigue strength of the repaired longitudinal stiffeners has been increased compared to the as-welded condition, since the weld angle was smaller after the repair.

Different, but few studies have shown that the fatigue life of welded joints could be significantly extended by gouging and re-welding, subsequently named as repair welding [4,13,18–20]. However, these experimental studies were not included in current design codes from the International Institute of Welding [21], Eurocode 3 [22] or the German FKM guideline [23]. For this reason, a validated approach for the repair of welded structures is needed to assure that a maximum possible fatigue life by re-welding could be reached. Furthermore, the fatigue classes (FAT) of such repaired weld details needs to be reported so they can be included for the design and fatigue life estimation of repaired welded steel constructions.

The aim of this work was to develop a validated approach for repair welding of fillet welds. Double sided transverse stiffeners were chosen as corresponding weld detail. For this, a stepwise procedure containing the fatigue test in original condition, crack detecting by NDT methods, gouging, multi pass re-welding and fatigue test in repaired condition were performed.

2. Materials and Specimens Detail

Two commonly used steel grades were investigated with significant different mechanical properties. As mild, structural steel S355J2+N in normalized condition with a yield of 402 MPa was used. Additionally, the high strength, quenched and tempered steel S960QL with a yield of 1011 MPa was used for this work. Fatigue tests were performed on 88 specimens. The chemical compositions of the base materials are shown in Table 1 and the mechanical properties are given in Table 2. The chemical composition was measured by spectral analysis.

Table 1. Chemical composition of the investigated base materials.

Materials	Elements (wt.%) (Fe = bal.)													
Element	C	Mn	Si	P	S	Cr	Ni	Mo	V	W	Cu	Al	Ti	CEV *
S355J2+N	0.161	1.47	0.17	0.0107	0.0053	0.040	0.035	0.007	0.008	0.004	0.015	0.032	0.0125	0.42
S960QL	0.155	1.23	0.20	0.0095	0.0017	0.194	0.084	0.599	0.046	0.007	0.013	0.057	0.003	0.53
G4Si1 **	0.08	1.65	1.0	-	-	-	-	-	-	-	-	-	-	-
Mn2NiCrMo **	0.10	1.80	0.80	-	-	0.350	2.300	0.600	-	-	-	-	-	-

* According to DIN EN 10025-2, ** data sheet.

Table 2. Mechanical properties of the investigated base materials.

Materials	Yield Strength (MPa)	Ultimate Strength (MPa)	Elongation (%)	Hardness (HV10)	Generic Name
S355J2+N	402	538	25 *	169	-
S960QL	1011	1060	14 *	316	Strenx S960E
G4Si1 *	390–490	510–610	≥25	-	SG3
Mn2NiCrMo *	880–920	940–980	16–20	-	Union X90

* data sheet.

In this work, the investigated fillet welds were manufactured with the gas metal arc welding (GMAW) process. The weld detail was a double-sided transverse stiffener. The specimens drawing is shown in Figure 1 and the cross section of the actual welds are shown in Figure 2. In total, three sections per material were extracted around the center of the welded sheets. Samples were hot mounted at 180 °C under 30 kN using an automatic metallurgical mounting press Struers LaboPress-3. Heating and cooling times lasted respectively 8 and 5 min. The polishing is realized on a Struers Tegramin-30 polishing machine using the following felts: Metall MD-Allegro 9 µm, Metall MD-DAC 3 µm, Metall MD-Chem 0.1 µm for further investigations. Some higher differences of the welds shape are visible for S960QL compared to S355J2+N but this could not be observed for every extracted section. The length of the welded joint was 1250 mm for S355J2+N and 1500 mm for S960QL. The manufacturing of each welded joint was performed by the aid of a welding robot. For the S960QL, the material was preheated to more than 90 °C but less than 150 °C, according to [24]. G4Si1 with a wire diameter of 1.2 mm was used as filler material for the welding process of S355J2+N and for S960QL the filler material Mn2NiCrMo was used with a wire diameter of 1.0 mm, according to DIN EN 757:1997-95. In both cases, M21-ArC-18 was used as inert gas with a flow rate of 15–18 L/min. The welding parameters for all cases are given in Table 3. For each material, the same welding parameters for each weld were used. For all welds the quality class B according to ISO 5817:2014-6 was reached.

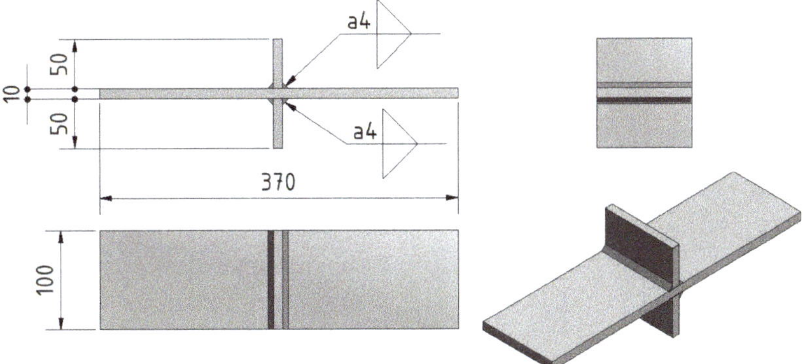

Figure 1. Technical drawing of welded specimen (mm).

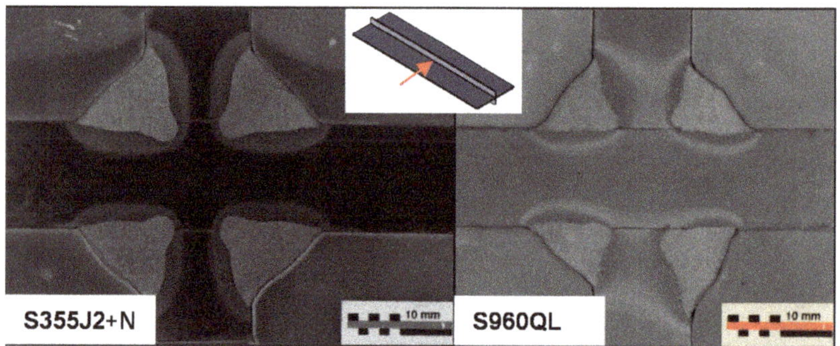

Figure 2. Macro graph of the weld cross sections.

Table 3. Process parameter for initial welding.

Parameter	Voltage (V)	Currency (A)	Heat Input (kJ/mm)	Welding Speed (mm/s)	Efficiency (-)	Wire Feed Speed (m/min)
S355J2+N	247	29.4	0.873	6.65	0.8	8.5
S960QL	216	29.4	1.016	5	0.8	9

Additionally, it should be mentioned that high-strength steels with yield strength of 960 MPa are susceptible to hydrogen-assisted cracking (HAC) during welding processing [25]. This should be considered in practicable applications but was not further investigated in this work.

In-situ temperature measurements were performed to determine the temperature-time-profile at the weld toe for all materials. For this, thermocouples type K with a wire diameter of 0.08 mm were used. The thermocouples were spot-welded in a distance of 0 mm, 0.5 mm, 1 mm, 2 mm, and 3 mm of the theoretical position weld toe. In total, six measurements with five thermocouples were performed in this work. Two measurements in a distance of 0.5 mm were chosen for the subsequent analyses of cooling and heating time. Figure 3 shows these temperature-time profiles.

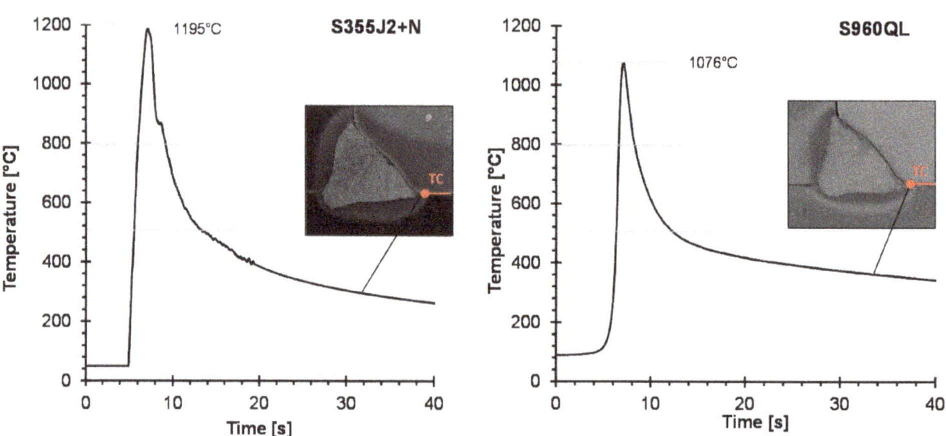

Figure 3. Temperature history from temperature measurement at the weld toe.

In practice, the cooling time represents an important parameter in joining of steels and enables the first verification of the welding process. To describe the cooling time, the so-called t85-time is used in fine-grained steels. This cooling time allows a qualitative statement on the microstructure of the HAZ. Besides of the performed in-situ measurement in this work, the t85-time can be calculated according to the German guidelines SEW 088 [26] and the standard DIN 1011 [27]. If the heat can dissipate from the molten pool on either side laterally in the parent material and in thickness direction the plate thickness is not negligible. In that case of 3D-dimensional heat flux, the t85-time is defined as

$$t_{85} = K_3 \, \eta \, E_1 \left(\frac{1}{500 - T_0} \right) - \left(\frac{1}{800 - T_0} \right) F_3 \qquad (1)$$

with

$$K_3 = \left(0.67 - 5 \times 10^{-4} T_0 \right) \qquad (2)$$

according to SEW 088 standard, where K_3 is the so-called pre-heating factor, T_0 is the initial temperature in °C, η is the thermal efficiency that is 0.8 for GMAW processes and F_3 is the heat dissipation factor that is 0.67 for fillet welds. Table 4 shows the comparison of t85-time from the measured temperature-time-profiles and the calculation according to Equation (1). The t85-time from the measurements was higher in both cases compared to the calculation.

Table 4. t85-time determined by temperature-measurement and analytical calculation.

t85-Time (s)	Measurement	SEW 088
S355J2+N	3.79	2.47
S960QL	4.21	3.61

3. Repair Procedure

The basis for the repair procedure is a publication by the Federal Highway Administration of the United States, derived by a workshop of experts under the supervision of Dexter and Ocel [3]. The main goal of this publication is the repair and retrofit of fatigue cracks in steel bridges. Besides the repair welding, bolted doubler plates and stop-hole-techniques are presented as possibility to retrofit structures. The repair procedure by welding uses several steps, illustrated in Figure 4: At first, the crack detection is necessary by visual testing (VT). The NDT-personnel should look for signs of rust and cracks in the coating surface. After that, penetrant testing (PT) or magnetic testing (MT) can be used to determine the specific length and orientation of the crack. Another possibility which is not clearly stated in the repair manual [3] is the use of ultrasonic testing (UT). The advantage is, that the depth of the crack can also be determined, as long as the crack path is only in one plate member. After the detection, the removal of the crack is performed by grinding or air arc gauging. Grinding can be used in general, whereas air arc gauging should only be used at thick plates. Between the removal passes, PT or MT should be repeated in order to follow the crack path and to obtain the information of the actual crack depth. In the case of the crack depth being below one half of the plate thickness, a one-sided repair should be made. Care should be taken by a final NDT, whether the crack has been completely removed. If this is not the case, the material removal should be made to a depth of three-quarter of the plate thickness. The first side welding should then be done according to the recommended specification of the base material manufacturer. In order to avoid a high magnitude of welding residual stresses, pre-heating in the vicinity of the repair weld is recommended. In case the necessary repair depth exceeds $0.5 \times t$, the repair also has to be done from the opposite side, again by grinding or air arc gouging. After performing the counterside welding, the weld toes should be grinded smooth to reduce the amount of sharp notches.

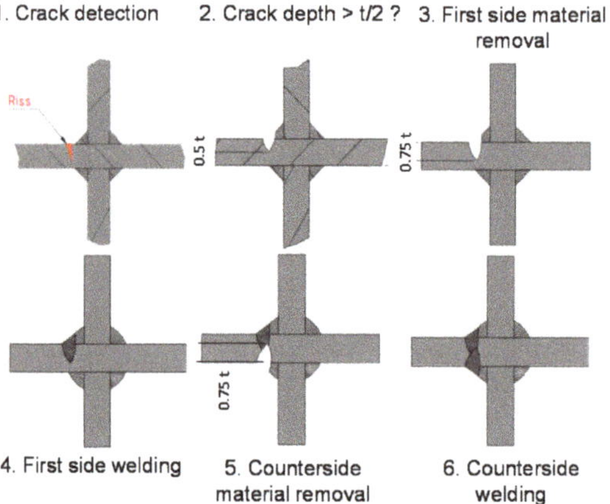

Figure 4. Principal procedure of the weld repair process.

Different crack depths are needed to be available for the two repair cases. For this reason, the specimens were loaded on a resonance-frequency machine RUMUL 150K (Russenberger Prüfmaschinen AG, Neuhausen am Rheinfall, Switzerland). A frequency decrease Δf was detected caused by propagating cracks associated with a stiffness decrease of the cyclic loaded specimen. A sinusoidal 4-point bending load was used in this work. Specimens with the width of 130 mm could be loaded without the crack growing through the complete specimen thickness and a comparably high weld length was available for the repair procedure. The test set-up is shown in Figure 5. Four specimens were used for a so-called beach-mark test to analyze the $\Delta f \sim$ a-correlation for different load levels. As illustrated in Figure 5, a shut-down criterion of $\Delta f = 0.2$ Hz was used for one-sided repair (a < t/2) and a shut-down criterion of $\Delta f = 1.2$ Hz was used for double-sided repairs (a > t/2). To assure that only cracks initiate at one specific weld toe, all other weld toes were treated by High Frequency Mechanical Impact (HFMI) treatment.

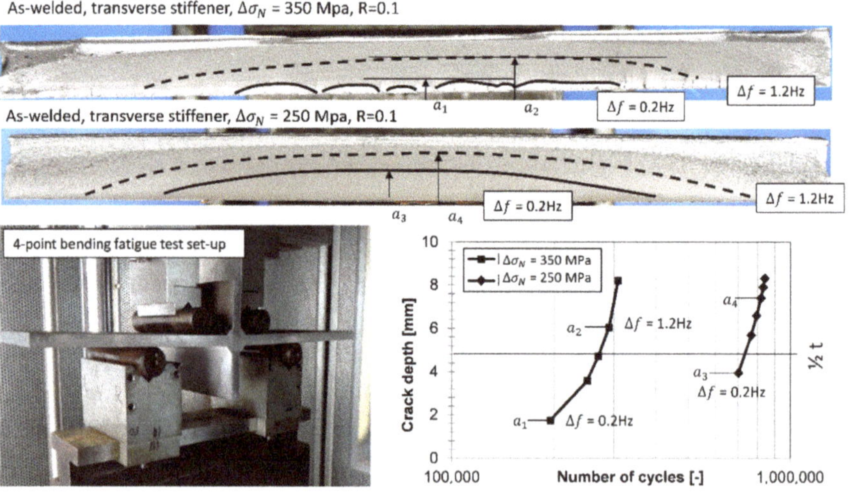

Figure 5. Fatigue test set-up and crack generation until a crack length of a < t/2 (one-sided repair) and a > t/2 (double-sided repair).

3.1. Non-Destructive Testing

The reliability of the crack detection with non-destructive testing methods is strongly dependent on the surface conditions in-situ. For this reason, the NDT-standards for MT (ISO 9934-1:2016 in conjunction with DIN EN ISO 17368:2019-05) and PT (DIN EN ISO 3452-1:2014-09 in conjunction with DIN EN ISO 23277:2015-06) give general guidelines on the necessary surface conditions for the inspection. The surface must be free from dirt, scale, rust, weld spatter, grease, oil, and other impurities and it has to be prepared in the way that relevant indications can be distinguished from false indications. In order to clarify the reason for the indication, it can be necessary to improve the surface condition by manual sanding or local grinding. For PT, precautions have to be taken not to smear surface defects by grinding. In addition, the examiner needs to consider deep grinding striations, which can lead to false indications. As the NDT-standards do not give a distinct threshold value for the surface roughness, it can be concluded, that the NDT-personnel needs to be as precise as possible in order to detect the defects. In Figure 6, an example is given on how dependent the indications are on the surface conditions.

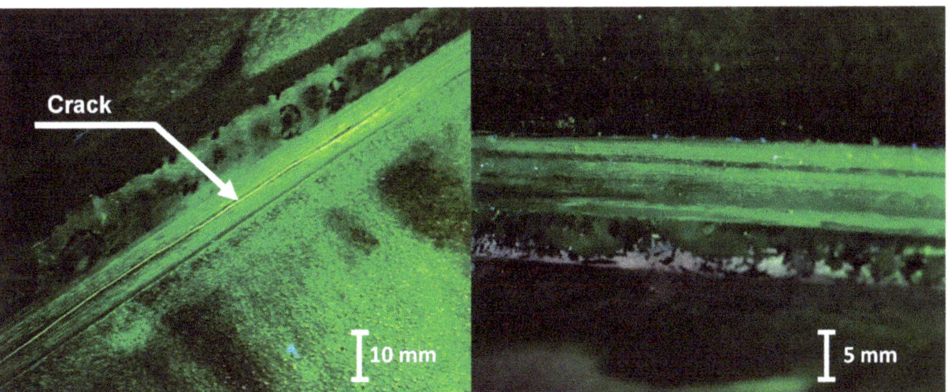

Figure 6. Detail of MT inspection.

On the left-hand side, a gently grinded U-groove is shown during an MT-inspection with fluorescent slurry with a clearly visible crack, which has not been fully removed. On the right-hand side, the striations introduced by strong grinding, preventing clear crack detection. A second gentle grinding pass would facilitate the interpretation.

All potentially cracked specimens were tested with PT after loading until the mentioned shut-down criterions were reached. In that way, it was assured that cracks really exists at the designated weld toe for the repair procedure and did not initiate at the other, HFMI-treated weld toes. Furthermore, these procedures proof, if such cracks are easily detectable in practical applications. However, in multiple cases, the PT lead to slightly visible or no visible cracks, illustrated in Figure 7. The percentage, were clearly or at least slightly detectable cracks were observed at the investigated specimen, are shown in Table 5. As displayed, this success rate is low for PT testing for specimen with crack depths < 0.5 t. However, PT was successfully applied for specimen with cracks depths > 0.5 t for all tested specimen. An increase of the exposure time from 5 min to the maximum of 60 min according to EN DIN ISO 3452-1:2013 leaded to no other results for PT. The specimen, where no crack was detectable with PT, MT was applied and showed for nearly all specimen clearly visible cracks. For this, is should be mentioned that all NDT methods were applied at non-loaded specimen. PT is most likely to deliver better results for specimen under mean stress or maximum stress because this leads to crack opening and to a better penetration and extraction of the PT-penetrant.

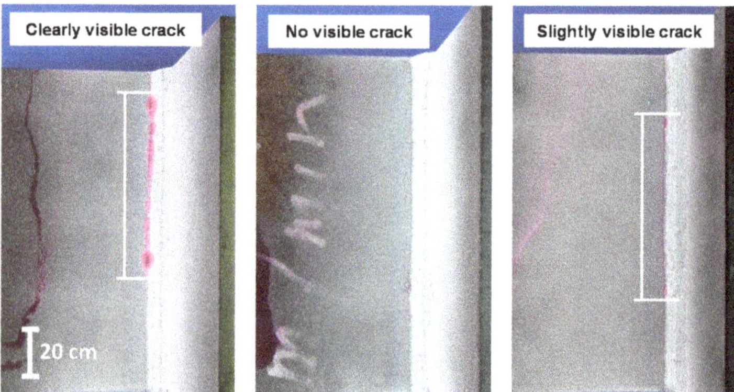

Figure 7. Detail of PT inspection.

Table 5. Percentage of detectable cracks for tested specimen.

Success Rate (Total Number of Tested Specimen)	S355J2+N		S960QL	
Crack depth	$>\frac{1}{2}$ t	$<\frac{1}{2}$ t	$>\frac{1}{2}$ t	$<\frac{1}{2}$ t
PT	5.2% (19)	100% (17)	14.2% (21)	100% (21)
MT	83% (12)	-	91% (12)	-

3.2. Repair Welding

The repair welding was performed by gas metal arc welding (GMAW). Before welding, the material on the complete length on the specimen was removed by manual angle grinding up to a certain depth, displayed in Figure 8. Additionally, run-off-tabs were spot-welded at both sides of the specimen to avoid run-out effects and to ensure quality class B according to ISO 5817:2014-6. Crack detection by MT was performed for every specimen before the actual repair welding process, as illustrated in Figures 6 and 8. In this way, it was assured that criterion for one side repairs (crack depth < t/2) was fulfilled. Furthermore, in a depth of 0.75 × t (both sided repair specimens) no cracks could be detected. The same filler material and preheat conditions are used as for the initial welding. The welding parameters are summarized in Table 6.

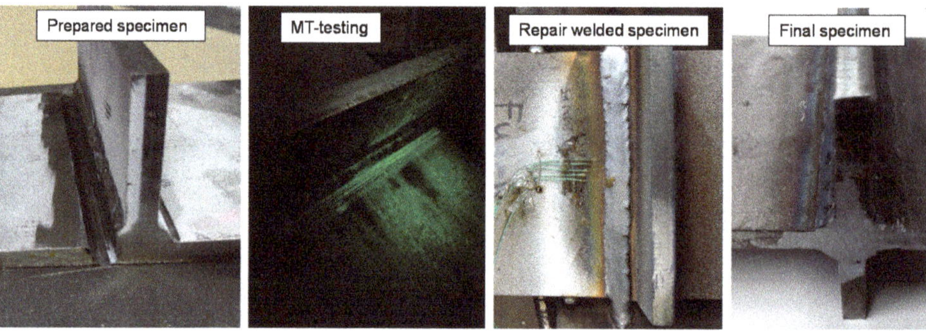

Figure 8. Repair procedure for investigated specimen.

Table 6. Process parameter for repair welding.

	Currency (A)	Voltage (V)	Energy per Length (J/mm)	Speed (mm/s)	Efficiency (-)	Wire Rate (m/min)
S355J2+N (1.pass)	212	42	1508	4.72	0.8	8.5
S355J2+N (2.pass)	222	42	1579	4.72	0.8	8.5
S960QL (1.pass)	215	42	1529	4.72	0.8	8.5
S960QL (2.pass)	220	42	1565	4.72	0.8	8.5

The cross sections of the different repair cases are summarized in Figure 9. As illustrated, the penetration of the molten zone was 1 mm and 2 mm deeper than the removed material. In such cases, it may be possible that short cracks, which have not been removed by disc grinding, are molten and therefore removed. Temperature measurements were also performed during the repair welding process. The results are shown in Figure 10. As displayed the t8/5-times are higher than for the initial welding process. This can be explained by the higher heat input and the smaller welded samples which leaded to less heat dissipation during the repair welding process. The t8/5 times are compared to the guideline SWE088. The results are summarized in Table 7.

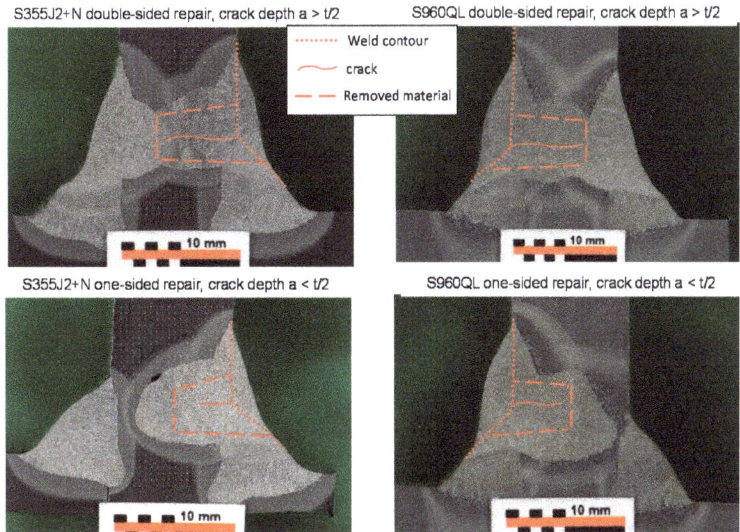

Figure 9. Cross section of the repaired specimen, illustrated with the original weld contour and repaired crack.

Table 7. t85-Time determined by temperature-measurement and analytical calculation.

t85-Time (s)	Measurement	SEW 088
S355J2+N (2.pass)	5.66 (0.87)	4.47
S960QL (2.pass)	6.81 (1.21)	5.94

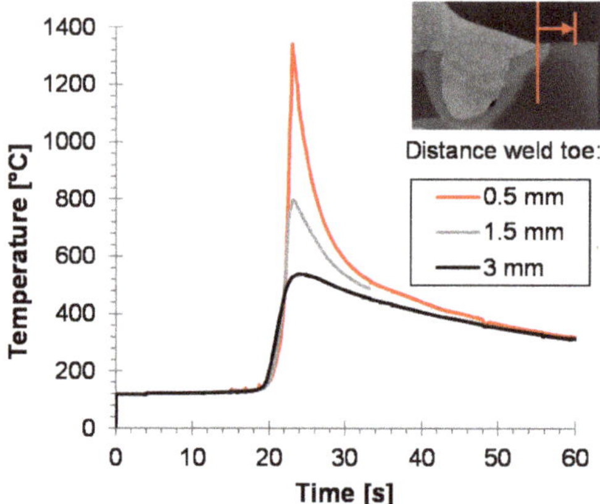

Figure 10. Temperature history from temperature measurement at the weld toe for the repair welding process.

4. Investigation of Weld Properties

4.1. Geometrical Properties

Recent studies showed a clear correlation between local, geometrical parameters of welded joints and their fatigue strength [28–30]. These investigations were transferred in the VOLVO quality standard that recommends specific values for weld toe radii for the first time. Besides this, recommendations on flank angles are given in standard ISO 5817:2014-06. Thus, the geometrical parameters of the weld flank angle θ_t and the weld toe radius ρ according to the definition of [31] were analyzed as part of study [32]. All measurements were performed with a 3D-laser scanner and evaluated according to the so-called curvature-method [33]. Furthermore, the angle of distortion was evaluated for every specimen. Mean values and Gaussian-standard deviations are given in Table 8. As shown, similar values of ρ were evaluated for every type of the investigated welded joints. However, the values of θ_t are significantly lower for the repaired (RP) conditions than for the as-welded (AW) conditions. The values of β are similar for every investigated welded joint, but slightly higher for the RP-condition. However, compared to ISO 5817:2014-6 the angles of distortion are less than $\beta = 1°$.

Table 8. Mean value (μ) and standard deviation (σ) for the geometrical parameters of the investigated welds.

System	Weld toe Radius (mm) [32]		Flank Angle (°) [32]		Angle of Distorsion (°)	
Weld	μ	σ	μ	σ	μ	σ
S355J2+N (AW)	0.915	0.356	39.93	11.28	0.12	0.051
S960QL (AW)	1.286	0.689	45.89	7.098	0.14	0.041
S355J2+N (RP)	0.802	0.209	18.54	7.170	0.82	0.125
S960QL (RP)	0.879	0.134	19.08	5.212	0.86	0.114

The assessment of the local, geometrical parameters was mainly motivated by the calculation of the stress concentration factors (SCFs) of the investigated welded joints. For the SCF calculation an approximation formulae proposed by [34] for fillet welds under tensile load and under bending load were used

$$\mathrm{SCF}\left(\frac{t}{\rho}, \theta_t, b, s\right) = m_0 + \left(1 + m_1\left(\frac{b}{\rho}\right)^{P_1}\left(\frac{s}{t}\right)^{P_2} + m_2\left(\frac{t}{\rho}\right)^{P_3} + m_3\sin(\theta_t)^{P_4}\right)\sin(\theta_t)^{P_5}\left(\frac{t}{\rho}\right)^{P_6} \tag{3}$$

where the SCF depends additionally on the throat thickness b and the root gap s, with parameters for bending load according to the Table 9.

Table 9. Geometrical parameters of investigated welds.

Load	m_0	m_1	m_2	m_3	p_1	p_3	p_4	p_5	p_6
tension	1.538	0.621	1.455	−2.933	−1.655	0.208	1.213	2.086	0.207
bending	1.256	0.023	2.153	−3.738	−3.090	0.154	0.481	1.723	0.172

Furthermore, the SCFs were determined by direct evaluation of the ratio of the highest principle stress and nominal stress determined by 2D-Finite Element calculation according to [33]. For this, around 3000 2D-profiles were evaluated from 3D-scans and transferred into a 2D-mesh with the commercial software package HYPERMESH. Then, all profiles were automatically calculated with the ABAQUS Standard Solver. The results are summarized in Table 10. As shown, the SCFs are between 20% and 30% higher in as-welded condition than in repaired condition dependent on the evaluation method. Because of similar weld toe radii for AW and RP-condition, this difference is mainly attributed to a significantly smaller flank angle.

Table 10. Stress concentration factors of investigated welds [32].

Method	Approximation [34]		2D-FEM [32]	
Weld type	μ	σ	μ	σ
S355J2+N (AW)	2.19	0.218	1.921	0.170
S960QL (AW)	2.28	0.164	2.035	0.112
S355J2+N (RP)	1.82	0.161	1.543	0.173
S960QL (RP)	1.71	0.099	1.492	0.113

4.2. Hardness

The toughness of the heat affected zone (HAZ) is also an important factor in order to secure that neither a brittle fracture under load nor cold cracks occur [35]. For that, the maximum hardness of the HAZ should be 320 HV, 380 HV and for some special applications at a maximum of 450 HV depending on the material class. These values are also established in the standards DIN EN ISO 15614-1, API Standard 1104, and CSA Z662-07. In the case of the investigated S355J2+N steel, a maximum hardness of 320 HV10 is allowed. According to DIN EN ISO 15614-1 the maximum hardness for fine grain steels with yield strength > 890 MPa needs to be determined separately. According to [36] maximum hardness of around 433 HV could be reached for S960QL. Figure 11 displays the cross section and the hardness distribution of the weld details from each base material. All hardness measurements were performed according to DIN EN ISO 6507-1:2018-07. The hardness scale for the measurements was HV1 with an indentation point distance of 0.5 mm. The size of the indentations is less than 50 μm. The indentation distance fulfills the requirement according to ISO 6507-1:2018-07.

The HAZ of the initial weld of S355J2+N shows a hardness increase of 255 HV1 (6.2 HV1) to 275 HV1 (6.6 HV1) compared to the BM of 196 HV1 (4.1 HV1). The hardness in brackets corresponds to the standard deviation. In the case of S960QL, the hardness of the HAZ was 396 HV1 (6.9 HV1) to 420 HV1 (7.5 HV1) compared to the BM of 342 HV1 (4.1 HV1). A very similar maximum hardness with a comparable low standard deviation was measured in the HAZs in the repaired condition. These values fulfill the general requirements.

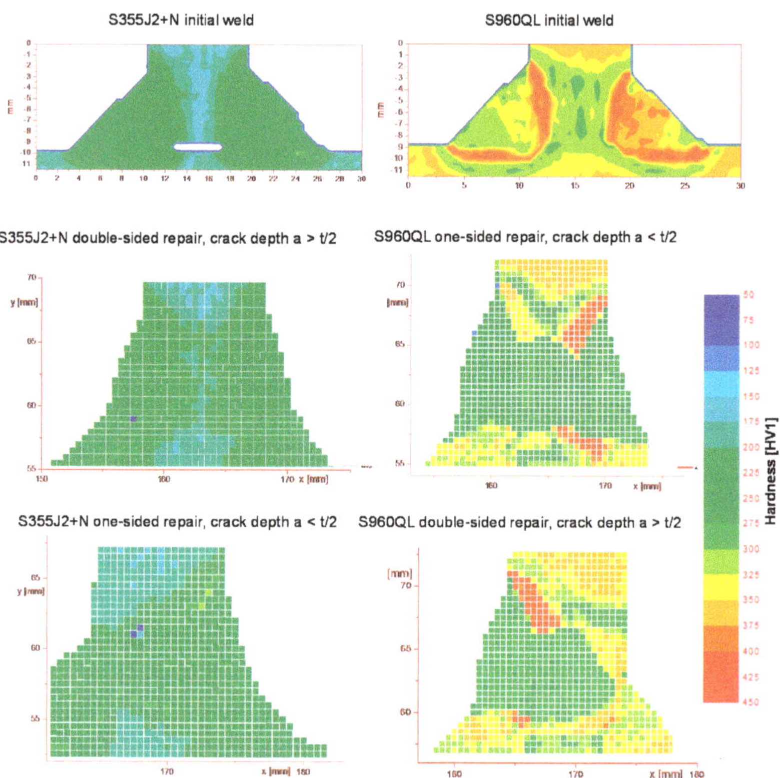

Figure 11. Hardness mappings for the investigated welded joints.

4.3. Residual Stress

It is strongly assumed that residual stresses caused by shrinking- and phase transformation effects during heating and cooling have a high influence on the fatigue behavior of welded joints. Usually, these residual stresses are handled as mean stresses in the current guidelines [23,37] or are not taken further into account for design recommendations [22]. However, low or medium residual stresses could be expected for the here-investigated, unconstrained and comparably short-width specimen according to the FKM-guideline [23] or IIW-recommendation [21]. However, this low or medium residual stress level leads to higher mean stress sensitivity and is covered by different enhancement factors. Thus, the residual stresses at the most-critical location, the weld toe was investigated experimentally for the specimen in the as-welded and the repaired condition.

The residual stresses were measured with X-ray diffraction techniques at the {211}-lattice plane with a Ca-Kr radiation. The collimator diameter for the measurement was 1 mm. For measurement preparation, the first 0.3 mm of the surface layer was removed electro-chemically to avoid measuring of overlaid compressive residual stress induced by blast-cleaning of the metal sheets before welding. The stresses in the transverse and longitudinal directions were evaluated by the $\sin(\psi)^2$-method assuming an even stress state at the surface layer.

Figure 12 illustrates the residual stress distribution from the weld toe to the base material for the as-welded (AW) and the repaired condition (RP) for both investigated materials. Furthermore, the residual stress redistribution by the separation of single specimen from the base plate was investigated. As displayed, the residual stresses in longitudinal direction are significantly higher than in transverse direction. The transverse residual stresses close to the weld toe range from 0 MPa to 100 MPa for all investigated conditions. It is strongly assumed that negative residual stress values at a distance of >3 mm from the weld toe are related to pre-welded abrasive blast cleaning or rolling processes. Furthermore, it is shown that a higher heat input of the repair-welding process does not lead to a significantly higher residual stress level than for the original welding process.

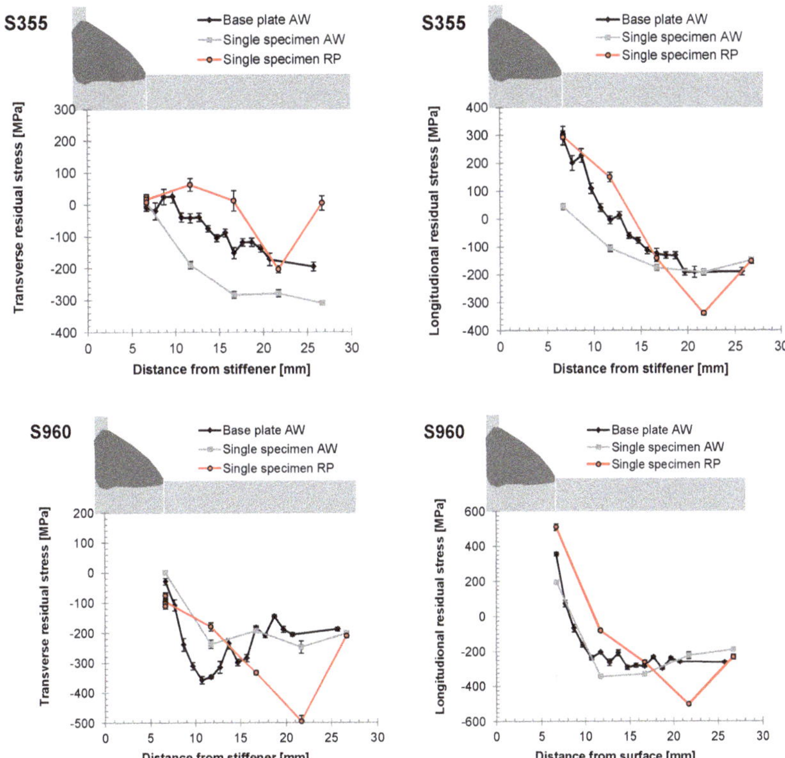

Figure 12. Residual stress distribution for initial welded joint and repaired welded joints.

5. Fatigue Analysis

The main focus of this work was the comparison of the fatigue performance of the initial-welded transverse stiffener (AW) with the repaired condition (RP). To cover both repair cases two shut down -criterions (frequency decrease) were used. $\Delta f = 0.2$ Hz for a crack depth of a < t/2 (one-sided-repair) and $\Delta f = 1.2$ Hz for crack depth a > t/2 (double-sided repair), as mentioned before. The corresponding fatigue results are displayed as S/N-diagramm in Figure 13 for the S355J2+N base material and in Figure 11 for the S960QL base material. No root cracks were observed at any of the investigated specimen. A clear tendency of higher fatigue life in RP-condition compared to the AW-condition is shown for both materials, but especially for the specimens made of S355J2+N. Furthermore, the fatigue test data is given in Tables 11 and 12. Specimens which did not fail from the repaired weld toe are included in this data but excluded from further evaluation. Also, run-out specimens were not included for the evaluation.

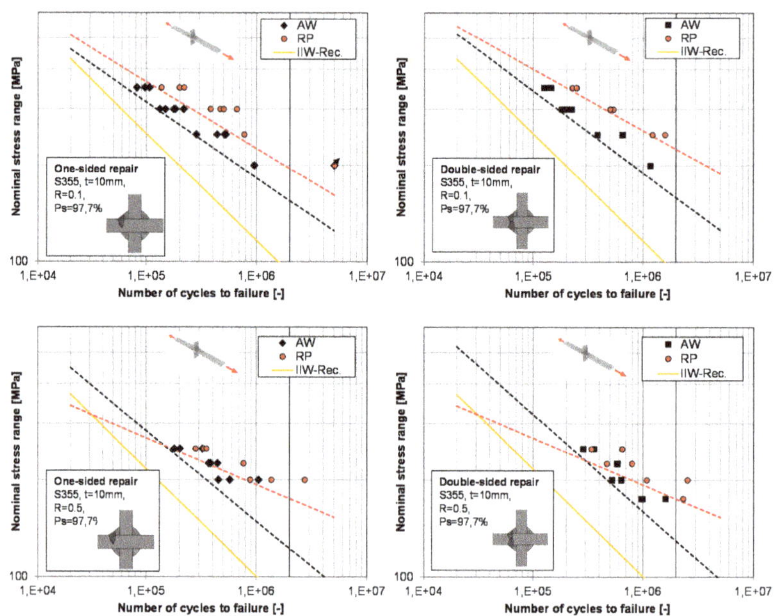

Figure 13. Fatigue test results for S355J2+N specimen for as-welded- (AW) and repaired (RP)-condition.

Table 11. Fatigue test data for S355J2+N specimen.

	R = 0.1					R = 0.5			
ID	$\Delta\sigma_N$ (MPa)	Δf (Hz)	N (-) AW	N (-) RP	ID	$\Delta\sigma_N$ (MPa)	Δf (Hz)	N (-) AW	N (-) RP
33	300	0.2	217,200	661,100	27	250	0.2	202,700	351,500
31	300	0.2	177,100	466,200	4	200	0.2	573,800	883,300
46	350	0.2	82,000	219,100	21	250	0.2	180,400	331,400
45	350	0.2	157,900	199,300	44 *	250	0.2	323,000	280,200
43	200	0.2	1,248,800	-	37 *	200	0.2	1,053,900	1,375,900
39	250	0.2	510,100	-	14	200	0.2	451,700	2,724,700
11	350	0.2	113,600	136,500	34	225	0.2	443,600	766,200
12	300	0.2	145,000	500,100	35	225	0.2	382,900	-
2	250	0.2	436,500	772,300	42	225	0.2	369,000	-
22	350	0.2	117,400	156,700	30	250	0.2	209,000	255,700
22	350	1	145,200	69,200	30	250	1.0	285,500	83,800
7	350	0.2	106,700	199,100	23	200	0.2	477,400	689,100
7	350	1	139,500	49,500	23	200	1.0	636,600	392,500
25	350	0.2	97,000	169,100	6 *	250	0.2	271,000	484,200
25	350	1	125,100	76,300	6 *	250	1.0	358,100	164,600
28	300	0.2	148,500	407,400	13	225	0.2	458,300	615,700
28	300	1	203,100	100,000	13	225	1.0	591,300	178,500
26	300	0.2	132,400	329,100	20 *	225	0.2	420,400	292,900
26	300	1	181,200	170,900	20 *	225	1.0	582,900	176,400
18	300	0.2	182,900	380,400	9	200	0.2	376,200	1,913,100
18	300	1	223,200	156,000	9	200	1.0	524,800	627,800
3	200	0.2	955,000	5,000,000	17	175	0.2	686,000	-
3	200	1	1,169,500	-	17	175	1.0	970,700	-
36	250	0.2	524,600	1,170,900	19	175	0.2	1,245,700	1,963,300
36	250	1	652,600	427,600	19	175	1.0	1,614,700	360,100
15	250	0.2	282,200	79,400	10	175	0.2	728,600	-
15	250	1	652,600	290,100	10	175	1.0	995,500	-

* failure from the HFMI-treated weld toe in RP condition.

Table 12. Fatigue test data for S960QL specimen.

	R = 0.1					R = 0.5			
ID	$\Delta\sigma_N$ (MPa)	Δf (Hz)	N (-) AW	N (-) RP	ID	$\Delta\sigma_N$ (MPa)	Δf (Hz)	N (-) AW	N (-) RP
85	500	0.2	84,100	37,400	48 *	300	0.2	199,400	124,800
61	400	0.2	192,700	71,000	43	250	0.2	179,600	290,200
22	250	0.2	2,000,000	598,200	44	200	0.2	585,500	754,800
45	400	0.2	-	80,700	23 *	300	0.2	147,400	104,000
4	500	0.2	30,500	45,200	10	250	0.2	223,500	585,600
6	400	0.2	214,800	111,300	11	200	0.2	360,800	615,800
17	250	0.2	5,000,000	798,600	30	300	0.2	97,200	391,000
12	400	0.2		156,000	15	250	0.2	477,200	255,000
35	300	0.2	828,200	263,300	16	200	0.2	544,700	1,373,300
24	500	0.2	51,200	39,700	13	175	0.2	5,000,000	-
33	400	0.2	61,100	136,900	13	300	0.2	148,200	
12	300	0.2	980,000	374,800	19	300	0.2	98,900	353,900
8	500	0.2	30,900	50,000	19	300	1.0	138,800	69,800
25	500	1.0	44,800	9,700	34	250	0.2	538,600	355,000
25	400	0.2	170,800	114,800	34	250	1.0	715,700	161,100
28	400	1.0	201,100	30,000	20	200	0.2	649,100	2,040,800
26	300	0.2	817,800	800,000	20	200	1.0	800,800	428,300
26	300	1.0	880,800	82,500	36	300	0.2	233,500	145,400
7	500	0.2	26,900	42,700	36	300	1.0	301,100	42,700
18	500	1.0	38,300	12,500	5 *	250	0.2	286,800	167,300
21	400	0.2	53,700	97,600	5 *	250	1.0	413,100	78,400
3	400	1.0	78,600	41,000	29	200	0.2	735,000	5,000,000
9	300	0.2	273,900	802,200	29	200	1.0	1,077,200	-
36	300	1.0	368,000	135,300	40	300	0.2	239,700	394,400
31	500	0.2	23,500	91,000	40	300	1.0	289,600	79,700
15	500	1.0	34,200	7,500	37	250	0.2	262,800	1,041,100
39	400	0.2	126,700	120,900	37	250	1.0	339,400	193,000
39	400	1.0	156,700	23,500	18	200	0.2	673,800	2,051,600
38	300	0.2	204,000	-	18	200	1.0	858,000	247,800
38	300	1.0	226,800	-	28	300	0.2	119,600	90,500
32	500	0.2	60,000	92,300	28	300	1.0	162,000	28,400
32	500	1.0	75,600	17,000	42	250	0.2	230,700	5,000,000
41	400	0.2	243,300	-	42	250	1.0	315,800	-
41	400	1.0	272,000	-					

* failure from the HFMI-treated weld toe in RP condition.

The evaluation of the fatigue test results was performed according to DIN EN 50100-2016 to assess the FAT values of each welded joint that corresponds to the nominal stress range at 2×10^6 cycles and a survival probability (P_s) of 97.7% according to the IIW-recommendation. The corresponding S/N-curves are shown in Figures 13 and 14. According to Eurocode 3 [22] and IIW-Recommendation [21] a basic FAT class of 80 was proposed for the here-investigated case of a transverse stiffener without further post-weld treatment. However, in this case, an enhancement factor of f(R) = −0.4R + 1.2 according to [21] was taken into account that takes into account the comparably low residual stress level of 0.2× base material yield. For a stress ratio of R = 0.1, this leads to a FAT value of 93. The design S/N-curves evaluated with variable slope k are also displayed in Figures 13 and 14. As illustrated, in all cases, the evaluated FAT-classes lie higher than in the recommendations.

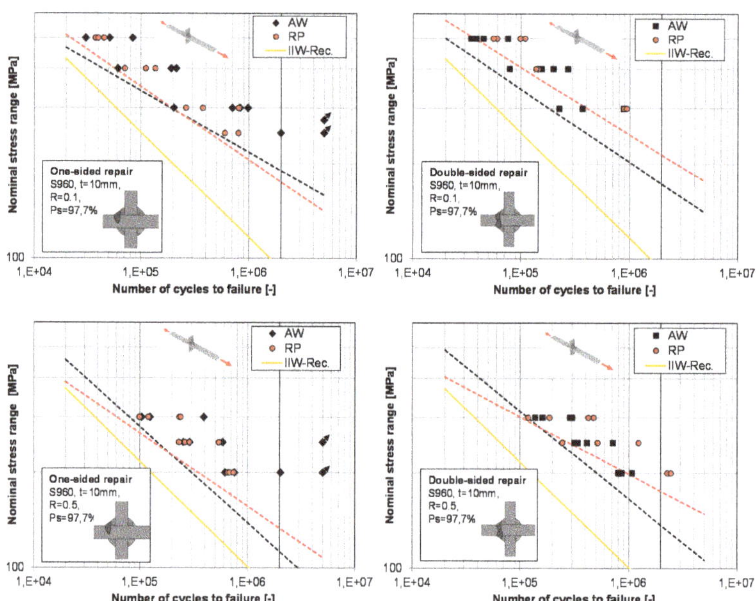

Figure 14. Fatigue test results for S960QL specimen for as-welded- (AW) and repaired (RP)-condition.

Furthermore, the evaluation of the FAT-classes was also performed with a fixed slope of k = 3 according to the design codes [10,11]. In this case, the FAT-classes for AW and RP conditions show lower differences of −3% to 15%. A higher difference of 25% was only observed for the base material S355J2+N and specimens loaded with R = 0.1. The comparison for both evaluations is illustrated in Figure 15.

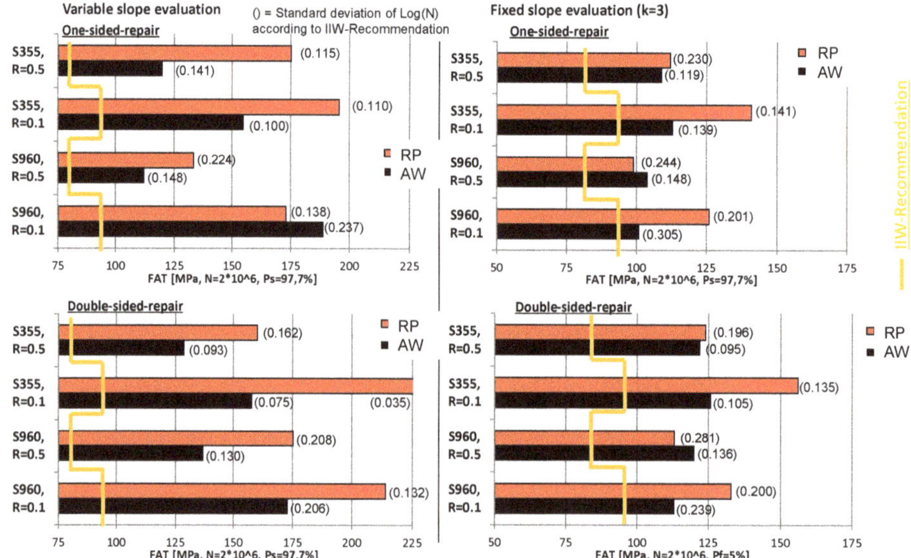

Figure 15. Evaluation of fatigue strength with variable and fixed slope.

Additionally, the fatigue analysis was performed according to the effective notch stress approach based on the micro support effect of Neuber [38] and interpreted on the basis of Radaj [39]. The effective notch stress was determined by Finite Element Analysis (FEA). For modelling a 2D-halfsymmetric model with a fixed weld toe radius of $\rho = 1$ mm was used, shown in Figure 16. Mesh size was chosen according to the recommendation of the German Welding Society [40]. The notch radius at the weld toe was meshed with three CPS8 elements (ABAQUS software package) with quadratic shape functions and a minimum size of 0.25 mm. The measured mean flank angle α, which are presented in Table 6, were used for modelling. A maximum angle of distortion of $\beta = 1°$ was taken into account for the calculation of the repaired case. This represents the maximum measured distortion angle, see Table 6.

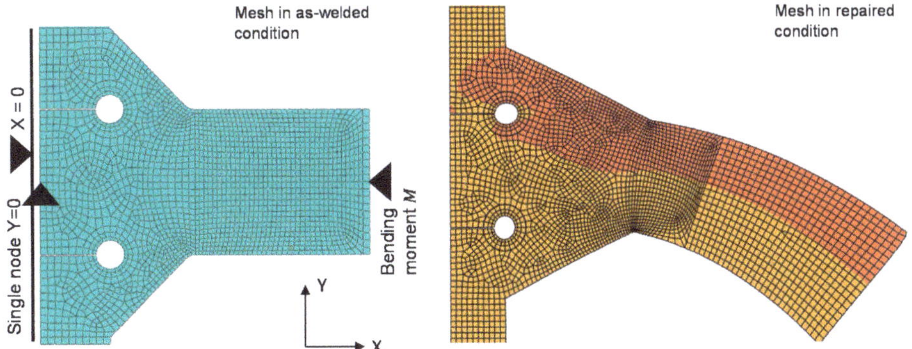

Figure 16. FE-mesh for the fatigue assessment according to the effective notch stress approach.

The FAT values according to the effective notch stress approach are given in Table 13. For this case, the IIW-recommendation recommends a value of FAT = 225 MPa. As shown, the determined FAT values are around 17% to 85% higher for R = 0.1 and around 5% to 51% higher for R = 0.5.

Table 13. FAT-classes determined according to the effective notch stress approach [39].

	S355J2+N		S960QL		IIW-Rec. [21]
	R = 0.1	R = 0.5	R = 0.1	R = 0.5	-
GZ	263	251	345	238	
RZ	365	326	399	250	225
RZ *	380	340	417	262	

* Taken a angle of distortion of 1° into account.

6. Discussion

The majority of the investigated specimen reaches at least their original life span in repaired condition. Similar results were published for similar fillet welds [13], butt joints [19] and complicated weld details like beams with gussets or weld repairs by cover plates [20] made of mild steel similar to S355 with a sheet thickness of 10 mm. A significantly higher fatigue life could be reached with a combination of weld repair and the TIG post weld treatment method [20].

The tendency of higher scatter of the fatigue tests results is shown in repaired condition, cp. standard deviation displayed in Figure 13. It is assumed that this is based on the manual repair-welding compared to the automatic original welding process. However, this trend was not observed in every test series.

The evaluated fatigue test values are significantly higher than the corresponding FAT-class of 80 MPa mentioned in the IIW recommendation [10,28]. Moreover, higher fatigue strengths in both investigated conditions are reached than in the investigation [13]. Similar transverse stiffener of S355 structural steel at R = 0.1 were tested in the AW condition and show a FAT-value of 98 MPa [41,42] and 93 MPa [15]. However, in that case, the specimens were tested under tensile load. It is assumed, that the comparably high fatigue strength in the as welded and repaired condition is based on the bending load type. Fatigue tests of identical welded joints under tensile loading confirm this assumption [43].

Similar investigations at transverse stiffeners made of S355 with the effective notch stress approach evaluate a FAT-class of 233 MPa in original condition and of 290 MPa in repaired condition [15]. It is assumed that the higher FAT-class of the here-investigated specimens is also based on the bending load type.

The residual stress level at the weld toe was low for all investigated conditions. Significant higher residual stresses are reported for GMAW welded butt joints made of S355J2 and S960QL [44]. However, no significant transverse residual stress in fillet welds was determined for SAE 1020 steel [45]. Therefore, a comparably low influence of the residual stress state on the fatigue behavior was assumed for the investigated welded joints.

A significant lower SCF was determined for the investigated welded joints in repaired condition caused by significant smaller flank angles [32]. Investigations at T-joints made of S355 show a high influence of the local weld geometry on the fatigue performance [28]. Furthermore, analytical formulae [34] for the determination of the SCF of fillet welds show a significant influence in case of constant weld toe radius. This leads to the assumption that the flank angle is the main factor for higher fatigue life of the repaired specimen in the investigated cases.

7. Conclusions

The aim of this work was to develop a comparably simple and reliable approach for the retrofit of fillet welds. For this, fatigue tests of transverse stiffeners made of S355J2+N and S960QL under four-point bending load under stress ratio of R = 0.1 and R = 0.5 were carried out. A crack depth of a < t/2 and of a > t/2 was controlled by a corresponding decrease of the machine test frequency. For a < t/2 an one-sided repair and for a > t/2 a repair procedure for both sides was performed. The repair contains non-destructive testing (NDT), removal of the material up to 0.75 t and a subsequent two-layer repair welding process followed by another NDT testing to assure that the specimen was crack-free. Furthermore, residual stress state, hardness, and weld geometry was investigated for the initial welded (AW) and repaired (RP) condition. The following conclusions could be made:

- No tensile residual stresses were determined at the weld toe in transverse direction for all investigated conditions.
- Hardness and microstructure are quite similar for AW and RP condition.
- The SCF is significantly lower for RP condition. This is related to a smaller flank angle of α = 40–45° compared to AW condition of α = 18–19° even if the angle of distortion is higher for RP condition.
- The majority of the repaired specimen (RP) reaches at least the fatigue life span in original condition (AW).
- All evaluated FAT values at R = 0.1 are higher for RP condition (for a fixed slope k = 3). For R = 0.5 higher FAT values in RP condition could be reached for S355 and slightly lower FAT values are reached for S960 (k = 3).
- In all cases, at least the FAT 100 according to the IIW-recommendation was reached.

According to these results, a life span for retrofitted fillet welds, repaired according to this approach, of at least the life span of the weld in original condition could be expected.

Author Contributions: Conceptualization, J.S.; methodology, P.L.; investigation, J.S. and P.L.; original draft preparation, J.S., P.L. and A.S.; writing—review and editing, P.K. and M.F.; project administration, P.K. and M.F.; All authors have read and agreed to the published version of the manuscript.

Funding: Financial support for this project was provided by IGF (German Federation of Industrial Research Associations) within the project IGF 18988 N.

Conflicts of Interest: The authors declare no conflict of interest and the funders had no role in the design of the study; in the collection, analyses, or interpretation of data; in the writing of the manuscript, or in the decision to publish the results.

References

1. Ladendorf, P.; Knödel, P.; Ummenhofer, T.; Schubnell, J.; Farajian, M. "Retrofit Engineering": Entwickeln und Validieren einer Prozedur zur schweißtechnischen Instandsetzung von Großbauteilen. In Proceedings of the Congress of German Welding Society, Düsseldorf, Germany, 7–8 May 2019.
2. über Die Qualität, B. *Dauerhaftigkeit und Sicherheit von Spannbetonbrücken*; Bau und Stadtentwicklung des Deutschen Bundestages: Berlin, Germany, 2006.
3. Dexter, R.J.; Ocel, J.M. *Manual for Repair and Retrofit of Fatigue Cracks in Steel Bridges, FHWA-IF-13-020*, 2nd ed.; University of Minnesota: Minneapolis, MN, USA, 2013.
4. Mikki, C. *Retrofitting Engineering for Fatigue Damaged Steel Structures*; International Institute of Welding: Paris, France, 2009.
5. Miki, C.; Hanji, T.; Tokunaga, K. Weld Repair for Fatigue-Cracked Joints in Steel Bridges by Applying Low Temperature Transformation Welding Wire. *Weld. World* **2012**, *56*, 40–50. [CrossRef]
6. Cruz, P.J.S.; Frangopol, D.M.; Neves, L.C. Bridge Maintenance, Safety, Management, Life-Cycle Performance and Cost. In Proceedings of the Third International Conference on Bridge Maintenance, Safety and Management, Porto, Portugal, 16–19 July 2006; Taylor&Francis: Porto, Portugal, 2006.
7. Al-Salih, H.; Bennett, C.; Matamoros, A.; Collins, W.; Li, J. Repairing Distortion-Induced Fatigue in Steel Bridges Using a CFRP-Steel Retrofit. In Proceedings of the Structures Congress 2020—Selected Papers from the Structures Congress 2020, St. Louis, MO, USA, 5–8 April 2020; American Society of Civil Engineers (ASCE): Reston, VA, USA, 2020; pp. 273–284.
8. Siwowski, T.; Kulpa, M.; Janas, L. Remaining Fatigue Life Prediction of Welded Details in an Orthotropic Steel Bridge Deck. *J. Bridg. Eng.* **2019**, *24*, 05019013. [CrossRef]
9. Alencar, G.; de Jesus, A.; da Silva, J.G.S.; Calçada, R. Fatigue cracking of welded railway bridges: A review. *Eng. Fail. Anal.* **2019**, *104*, 154–176. [CrossRef]
10. Teixeira de Freitas, S.; Kolstein, H.; Bijlaard, F. Fatigue Assessment of Full-Scale Retrofitted Orthotropic Bridge Decks. *J. Bridg. Eng.* **2017**, *22*, 04017092. [CrossRef]
11. Yu, Q.Q.; Wu, Y.F. Fatigue retrofitting of cracked steel beams with CFRP laminates. *Compos. Struct.* **2018**, *192*, 232–244. [CrossRef]
12. Walbridge, S.; Liu, Y. Fatigue Design, Assessment, and Retrofit of Bridges. *J. Bridg. Eng.* **2018**, *23*, 02018001. [CrossRef]
13. Wylde, J.G. *The Fatigue Performance of Repaired Fillet Welds*; Welding Institute: Cambridge, UK, 1983.
14. Pasternak, H.; Chwastek, A. Zur Entwicklung eines Bemessungskonzeptes für die Lebensdauer von Reperaturschweißungen, Teil 1—Stand der Technik und Versuche. *Bauingenieur* **2015**, *90*, 47–54.
15. Pasternak, H.; Chwastek, A. Zur Entwicklung eines Bemessungskonzeptes für die Lebensdauer von Reperaturschweißungen, Teil 2—Ermittlung der Schwingfestigkeitsklassen. *Bauingenieur* **2015**, *90*, 272–277.
16. Pasternak, H.; Chwastek, A. A current Problem for Steel Bridges: Fatigue Assesment of Seams Repair. *Int. J. Civ. Environ. Eng.* **2016**, *10*, 1269–1276.
17. Ummenhofer, T.; Weidner, P.; Zinke, T.; Mehdianpour, M.; Rogge, A. *Schlussbericht P864: Fertigungs- und Instandhaltungsoptimierung bei Tragstrukturen von Offshore-Windenergie-Anlagen/Optimization of Fabrication and Maintance for Supporting Structures of Offshore Wind Energy Plants*; Verlag und Vertriebsgesellschaft GmbH: Düsseldorf, Germany, 2016; ISBN 9783942541992.
18. Yamada, K.; Sakai, Y.; Kondo, A.; Kikuchi, Y. Weld repair of cracked beams and residual fatigue life. *Proc. JSCE* **1986**, *1986*, 373–382. [CrossRef]
19. Kelly, B.A. *Fatigue Performance of Repair Welds*; Lehigh University: Bethlehem, PA, USA, 1997.
20. Miki, C.; Takenouchi, H.; Mori, T.; Ohkawa, S. Repair of fatigue damage in cross bracing connections in steel girder bridges. *Doboku Gakkai Ronbunshu* **1989**, *1989*, 53–61. [CrossRef]
21. Hobbacher, A.F. *Recommendations for Fatigue Design of Welded Joints and Components*, 2nd ed.; Mayer, C., Ed.; Springer: Berlin/Heidelberg, Germany, 2016; ISBN 78-3-319-23757-2.
22. Europeon Commitee of Standardization. *Eurocode 3: Design of Steel Structures—Part 1–9: Fatigue, 1993-1-9:2005*; Europeon Commitee of Standardization: Brussels, Belgium, 2009.
23. Rennert, R.; Kullig, E.; Vormwald, M.; Esderts, A.; Siegele, D. *FKM-Richtlinie: Rechnerischer Festigkeitsnachweis für Maschinenbauteile*, 6th ed.; VDMA Verlag: Frankfurt, Germany, 2012; ISBN 3816306055. (In German)
24. SEW 088 Guidline. *Supplementary Sheet 1 to SEW 088: Weldable Fine Grain Steels: Guidelnes for Processing, Particulary for Fusion Wleding, Cold Cracking During Welding, Determining Appropriate Minimum Preheating*; DIN German Institute for Standartization: Berlin, Germany, 2017.

25. Schaupp, T.; Ernst, W.; Spindler, H.; Kannengiesser, T. Hydrogen-assisted cracking of GMA welded 960 MPa grade high-strength steels. *Int. J. Hydrog. Energy* **2020**, *45*, 20080–20093. [CrossRef]

26. SEW 088 Guideline. *Supplementary Sheet 2 to SEW 088: Weldable Fine Grain Steels: Guidline for Processing, Particulary for Fusion Welding, Determine the Cooling Time t8/5 for the Indentificatrion of Welding Thermal Cycles*; DIN German Institute for Standartization: Berlin, Germany, 2017.

27. DIN EN 1011-2. *Welding-Recommendations for Welding of Metallic Materials—Part 2: Arc Welding of Ferritic Steels*; DIN German Institute for Standartization: Berlin, Germany, 2002.

28. Barsoum, Z.; Jonsson, B. Fatigue Assesment and LEFM Analysis of Cruciform Joints Fabricated with Different Welding Processes. *Weld. World* **2007**, *52*, 93–105. [CrossRef]

29. Jonsson, B.; Samuelsson, J.; Marquis, G.B. Development of Weld Quality Criteria Based on Fatigue Performance. *Weld. World* **2011**, *55*, 79–88. [CrossRef]

30. Nykänen, T.; Marquis, G.B.; Björk, T. A simplified fatigue assessment method for high quality welded cruciform joints. *Int. J. Fatigue* **2009**, *31*, 79–87. [CrossRef]

31. Berge, S. On the effect of plate thickness in fatigue of welds. *Eng. Fract. Mech.* **1985**, *21*, 423–435. [CrossRef]

32. Schubnell, J.; Jung, M.; Le, C.H.; Farajian, M.; Braun, M.; Ehlers, S.; Fricke, W.; Garcia, M.; Nussbaumer, A.; Baumgartner, J. Influence of the optical measurement technique and evaluation approach on the determination of local weld geometry parameters for different weld types. *Weld. World* **2020**, *64*, 301–316. [CrossRef]

33. Jung, M. Entwicklung und Implementierung Eines Algorithmus zur Approximation und Bewertung von Kerbfaktoren an Kehlnähten auf Basis Berührungsloser 3D-Vermessung. Master's Thesis, Karlsruhe Institut of Technology, Karlsruhe, Germany, 2018. (In German).

34. Anthes, R.; Köttgen, V.; Seeger, T. Kerbformzahlen von Stumpfstößen und Doppel-T-Stößen. *Schweißen Schneid.* **1993**, *45*, 685–688.

35. Tomków, J.; Janeczek, A. Underwater In Situ Local Heat Treatment by Additional Stitches for Improving the Weldability of Steel. *Appl. Sci.* **2020**, *10*, 1823. [CrossRef]

36. Seyffarth, P.; Meyer, B.; Scharff, A. *Großer Atlas Schweiss-ZTU-Schaubilder*, 2nd ed.; DVS Media: Düsseldorf, Germany, 2018; ISBN 9783961440108. (In German)

37. Hobbacher, A. *Recommendations for Fatigue Design of Welded Joints and Components*; Welding Research Council: New York, NY, USA, 2009.

38. Neuber, H. Über die Berücksichtigung der Spannungskonzentration bei Festigkeitsberechnungen. Konstruktion 20. *Konstruktion* **1968**, *20*, 151–245. (In German)

39. Radaj, D.; Sonsino, C.; Fricke, W. *Fatigue Assessment of Welded Joints by Local Approaches*, 2nd ed.; Woodhead Publishing Ltd.: Cambridge, UK, 2006; ISBN 1845691881.

40. Society, G.W. *Merkblatt 0905: Industrielle Anwendung des Kerbspannungskonzeptes für den Ermüdungsfestigkeitsnachweis von Schweißverbindungen*; DVS Media GmbH: Düsseldorf, Germany, 2017. (In German)

41. Dürr, A. *Zur Ermüdungsfestigkeit von Schweißkonstruktionen aus Höherfesten Baustählen bei Anwendung von UIT-Nachbehandlung, Fatigue Atrength of Welded High Strength Steels by Application of UIT-Post-Weld Treatment*; Univercity of Stuttgart: Stuttgart, Germany, 2007. (In German)

42. Kuhlmann, U.; Gunther, H. *Experimentelle Untersuchungen zur ermüdungssteigernden Wirkung des PIT-Verfahrens (in German)*, *Versuchsbericht*; University of Stuttgart: Stuttgart, Germany, 2009.

43. Schubnell, J.; Gkatzogiannis, S.; Farajian, M.; Luke, M.; Ummenhofer, T. *Schlussbericht IGF-Vorhaben IGF Nr. 19.227 N / DVS-Nr.: 09.080: Rechnergestütztes Bewertungskonzept zum Nachweis der Lebensdau-Erverlängerung von Mit Dem Hochfrequenz-Hämmerverfahren (HFMI) Behandelten Schweißverbindungen aus Hochfesten Stählen*; Karlsruher Institut für Technologie: Karlsruhe, Germany, 2020. (In German)

44. Hensel, J.; Nitschke-Pagel, T.; Tchoffo Ngoula, D.; Beier, H.-T.; Tchuindjang, D.; Zerbst, U. Welding residual stresses as needed for the prediction of fatigue crack propagation and fatigue strength. *Eng. Fract. Mech.* **2018**, *198*, 123–141. [CrossRef]

45. Teng, T.-L.; Fung, C.-P.; Chang, P.-H.; Yang, W.-C. Analysis of residual stresses and distortions in T-joint fillet welds. *Int. J. Press. Vessel. Pip.* **2001**, *78*, 523–538. [CrossRef]

Article

Damage-Based Assessment of the Fatigue Crack Initiation Site in High-Strength Steel Welded Joints Treated by HFMI

Yuki Ono [1,*], Halid Can Yıldırım [2], Koji Kinoshita [1] and Alain Nussbaumer [3]

[1] Department of Civil Engineering, Gifu University, Gifu 501-1193, Japan; kinosita@gifu-u.ac.jp
[2] Department of Civil and Architectural Engineering, Aarhus University, 8000 Aarhus, Denmark; halid.yildirim@cae.au.dk
[3] ENAC-RESSLab, Ecole Polytechnique Fédérale de Lausanne (EPFL), CH-1015 Lausanne, Switzerland; alain.nussbaumer@epfl.ch
* Correspondence: yono@gifu-u.ac.jp; Tel.: +81-58-293-2414

Abstract: This study aimed to identify the fatigue crack initiation site of high-frequency mechanical impact (HFMI)-treated high-strength steel welded joints subjected to high peak stresses; the impact of HFMI treatment residual stress relaxation being of particular interest. First, the compressive residual stresses induced by HFMI treatment and their changes due to applied high peak stresses were quantified using advanced measurement techniques. Then, several features of crack initiation sites according to levels of applied peak stresses were identified through fracture surface observation of failed specimens. The relaxation behavior was simulated with finite element (FE) analyses incorporating the experimentally characterized residual stress field, load cycles including high peak load, improved weld geometry and non-linear material behavior. With local strain and local mean stress after relaxation, fatigue damage assessments along the surface of the HFMI groove were performed using the Smith–Watson–Topper (SWT) parameter to identify the critical location and compared with actual crack initiation sites. The obtained results demonstrate the shift of the crack initiation most prone position along the surface of the HFMI groove, resulting from a combination of stress concentration and residual stress relaxation effect.

Keywords: high-strength steel; HFMI; residual stress relaxation; X-ray diffraction; neutron diffraction; damage assessment

Citation: Ono, Y.; Yıldırım, H.C.; Kinoshita, K.; Nussbaumer, A. Damage-Based Assessment of the Fatigue Crack Initiation Site in High-Strength Steel Welded Joints Treated by HFMI. *Metals* **2022**, *12*, 145. https://doi.org/10.3390/met12010145

Academic Editor: Francesco Iacoviello

Received: 6 November 2021
Accepted: 27 December 2021
Published: 12 January 2022

Publisher's Note: MDPI stays neutral with regard to jurisdictional claims in published maps and institutional affiliations.

1. Introduction

For welded joints, fatigue strength improvement can be achieved by using post-weld treatment in which the aim is to modify the weld toe regions to avoid fatigue crack development. Among others, high-frequency mechanical impact (HFMI) treatment has received much attention in the last two decades [1–11]. The application of the HFMI treatment introduces compressive residual stress in the weld toe and material hardening in the surface layer, simultaneously improves the local weld geometry, and removes typical weld imperfections. The degree of the fatigue strength improvement of HFMI has been related with the material yield strength (f_y), namely, it increases together with the steel grade [9–11]. Therefore, the use of HFMI treatment for high-strength steel welded joints may lead to a superior fatigue performance [9–11]. The primary reason for this is the extended fatigue crack initiation and propagation periods within short crack lengths [12]. The extension in these periods is strongly related to the local parameters such as material property, residual stress, imperfection, and notch geometry. Thus, theoretical modeling with all the information for the life estimation of high-strength steel welded joints treated by HFMI is challenging.

In this context, some studies have included analytical calculations of crack initiation life for HFMI-treated joints in S355 or S960 steel grades based on the local strain approaches [13–16].

In general, these approaches consider the experimental measurements of influencing parameters into finite element (FE) analysis. The simulated local strains/mean stresses at the crack initiation site have been used for calculating the crack initiation life from strain–life relationships of unnotched materials. The results presented in the studies have shown that the local strain approaches have the possibility for the improved life estimation of HFMI-treated joints and the effectiveness in knowing the contribution of local parameters on the improvement effect. In real situations, a matter of concern is residual stress relaxation by high peak stress, either a single overload or a part of variable amplitude loading [17–20]. This is because, for HFMI-treated joints, the beneficial compressive residual stress can be partially or fully reduced if the high peak stress results in significant local yielding [21–23]. Thus, a study by Mikkola et al., 2017 [24] assessed the contribution of the residual stress relaxation to fatigue damage in a HFMI-treated joint made of S700 steel grade. Relative comparisons of local strain-based damage parameters that correlate with the crack initiation life were conducted under various loading scenarios. With the same concept as Mikkola et al., 2017 [24], an extended study including different welded details and loading scenarios was executed by Nazzal et al., 2021 [25]. The results have shown that the residual stress relaxation has a high influence on the fatigue damage. Due to this, the contribution of each benefit—compressive residual stress, improved weld geometry, and material hardening—was shown to vary depending on the combination of high peak stresses and its R-ratios. However, the impact of the difference in the resulting damage parameter values on the actual fatigue phenomena such as the most prone sites of crack initiation, the crack initiation life at the site, and then the improvement level, has not been clarified experimentally.

To perform a better estimation for these phenomena by the damage parameter values, it is necessary to describe the residual stress states in the FE as accurately as possible. Recent numerical studies by considering thermo-mechanical welding and dynamic elastic-plastic analysis of HFMI process have been carried out by, e.g., Ruiz et al., 2019 [26] and Schubnell et al., 2020 [22]. The simulations were utilized for the analysis of the residual stress relaxation. Nevertheless, the experimental measurements are still essential and indispensable, being the only way to at least calibrate and often validate the numerical simulation results. Only limited experimental measurements are available and show these changes in the residual stress state due to the high peak stresses in HFMI-treated joints, particularly, for the steels with $f_y > 690$ MPa.

Based on the available literature, this study aimed to investigate the actual crack initiation site of the HFMI groove surface. In order to develop better tools for the estimation of improved fatigue life, the local strain-based damage parameter and the residual stress relaxation were taken into account.

Firstly, the residual stress states and crack initiation sites were experimentally characterized. Specifically, residual stress measurements were carried out on the specimens made of S690QL steel grade by means of X-ray diffraction (XRD) and, more extensively, by neutron diffraction (ND) methods. Secondly, fracture observations were performed on previously tested HFMI specimens subjected to high peak stresses as a part of the variable amplitude loading history [27]. Thirdly, following the experimental observations, the local stress–strain response was studied with the FE analyses to clarify the level of residual stress relaxation. The FE model included the initial residual stress state, improved weld geometry, high peak loading, and non-linear material behavior, as proposed by Mikkola et al., 2017 [24]. Finally, the fatigue damage required to initiate a crack was evaluated using the Smith–Watson–Topper (SWT) parameter along the surface of the HFMI groove, to identify and compare the critical location with the results of the failure observation.

2. Materials and Experimental Methods

2.1. Material Property and Specimen Detail

This study used the high-strength, quenched and tempered steel, S690QL. The S690QL steel has a thickness 6 mm. The mechanical property and chemical composition of S690QL are shown in Table 1. The constructional detail and data investigated in this paper consists of a plate with transverse non-load-carrying attachments and with fillet welds in both as-welded (AW) and HFMI-treated states [27]. The HFMI treatment was applied to the weld toe regions by using indenters with a round tip of 1.5 mm radius. Figure 1 shows the configuration of the specimens. For these specimens, the previous experimental study provides the microstructure analysis and hardness measurement [27]. The hardness was measured over weld metal, hear-affected zone (HAZ), and base plate at 0.3 mm from the surface. For the as-welded state, the hardness of base plate and weld metal were around 282 Hv and 306 Hv, respectively. Three HAZ observed were intercritical, fine-grained, and coarse-grained zones, and the peak value of 390 Hv was measured for the fine-grained zone. The HFMI-treated state showed nearly identical profiles and hardness values. Mikkola et al., 2016 [28] studied the influence of material hardening by HFMI treatment on the material properties for S700MC, similar steel grade to S690 in this study. The hardness of HFMI-treated material was about 6.0% higher than that of the base plate.

Table 1. Mechanical properties and chemical compositions of S690QL.

Steel	Mechanical Properties			Chemical Composition															
	Yield Strength f_y (N/mm^2)	Tensile Strength f_u (N/mm^2)	Elongation (%), Minimum	C	Si	Mn	P	S	Al	Nb	V	Ti	Cu	Cr	Ni	Mo	Ca	N	EW
S690QL	832	856	0	0.14	0.29	1.21	0.011	0.001	0.047	0.021	0.028	0.10	0.010	0.28	0.05	0.150	0.0	0.002	0.43

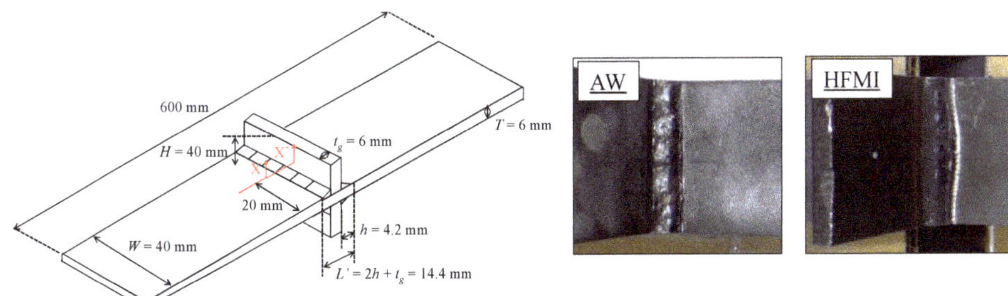

Figure 1. Configuration of the non-load-carrying attachments, Reprinted with permission from ref. [27]. 2021 International Journal of Fatigue (in axonometry and pictures showing weld geometries).

2.2. Residual Stress Measurement Methods

Residual stresses were measured for AW, HFMI, and HFMI-LC (HFMI after load cycles) states by X-ray diffraction (XRD) and Neutron diffraction (ND) methods (see Table 2). HFMI-LC were neutron-scanned after the application of a high peak load cycle corresponding to $\sigma_{max} = 0.8f_y$. This was followed by 20 load cycles of $\sigma_{max} = 0.2f_y$ at a constant amplitude of $R = -1$, as shown in Figure 2. This load cycle was based on a previous study by Mikkola et al., 2017 [24]. The first single cycle assumed an extreme event in which a high peak stress might be induced. For instance, it may correspond to transportation, erection, and mounting of steel structures, or just extraordinary large loading case in a part of service loading such as in an earthquake, storm, or heavy sea wave. To explore the effectiveness of the HFMI treatment even for such extreme cases, the relaxation at a load close to the material yield strength was considered. A similar approach by Mikkola et al., 2017 [24] showed that only the first high peak load cycle was critical with respect to the residual stress relaxation. The following smaller cycles intended to represent the stabilized mean stress behavior and the stresses due to daily live loads. The load corresponding to $0.2f_y$ was referred to typical equivalent stress levels used in variable amplitude loading in a previous study by Yıldırım and Marquis [17].

Table 2. Test matrix of residual stress measurements.

Condition Name	XRD	ND						Note
	Meas. Points	Meas. Paths		Meas. Paths		Meas. Paths		
	0 mm	Top	Bottom	Top	Bottom	Top	Bottom	
AW	-	1	1	1	1	1	1	Initial state
HFMI	4	1	1	1	1	1	1	Initial state
HFMI-LC	-	-	1	-	1	-	1	After load cycles in Figure 2

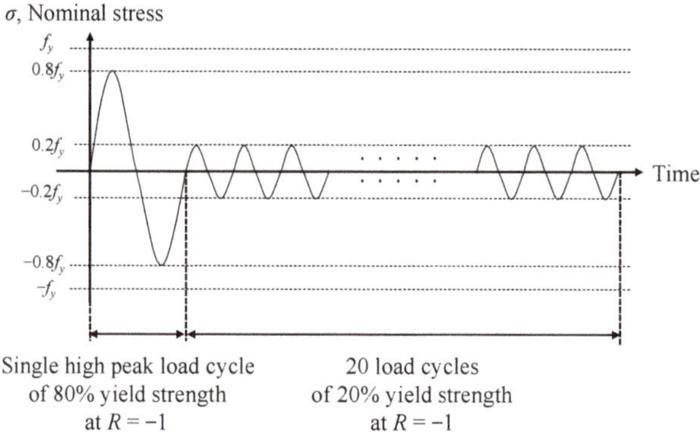

Figure 2. Applied load cycles at the constant amplitude of stress ratio $R = -1$.

Figure 3 shows the measurement points and paths considered in this study. The aim was to measure the center line of the main plate, i.e., 20 mm away from the specimen's edge (see X-X′ in Figure 1). The three main directions with respect to the weld line were chosen as the longitudinal direction (Ld), transverse direction (Td) and normal direction (Nd), in which the load was applied in the Ld. The measurement was done through the thickness around the weld toe or HFMI-treated region. In Figure 3, the coordinate origins ($x = 0$, $y = 0$) are located at the surface of the weld toe or at the bottom of the HFMI groove. In-depth measurements by ND were performed up to $y = 3$ mm for three paths at $x = 0$ mm and $x = \pm 2$ mm. For AW, HFMI states and for these three paths, the measurements were carried out at both the top and bottom sides (see Figure 3 and Table 2). XRD method was used to measure the residual stress state in Ld at the HFMI-treated surface. The measurements were repeated at four different HFMI grooves (see Figure 3). Neutron diffraction was used to measure residual strains at each point in three main directions, assumed to be the principal ones, required to calculate the three residual stress components of the tensor. The step sizes through the thickness were chosen as 0.07–0.11 mm close to the surface, 0.2, 0.4 or 1.0 mm away from the surface.

XRD measurements were carried out by the device Xstress 3000 G2R (StressTech, Vaajakoski, Finland), which is a portable system for in-lab and in-field use [29]. The measurements were performed with Cr-Kα radiation and a round collimator of 1 mm in diameter. The experiments by ND were conducted with the SALSA instrument (Stress–strain Analyzer for Large-Scale Engineering Applications) at the ILL (Institut Laue Langevin, Grenoble, France) located in Grenoble, France [30]. This instrument uses a monochromatic neutron beam with a wavelength of 1.71 Å and the 2θ measurement method. The measurements were performed with different rectangular-shaped gauge volumes, either $0.6 \times 2 \times 0.6$ mm^3 or $2 \times 2 \times 2$ mm^3, where the latter one was used for 3 mm depth only.

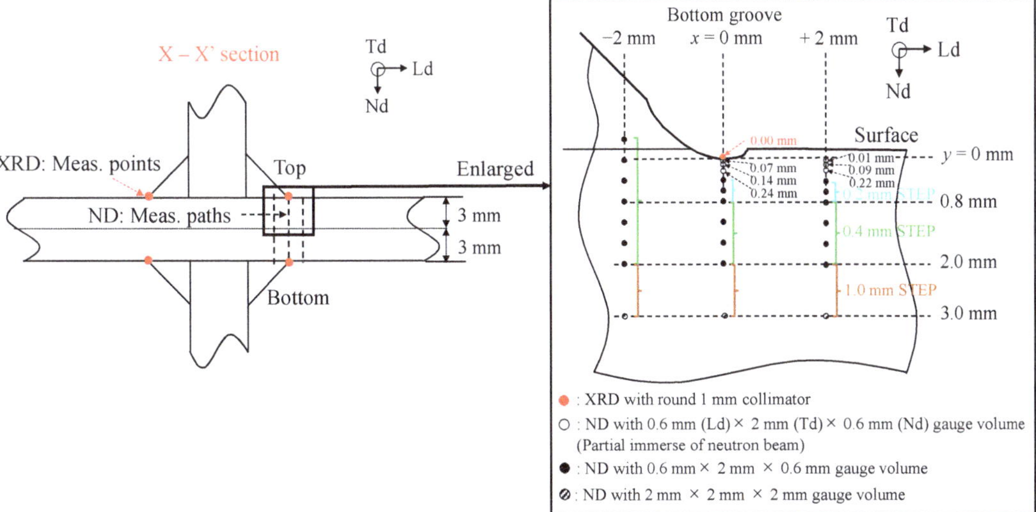

Figure 3. Measurement points and paths through the thickness for XRD and ND methods.

2.3. Fatigue Test Methods

The specimens were previously tested under high peak stresses as a part of variable amplitude loading [27]. The applied loading history was the cumulative amplitude distribution as shown in Figure 4. This distribution represents the occurrence frequency of relative load amplitude, nearly straight line on semi-log scale. The sequence length L_s was 2×10^5 cycles and the irregularity factor I was 0.99. The order of the individual cycles within this load sequence was randomly chosen, and the highest peak stress ranges had a stress ratio of $R = -0.43$. For the HFMI specimens, the three highest load cycles with respect to maximum stress, which are $\sigma_{max} = 1.00f_y$ and $\sigma_{max} = 0.80f_y$, $\sigma_{max} = 0.70f_y$, were applied to the specimens.

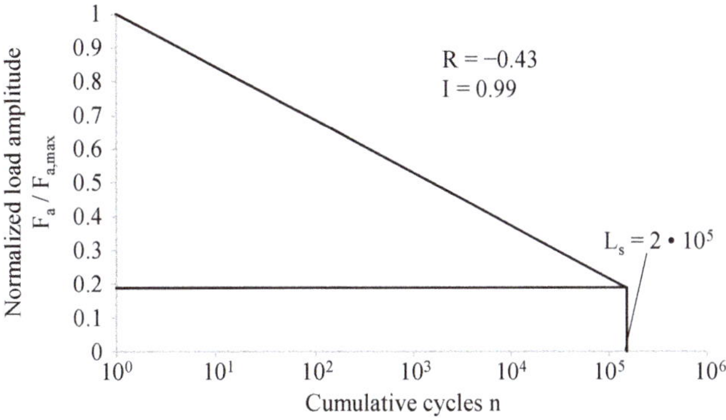

Figure 4. Applied load sequence of the cumulative amplitude distribution, Reprinted with permission from ref. [27]. 2021 International Journal of Fatigue.

3. Experimental Results

3.1. Results of the Residual Stress Measurement

Figure 5 shows the residual stress profiles as a function of depth at or close to the weld toe as well as at the HFMI groove for all cases: AW, HFMI, and HFMI-LC. The scatter bands of the residual stress distribution for the ND method are presented with the maximum and minimum values estimated statistically. All the results of XRD are plotted at $y = 0$ mm with asterisk symbols in Figure 5b. For the profiles of $x = 0$ and $+2$ mm, the residual stress gradients of the HFMI states were quite noticeable up to the mid-plane at $y = 3$ mm depth. The distribution features were very similar between $x = 0$ and $+2$ mm. The high compressive residual stresses induced by HFMI can be observed at the surface and subsurface. On the other hand, for the $x = -2$ mm profile, there was not a significant difference between AW and HFMI. Thus, for this particular case, the HFMI treatment did not sufficiently affect this region. In Figure 5b, one can observe a large dispersion in the measurement results of HFMI state for the top and bottom sides. The high compression state of about -200 to -700 MPa was found close to the surface, within $y = 0.5$ mm of depth, whereas XRD showed relatively lower values, corresponding to about -100 to -300 MPa at the surface only. Nevertheless, the compressive stresses were maintained up to depths of $y = 0.70$ and 1.60 mm at the top and bottom sides, respectively, and shifted to the tensile stresses at further depths. For the measurements of HFMI-LC specimens, the applied load cycle led to a great reduction of compressive stresses, resulting in almost 0 MPa close to the HFMI treatment and remained tensile stresses at further depths.

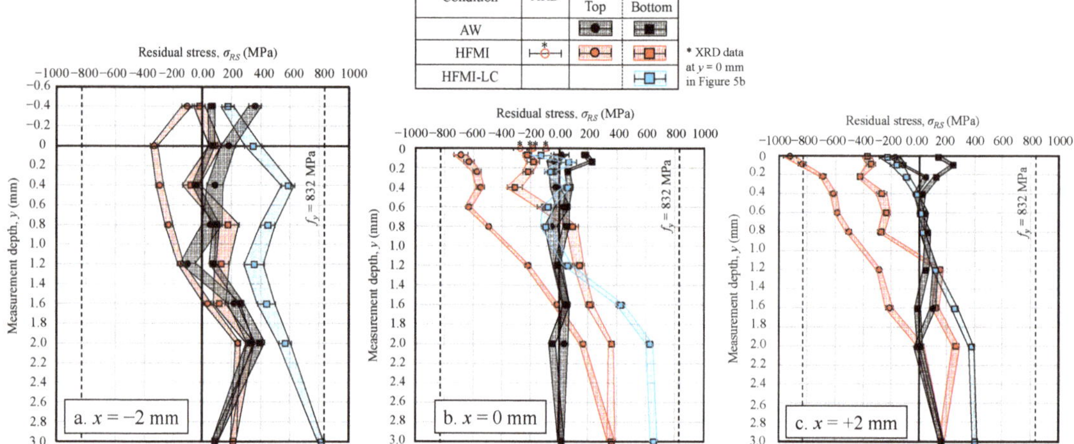

Figure 5. Residual stress profiles of AW, HFMI and HFMI-LC through thickness.

These measurement results of the HFMI cases were discussed together with the existing studies in the literature for in-depth distribution at $x = 0$ mm on the HFMI groove bottom. The data were collected for the same joint type with HFMI treatments and two types of steel grades [14,22,31–37]. These steels are S690 steel grade, which is the same as this study, and S355 steel grade to confirm the influence of material strength on the distribution. The detailed information of the literature is shown in Tables A1 and A2 in Appendix A, for each type of steel, the HFMI treatment conditions, specimen geometry, and residual stress measurement methods. Of those indications, mainly, the HFMI treatment condition is known to make a difference in the induced residual stress values and distribution through the thickness. Therefore, this paper only focuses on comparing the overall trend of distribution within the results obtained herein. A comparison of in-depth profiles is shown in Figure 6, in which the residual stress distributions were normalized with respect to the material yield strength. First, as illustrated in Figure 6, it must be acknowledged that the scatter of the residual stresses close to the surface is considerable, including the measurement results presented in this study. However, the scatter is almost the same regardless of the different steels. There is a slight difference in how deep the compressive residual stresses are maintained for the different steels. The experimental data of previous studies show that the depth of the compression layer is about 1.0 mm and 2.0 mm for S690 and S355 steel grade, respectively, where the first one agrees well with the data observed in this study. It means that the residual stress distribution of the S690 steel grade has a steeper gradient in depth. For this reason, and for S355 steel grade, the application of the HFMI treatment may have penetrated deeper, with large plastic deformations, in comparison to S690 steel grade. As a result, with respect to the depth of the compressive layer, different material properties can be seen for different layers. Consequently, it was confirmed that the dispersion observed in these study results was quite similar to the ones found in the existing literature. Therefore, the residual stress distribution of the HFMI conditions can be utilized for numerical simulations calibration.

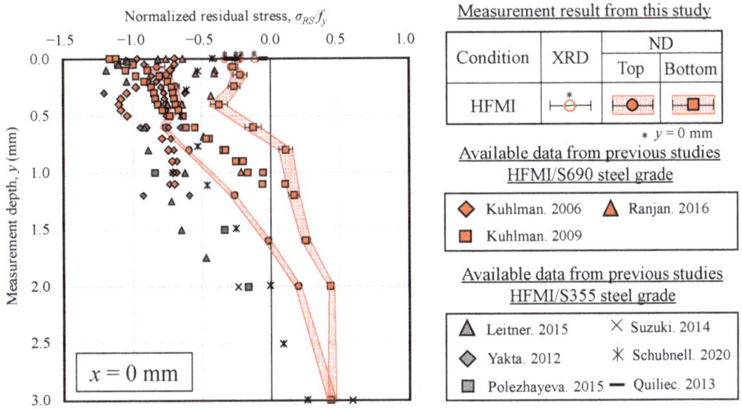

Figure 6. Comparison between measurement results and existing data for HFMI specimens and for two different steels at $x = 0$ mm.

3.2. Results of the Fatigue Tests and Fracture Observations

The fatigue test results are summarized in Figure 7. In Figure 7, the high improvement effect of 164% on the median fatigue strength (analyzed with a free *S-N* slope, *m*) was confirmed in spite of involving high peak stresses. The detailed information of the specimens is given in Table 3.

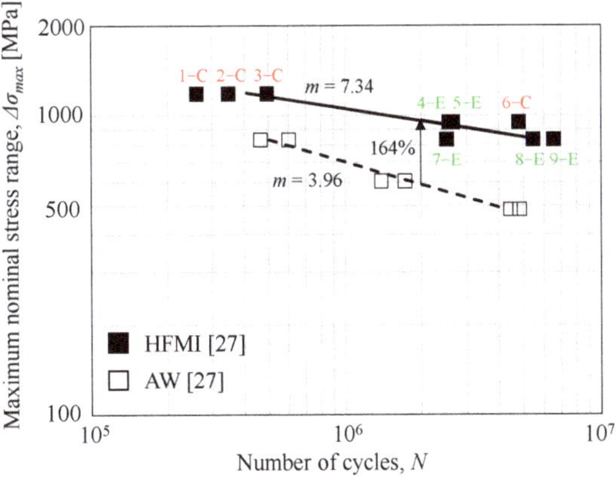

Figure 7. Fatigue test results of S690QL in AW and HFMI conditions.

Table 3. Fatigue test results and fracture observation of S690QL in HFMI condition.

Specimen Number	The Highest Peak Stress in a Part of Variable Amplitude Loading			The Number of Cycles to Complete Failure, N	Crack Location
	$\Delta\sigma_{max}$	σ_{max}	σ_{min}		
1−C	1191 MPa	$1.00f_y$ (833 MPa)	$-0.43f_y$ (358 MPa)	258750	Center
2−C	1191 MPa	$1.00f_y$ (833 MPa)	$-0.43f_y$ (358 MPa)	346500	Center
3−C	1191 MPa	$1.00f_y$ (833 MPa)	$-0.43f_y$ (358 MPa)	492000	Center
4−E	911 MPa	$0.80f_y$ (665 MPa)	$-0.37f_y$ (246 MPa)	2565750	Edge
5−E	911 MPa	$0.80f_y$ (665 MPa)	$-0.37f_y$ (246 MPa)	2637000	Edge
6−C	911 MPa	$0.80f_y$ (665 MPa)	$-0.37f_y$ (246 MPa)	4788750	Center
7−E	780 MPa	$0.70f_y$ (582 MPa)	$-0.34f_y$ (198 MPa)	2499000	Edge
8−E	780 MPa	$0.70f_y$ (582 MPa)	$-0.34f_y$ (198 MPa)	5466000	Edge
9−E	780 MPa	$0.70f_y$ (582 MPa)	$-0.34f_y$ (198 MPa)	6579000	Edge

To understand the influence of high peak stresses on the crack initiation behavior of HFMI-treated joints, the failure surfaces, crack initiation sites and types of the failed specimens were studied. The pictures on the first row in Figure 8 show the failure surfaces for fatigue tests at different load ranges. Different failure surfaces can be observed: the first crack initiated from the center for the case $\sigma_{max} = 1.0f_y$, and the second crack initiated at or near the plate edge for the case $\sigma_{max} = 0.7f_y$. In the former, several ratchet marks were observed on the fracture surface, corresponding to the blue lines in the schematic surface which is just below the pictures. This means that there were several crack initiation points along the HFMI groove surface. These cracks repeatedly coalesce and the finally resulting crack propagated with a very flat semi-elliptical shape. In the latter, no ratchet mark was observed, and the crack initiation was localized near the edge, which differs from the former. Of all the specimens in Table 3, the crack pattern tends to change from a center to an edge crack as the highest peak stress is lowered. It should be noted that the labeling of specimens such as 1-C and 4-E in Figure 7 and Table 3 are related to C (center) and E (edge), respectively.

To confirm the site change of the crack initiation, pictures of cross sections or specimen surfaces are presented in the middle row in Figure 8. Typically, cracks started from two main sites: the HFMI groove, and the boundary between the HFMI groove and the weld metal. On one hand, for specimen under $\sigma_{max} = 1.0f_y$, the cracks were initiated from the HFMI-treated zone. The failure plane was identified as the middle of the treatment width, where the severe stress concentration was present. On the other hand, for the specimen under $\sigma_{max} = 0.7f_y$, the cracks initiated from the boundary, i.e., away from the point with the highest stress concentration. As a result, the most prone site of crack initiation was variable from the middle of the treatment to the boundary of the treatment, depending on the magnitude of the highest peak stresses.

To reveal the crack initiation type, fatigue fractography was observed under scanning electron microscope (SEM). The analysis results are shown in the last row in Figure 8. According to Anami et al., 2000 [38], a steep ditch induced by peening at the weld toe may cause high stress concentrations and initiate cracks. In addition, Fisher et al., 1974 [39] and Marquis et al., 2016 [9] indicated that peening weld toe regions may leave a lap-type imperfection (or a crack-like lap imperfection), providing a site for the crack initiation. Therefore, this paper investigated if similar imperfections were present around the crack initiation site. In the SEM micrograph of the specimen tested under $\sigma_{max} = 1.0f_y$, the crack initiation was pointed in the area of the near-surface of the HFMI groove, even though the primary site was difficult to observe due to the multiple crack initiation points. The initiation site was located around a steep discontinuity induced by the HFMI treatment. On the other hand, the lap-type imperfections located near the boundary of the treatment were identified for the crack initiation point in the specimen tested under $\sigma_{max} = 0.7f_y$. These imperfections were found in almost all specimens with cracks initiating from the boundary. Such imperfections are believed to be typical for the treated welds due to the indenters.

Specifically, imperfections might be present as a result of pressing the material of the weld toe region toward the weld gusset sides.

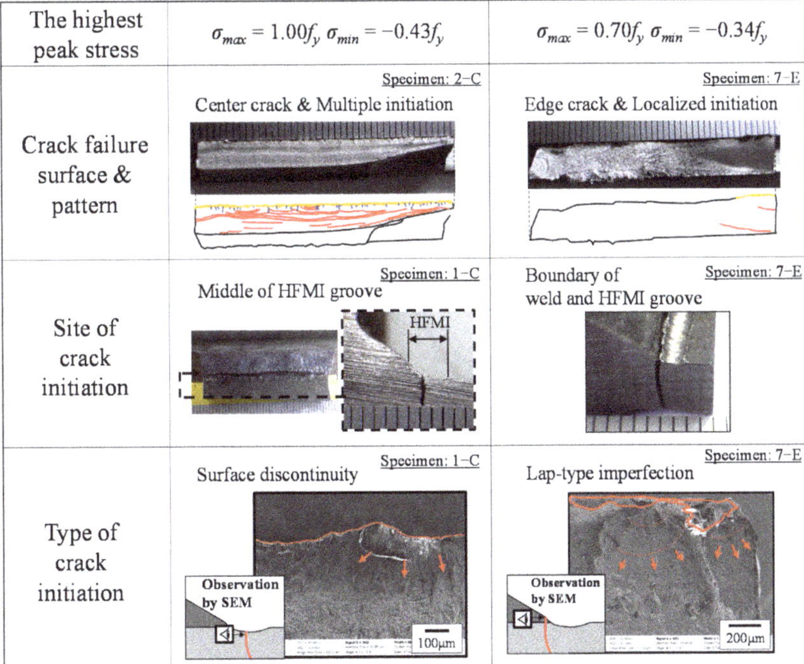

The highest peak stress	$\sigma_{max} = 1.00f_y, \sigma_{min} = -0.43f_y$	$\sigma_{max} = 0.70f_y, \sigma_{min} = -0.34f_y$
Crack failure surface & pattern	Specimen: 2–C Center crack & Multiple initiation	Specimen: 7–E Edge crack & Localized initiation
Site of crack initiation	Specimen: 1–C Middle of HFMI groove	Specimen: 7–E Boundary of weld and HFMI groove
Type of crack initiation	Specimen: 1–C Surface discontinuity	Specimen: 7–E Lap-type imperfection

Figure 8. Observation results of fatigue crack initiation sites under different high peak stresses in the HFMI condition.

4. Numerical Methods

Finite Element simulations were performed on the HFMI-LC specimens to validate the experimental behavior of residual stress relaxation. The global dimension of the transverse attachment considered in this study is represented as a one-fourth 2D model, in Figure 9. This analysis focuses on the local behavior at the HFMI-treated zone; thus, the weld root was not included in the model. The weld geometry modelled with the weld leg length (h_x/h_y) of 5.2 mm for the base plate side and 6.4 mm for the gusset plate side. The weld angle (θ) was 42 degrees. The geometrical values for the HFMI groove were 2.2 mm radius (ρ_H), 0.14 mm depth (d_H), and 2.8 mm width (w_H). These weld geometries were representative of the average values measured on an image of the cross section where the microstructure was revealed [27]. Thus, note that the scatter of the weld geometry was not considered in this work. Linear plane strain elements were used. Finite strain theory was applied to represent the large displacements and material non-linearity. The element size was set to about 0.1 mm around the HFMI groove, and then the size was gradually increased towards the other global parts.

Table 4 provides the material properties of S690 steel grade used in the FE simulation. Combined non-linear isotropic-kinematic hardening parameters, so-called Voce-Chaboche's (VC) parameters [40], were employed. In the FE simulation, an assumption was made that the material characterization considers only homogenous properties of the base materials with reference to [13]. Three kinds of VC parameters, BM-1, BM-2, and BM-3, were considered to compare the simulation results in between. However, as shown in Section 2.1, there are various material properties such as weld metal, HAZ, base metal, and hardened metal in and around HFMI-treated regions. As the hardened metal due to the HFMI has higher local yield strength compared to the base metal [24,25,28], it might limit the amount of residual stress relaxation. On the other hand, the local yield strength of coarse-grained HAZ has been found to be slightly lower than that of the base metal [24,28,41], thus probably changing the residual stress distribution around the HFMI-treated region. In this work, these material gradient effects on the residual stress relaxation were not considered. The VC parameters in Table 4 were extracted from a recent study by Garcia 2021 [41], in which the material parameters were obtained by calibrating the cyclic hysteresis loops of the base materials with a unique optimization algorithm. The procedure provided a better transition between the elastic and plastic behavior on the first cycle for high-cycle fatigue modelling with little plasticity. Since the residual stress relaxation of HFMI-treated joints occurs only during the very first cycle [24], the VC parameters by Garcia 2021 [41] were utilized in this study.

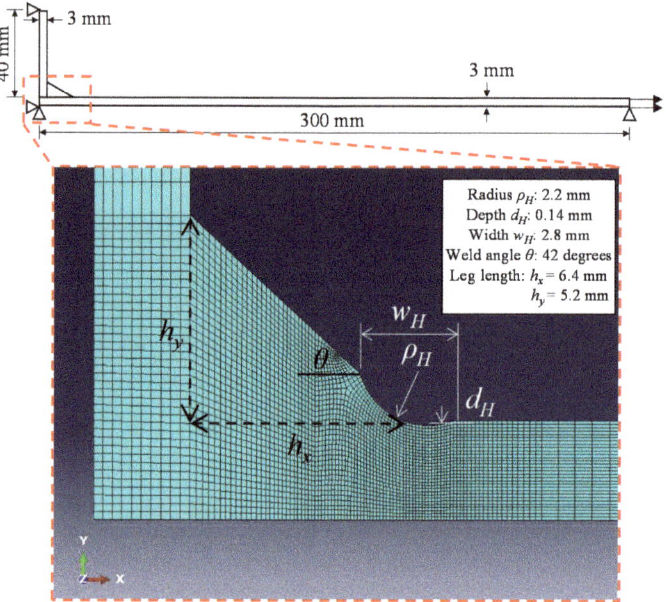

Figure 9. The global and local geometry for HFMI-treated detail.

Table 4. Voce-Chaboche's (VC) parameters for high-strength steels of S690 grade presented by Garcia, data from ref. [41].

Name	Author	Material	Linear-Elastic Behavior		Non-Linear Behavior					
			Elastic Properties		Isotropic Properties		Kinematic Properties			
			E [MPa]	v	Q	q	σ_y [MPa]	C_1	γ_1	C_2	γ_2
BM-1	Garcia	Base material	210000	0.3	0	0	578	1832	8	17421	88
BM-2	Castro e Sousa	Base material	206000	0.3	0	1	590	19018	83	771	5
BM-3	Mikkola	Base material	200000	0.3	1	1	772	11478	395	11478	395

E: Young modulus, v: Poisson ratio, Q: Maximum increase in size of yield surface due to hardening at saturation, q: How quickly the increase of yield surface approaches the saturation, σ_y: Yield stress at zero plastic strain, C: Initial kinematic hardening modulus, γ: Rate at which the kinematic hardening modulus decreases with increasing plastic deformation.

In-depth residual stress distributions shown in Figure 10 were implemented in the FE model. The distributions were defined as the initial mean stress state $x = 0$ in Figure 5. As a simplification of the profile, the compressive residual stress at the surface up to 0.5 mm depth was kept constant. Thus, a fixed value of the high or low compressive residual stress was chosen as an average of the residual stresses within the 0.5 mm depth. Above the depth, these values met at the crossing point of $\sigma_{RS}/f_y = 0$. The tensile residual stress at this meeting point and $y = 3$ mm depth were represented by straight lines. The intersection corresponds to $y = 1.15$ mm depth, which was calculated by averaging the compressive layer depth of distributions at the top and bottom sides. To satisfy the self-equilibrium, the areas above and below $\sigma_{RS}/f_y = 0$ were set to be equal. The two distributions were determined as HC (High compression) and LC (Low compression). These residual stress distributions were represented by means of "predefined temperature field" in Abaqus [42], as shown in Figure 11. The benefit of this approach is that the initial residual stresses can be modelled as a thermal step before applying any external load. Moreover, the FE software is able to calibrate the stress equilibrium in the area of interest, e.g., from $x = -3.5$ mm to $x = 3.5$ mm. To create the residual stress distribution, the change in the temperature (ΔT) was defined at the nodes for the three paths at $x = -3.5$ mm, $x = 0$ mm, and $x = 3.5$ mm. To obtain the thermal strains at the HFMI groove, ΔT values of HC and LC were divided by the elastic stress concentration factors (see Figure 11). Besides, the two-third and one-third of ΔT for HC and LC, without any stress concentration, were applied at $x = 3.5$ mm and $x = -3.5$ mm, respectively. These were adjusted by trial and error to find the best fit with the experimental data of this study. To compare the experimental observations for the residual stress relaxation, HFMI-LC was subject to the load history provided in Figure 2.

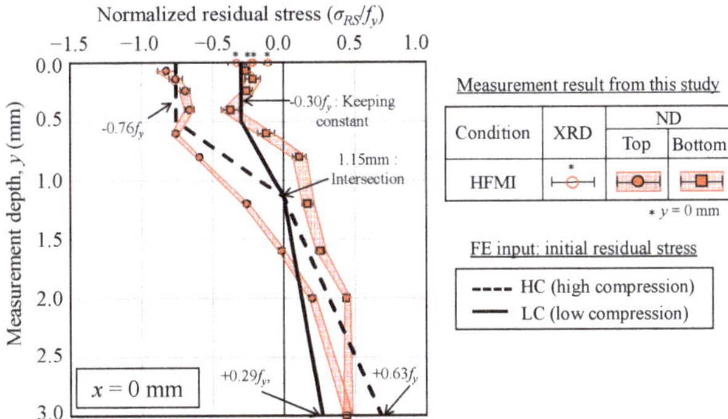

Figure 10. Experiment-based in-depth residual stress distribution at $x = 0$ mm.

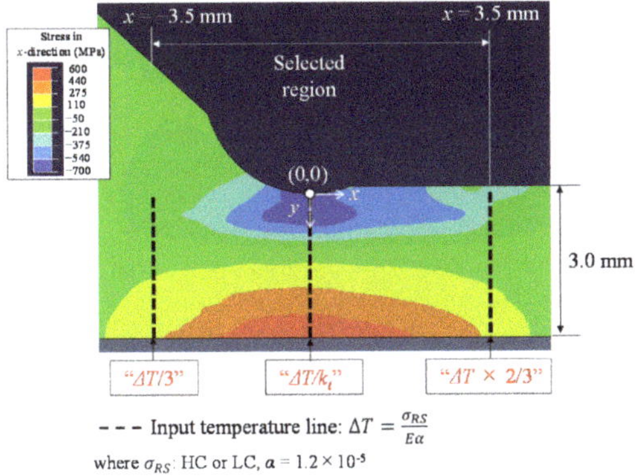

Figure 11. Implemented residual stress distribution in FE model.

5. Numerical Results

Comparisons between the residual stress distributions after the relaxation found in the FE simulations and experimental work using the ND method are presented in Figure 12. In this figure, six simulation results are presented as parameters: the residual stress either as HC or LC and the material properties as BM-1, BM-2 or BM-3. The experimental data showed that the stress state after the load cycles was almost zero below $y = 1.2$ mm and it was tensile in further depth. The FE simulations from any material properties had a similar tendency, as the experiment in that the applied load cycle reduced the compressive residual stresses near the surface. The change of residual stresses took place below $y = 0.5$ mm, giving the stresses towards zero or slightly tensile. The difference of the initial compressive stresses was small after the load cycles. The experimental results in the HFMI-treated local region below $y = 0.14$ mm, which is the most interesting region to study the crack initiation, were about $-0.24f_y$ to $+0.13f_y$. The simulation result for BM-3 showed that the near-surface residual stresses in that region agreed with the experiment. However, the overestimation could be observed in case of BM-1 and BM-2. To summarize, the FE models were able to reproduce the relaxed residual stresses; however, their accuracy was ensured only for the local zone of HFMI treatment, i.e., the near surface. Thus, the FE models are validated; they can be considered suitable for studying the crack initiation or short-crack propagation behavior. Different material properties may affect the values of relaxed residual stress at the near surface, and therefore BM-3 will be utilized in the subsequent investigation.

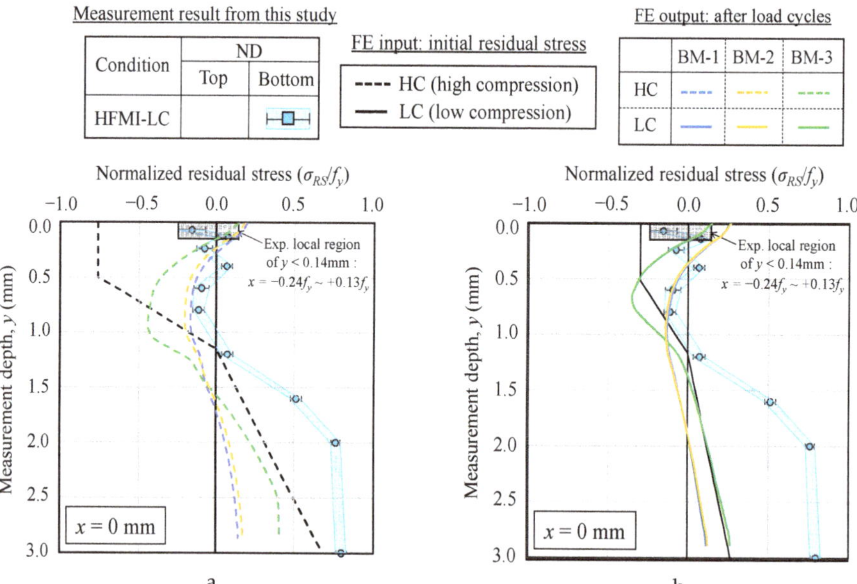

Figure 12. Numerical and experimental comparison of the residual stress distributions after the stress relaxation at $x = 0$ mm. (**a**) HC case as FE input; (**b**) LC case as FE input.

The change of the defined initial residual stress in relation to the applied load cycles is shown in Figure 13. From the figure, one can see that the tensile peak stress was not responsible for the relaxation of the residual stress state around the HFMI-groove. A dramatic change of stress field occurred after the load cycles, including compressive peak stress equal to $0.8f_y$. The beneficial compressive residual stress induced by the HFMI reduced to almost zero or even slightly tensile stress. Therefore, a compressive peak stress equal to the $0.8f_y$ prior to small load cycles proved to be detrimental, as it led to local yielding when under compression, and thus to large stress relaxation around the HFMI-groove. On the contrary, a tensile peak stress equal to the $0.8f_y$ did not reduce the residual stress around the HFMI-groove, even considering the stress concentration in this region.

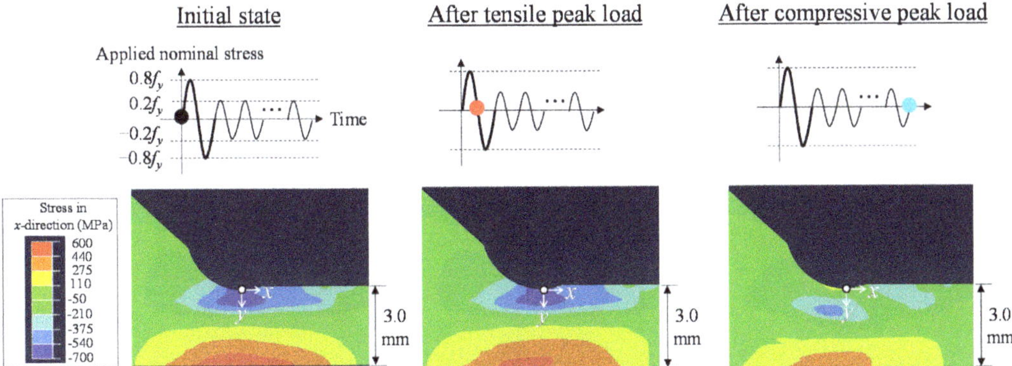

Figure 13. Change of residual stress state due to the applied load cycles in the FE model.

6. Fatigue Damage Assessment Considering Residual Stress Relaxation

After validating the FE models, the HFMI specimens of Figure 7 were simulated to assess the most prone site of the crack initiation. The applied load cycles are depicted in Figure 14. The highest peak stress and its *R*-ratio, according to the ones applied in the experiments, were studied in the simulations to understand the impact of loading conditions on the crack initiation site based on a local strain-based method. Two types of loadings were compared: (i) cycles with $\sigma_{max} = 1.0f_y$ and $R = -0.43$ as the higher peak stress and (ii) cycles with $\sigma_{max} = 0.7f_y$ and $R = -0.43$ as the lower peak stress. With the results from the preceding investigation, this part of the study concentrated on the effect of the first high peak load. The following load cycles, corresponding to the $0.2f_y$, were applied to calculate the relative damage after the stress relaxation.

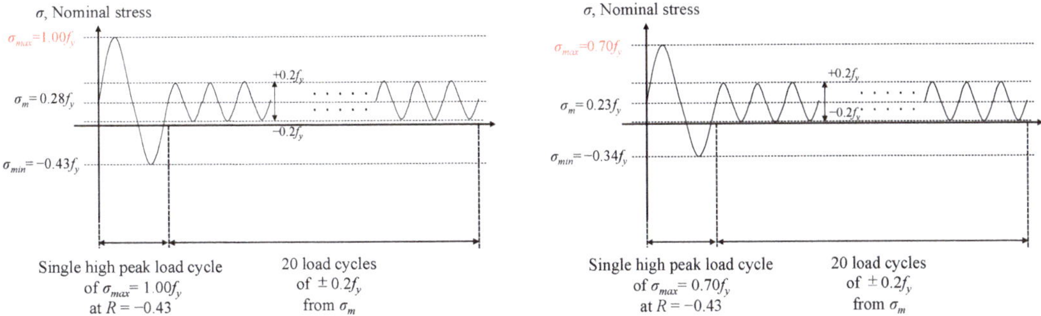

Figure 14. Applied load cycles for fatigue damage assessment.

Following the FE simulation of the residual stress relaxation, damage analysis of HFMI-treated joints was carried out using a local strain-based method. The Smith–Watson–Topper (SWT) parameter, as given in (1), was employed, as it allows for handling the mean stress/residual stress influence.

$$P_{SWT} = \sigma_{t,max}\frac{\Delta\varepsilon_{t,T}}{2} \tag{1}$$

where $\sigma_{t,max}$ is the true maximum stress and $\Delta\varepsilon_{t,T}$ is the total true strain range. The SWT parameter was originally derived from a combination of the Basquin–Coffin–Manson relationship and concept of strain energy density [43,44]. Thus, the calculated P_{SWT} from (1) represents the "fatigue damage" that is required to a crack initiation. In this study, the resulting fatigue damage at the surface of the HFMI groove was compared with the observation results for the actual crack initiation sites of Section 3.2. Figure 15 shows the SWT parameter fatigue damage distribution. The P_{SWT} was calculated using the maximum true stress and total true strain for each element along the surface of HFMI groove. As shown in Figure 15, the position of the groove bottom is defined as zero, meaning that the weld gusset side was chosen as positive. The stresses and strains used were the principle ones along the curvature of the HFMI groove. In the calculations, the closed hysteresis loop following the first high peak load cycle was used. It should be noted that the P_{SWT} was considered to provide a comparative assessment of the resulting fatigue damage based on the assumption of homogeneous material properties.

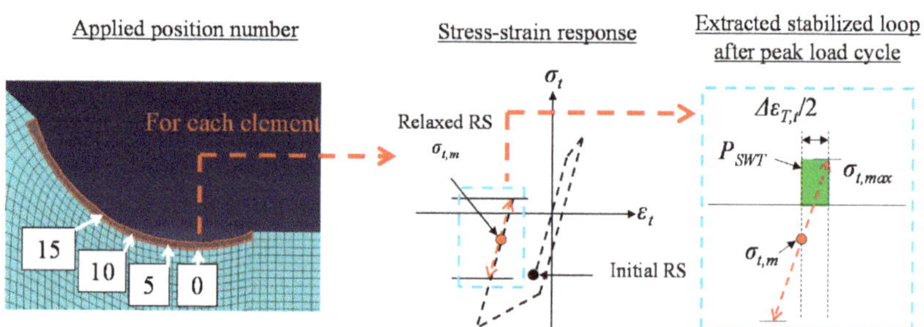

Figure 15. Positions subjected to SWT damage parameter.

Figure 16 shows the calculated P_{SWT} values of the load cycle with $\sigma_{max} = 1.0f_y$, together with the local mean stress ($\sigma_{t,m}$) and the actual crack initiation site. The applied loading cycle with $\sigma_{max} = 1.0f_y$ shifted the local compressive stresses to the tensile stresses for almost all positions, regardless of the level of initial residual stress. For the fracture surface in Figure 8, several ratchet marks existed on the surface, indicating there were multiple crack initiation points along the HFMI groove. Here, this can be explained by the full relaxation of the induced compressive residual stresses, thus that the stress concentration became the most critical factor for cracking. The P_{SWT} analysis showed that the highest fatigue damage appeared slightly to the left of HFMI groove bottom; the position was between the numbers 2 to 8, where also the more severe stress concentration occurred. The actual crack initiation site was almost at the location with the high P_{SWT} values.

Next, the results for the load cycle with $\sigma_{max} = 0.7f_y$ are similarly shown in Figure 16. The applied loading cycle with $\sigma_{max} = 0.7f_y$ provided the residual stress relaxation; nevertheless, the local mean stress remained between compressive to being close to zero. For the fractured surface in Figure 8, the crack developed close to the corner of the specimen. Here, it could be proven that the effect of the compressive residual stress was maintained, i.e., minor relaxation, and then the crack was localized around the corner because the stress concentration was slightly more aggressive at the edge of plate. For the case of HC, the large damage of P_{SWT} could be observed somewhat away from the position of the highest stress concentration. Those positions corresponded to numbers in the range of 12 to 15. Thus, the shape of the damage diagram was different from the former case (with $\sigma_{max} = 1.0f_y$). A similar tendency was confirmed for the case of LC, where numbers ranging from 5 to 13 corresponded to relatively high P_{SWT} values, i.e., the wider surface of the HFMI groove had higher damage values. In terms of the actual fracture site, the crack started from the boundary between the weld metal and the HFMI-treated zone. This site was about 1.4 mm away from the treated edge. Based on the above discussion, this behavior can be explained from the damage diagrams. In other words, the combination of stress concentration and relaxation effect makes the crack initiation site move close to the boundary side where a lap-type imperfection existed.

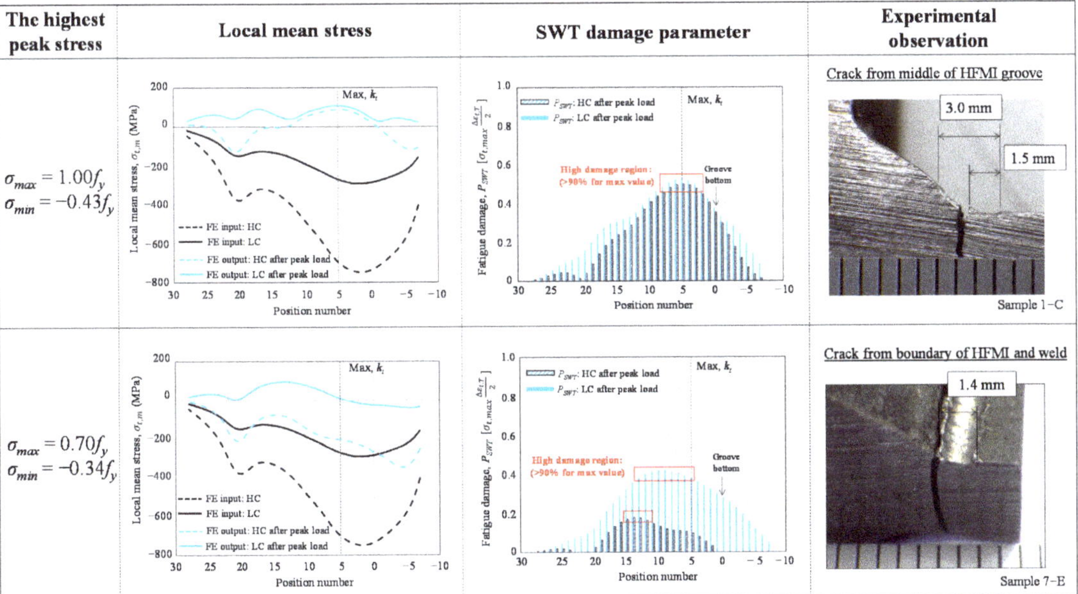

Figure 16. Comparison of fatigue damage distribution along the HFMI-treated surface between different high peak stresses.

To summarize, the SWT parameter considering the local mean stress after the relaxation of residual stress demonstrated the variability of crack initiation sites with some cracks from the HFMI groove and others from the boundary between the HFMI-treated zone and the weld metal. The combination of stress concentration and relaxation effect made a shift of the most prone crack initiation site. For example, in Figure 16, when the peak stress of $\sigma_{max} = 1.0f_y$ was modelled, a full relaxation of residual stress took place. This result explains that multiple cracks initiated from the surface discontinuity induced by HFMI treatment, where the stress concentration is dominant. Another example is shown in Figure 16, in which less relaxation occurred at the peak stress of $\sigma_{max} = 0.7f_y$. The effect of compressive residual stress remained; thus, the cracks were localized around the corner of the specimen and the boundary of the treatment became more critical for initiating the cracks. Therefore, the lap-type imperfections located around the boundary have the highest probability of initiating cracks.

At the end, the authors want to emphasize the evitable limitations of this study's modelling. The model mainly focused on the impact of the residual stress state and its relaxation on the crack initiation site. However, the variability of crack initiation may also depend on further parameters such as the level of stress concentration (i.e., improved weld geometry), microstructure and imperfection size. The fatigue damage analysis by considering the variations of these parameters is, however, out of the scope of the current work. Moreover, the relaxation effect should be investigated further for different cases of high peak loading and R-ratio.

7. Conclusions

This study aimed to identify the fatigue crack initiation site for high-frequency mechanical impact (HFMI)-treated joints made of S690QL by clarifying the relaxation effect of residual stress. At first, the residual stress measurements were conducted on the non-load-carrying transverse attachment specimens in the as-welded and HFMI-treated states, using advanced diffraction methods. Then, fracture observations were performed on the specimens tested under fatigue loading involving high peak stresses. After these experimental works, the relaxation behavior of HFMI-induced residual stress was simulated by FE models implementing the load cycles and high peak load, the characterized residual stress, the local weld geometry as well as non-linear material behavior parameters. Finally, the observation results for the fracture specimens were discussed through the damage-based assessment using the Smith–Watson–Topper parameter based on the local strain. The conclusions drawn from these investigations are summarized below.

For the experimental investigation:

- At the HFMI groove bottom, the in-depth profiles of residual stress showed the high compressive stress of about $-0.30f_y$ to $-0.76f_y$ within 0.5 mm of depth. The compressive stresses were maintained up to the depths in the range of 0.70 to 1.60 mm and were shifted to tensile stresses more deeply, which in-depth gradient was even steeper than that observed on available data of S355 steel grade. The high peak stress equal to $0.8f_y$ led to a significant reduction of the beneficial compressive stresses, which, after relaxation, were close to zero near the surface and up to 1.2 mm, and remain tensile more deeply.

- Different features on fracture surface, crack pattern, crack initiation site, and crack initiation type were observed according to different applied high peak stresses. Particularly interesting, as the applied peak stress was lowered from $1.0f_y$ to $0.7f_y$, the initiation site within the weld shifted from the HFMI groove to near the boundary between the HFMI-treated zone and the weld metal. In the latter case, the lap-type imperfections for the site near the boundary became the origin of crack initiation.

For the numerical investigation:

- The FE models developed, incorporating the measured in-depth residual stress profiles and applied load cycle with high peak load, was able to reproduce the residual stresses after relaxation; however, it was accurate only at the near surface with the HFMI treatment region.

- The simulation results with the FE models demonstrated that the significant reduction of compressive residual stress near the surface, observed in residual stress measurement, was mainly occurred by compressive peak stress, as it led to immediate local yielding on the compressive sides of the HFMI treatment region.

For the fatigue damage assessment:

- The damage-based assessment considering the local mean stress after high peak stress equal to $1.0f_y$ and $0.7f_y$ confirmed a shift of the crack initiation most prone position along the surface of the HFMI groove, resulting from a combination of stress concentration and relaxation effect of residual stresses.

- When the peak stress was equal to $1.0f_y$, full relaxation of the compressive residual stress took place, such that cracks initiated from the HFMI groove where the stress concentration was dominant; less relaxation occurred under the peak stress equal to $0.7f_y$. Thus, in the latter case, the lap-type imperfections located near the boundary of the treatment became more critical for initiating the cracks even though the stress concentration was smaller than that of the HFMI groove. The above explains and confirms the experimental observations.

Author Contributions: Conceptualization, Y.O., H.C.Y., K.K. and A.N.; methodology, Y.O., H.C.Y. and A.N.; software, Y.O. and H.C.Y.; validation, Y.O.; formal analysis, Y.O.; investigation, Y.O., H.C.Y., K.K. and A.N.; resources, Y.O., H.C.Y. and A.N.; data curation, Y.O.; writing—original draft preparation, Y.O.; writing—review and editing, H.C.Y., K.K. and A.N.; visualization, Y.O.; supervision, H.C.Y., K.K. and A.N.; project administration, H.C.Y.; funding acquisition, Y.O., H.C.Y. and K.K. All authors have read and agreed to the published version of the manuscript.

Funding: This research was mainly conducted during study abroad of the first author at EPFL and Aarhus University, supported by two scholarships: scholarship in the Japan Public–Private Partnership Student Abroad Program (Tobitate! Young Ambassador Program) from the Japanese

Student Services Organization (JASSO) and the Valor & V-drug Study Abroad Scholarship from Gifu University. A part of this research was funded by the European Project Hi-Life of Horizon 2020 with the Grant ID 702233, in which the second author is project manager.

Institutional Review Board Statement: Not applicable.

Informed Consent Statement: Not applicable.

Data Availability Statement: Not applicable.

Acknowledgments: The authors gratefully appreciated the valuable discussions and scientific input from Heikki Remes, from Aalto University. The authors would like to thank Sylvain Demierre from EPFL for his support in performing the fatigue tests and Gregoire Baroz from EPFL for SEM training, sample preparation, and micrograph imaging. Our special thanks go to ILL and Pirling Thilo, as well as Paul Lefevre from SONATS, for the residual stress measurements, and Salim Sleiman Nazzal from Aarhus University for sharing numerical modelling techniques.

Conflicts of Interest: The authors declare no conflict of interest.

Appendix A

Table A1. Detailed information of literature on specimen geometry and residual stress measurement method.

Ref	Author	Steel Grade (f_y)	Specimen Geometry			Method of Residual Stress Measurement
			$L \times W \times T$	$H \times t_g$	h	
[31]	Kuhlman. 2006	S690QL (813 MPa)	$23 \times 160 \times 12$	18×12	5.7	·Hole drilling (1.80 mm hole): 1.5 mm away from weld toe
[32]	Kuhlman. 2009	S690QL (830 MPa)	$26 \times 80 \times 12$	40×12	7.1	·Hole drilling (1.77 mm hole): 1.0 mm away from weld toe
[35]	Ranjan. 2016	A514 (793 MPa)	$19 \times 30 \times 9.5$	25×6.4	6.4	·Laser X-ray diffaction ($\sin^2\psi$) & Layer removal by electronic poslihing
[27]	Yildirim. 2020 & This study	S690QL (832 MPa)	$14 \times 40 \times 6$	40×6	4.2	·X-ray diffaction ($\sin^2\psi$, 1-mm collimator) ·Nutron diffraction at SALSA ($0.6 \times 2.0 \times 0.6$ mm^3 or $2.0 \times 2.0 \times 2.0$ mm^3 collimator)
[33]	Tehrani Yekta. 2012	350W (396 MPa)	$19 \times 30 \times 9.5$	25×6.4	6.4	·Laser X-ray diffaction ($\sin^2\psi$) & Layer removal by electronic poslihing
[14]	Quilliec. 2013	S355K2 (490 MPa)	$- \times - \times 15$	$- \times 15$	-	·X-ray diffraction ($2.5 \times 1.0 \times 0.006$ mm^3 collimator)
[34]	Suzuki. 2014	SM490 ($\geq$325 MPa)	$- \times 100 \times 16$	50×16	-	·X-ray diffraction ·Neutron-diffraction ($2.0 \times 2.0 \times 2.0$ mm^3 collimator)
[36]	Leitner. 2015	S355 ($\geq$350 MPa)	$- \times 90 \times 13$	40×16	-	·X-ray diffaction ($\sin^2\psi$, 1-mm collimator)
[37]	Polezhayeva. 2015	080A15 (560 MPa)	$46 \times 80 \times 20$	50×20	13	·Neutron-diffraction at UK's ISIS neutron source (1-mm collimator)
[22]	Schubnell. 2020	S355J2 + N (420 MPa)	$21 \times 50 \times 10$	50×10	5.7	·X-ray diffaction (2-mm collimator) ·Neutron diffraction ($2.0 \times 2.0 \times 2.0$ mm^3 or $2.0 \times 2.0 \times 5.0$ mm^3 collimator)

L: toe-to-toe length (mm), W: main plate width (mm), T: main plate thickness (mm), H: gusset height (mm), t_g: gusset thickness (mm), h: weld leg length (mm).

Table A2. Detailed information of literature on HFMI treatment condition.

Ref	Author	Method	Ultrasonic Frequency (kHz)	Ultrasonic Amplitude (μm)	Impact Frequency (Hz)	Indenter Diameter, D or Tip Radius, R (mm)	Travel Speed (m/mm)	Note
[31]	Kuhlman. 2006	UIT	27			3 (D)		Power consumption: 900 W
[32]	Kuhlman. 2009	PIT			90	2.0(R)8.0(D)/ 2.5(R)8.0(D)	2–3	Working pressure: 6 bars Angle for plate: 50–70 degree
[35]	Ranjan. 2016	UIT	20	50–60	220	3.0(R)		
[27]	Yildirim. 2020 & This study	UNP	20	30–60	100–400	1.5(R)		Angle for plate: 45 degree Angle for travel direction: 90 degree
[33]	Tehrani Yekta. 2012	UIT		27–29			6.0	Number of pass: 4 Angle for plate: 30–60 degree Groove radius: 1.69–2.37 Groove depth: 0.27–0.36
[14]	Quilliec. 2013	UIT	27			3.0(D)	4.0	Power consumption: 1200 W Number of pass: 1 Indenter: 3 pins Angle for plate: 67 degree
[34]	Suzuki. 2014	UIT	27	30		3.0(R)	6.0	Power consumption: 1000 W
[36]	Leitner. 2015	PIT				2.0(R)	0.6–1.8	Angle for plate: 30–60 degree Angle for travel direction: 90 degree
[37]	Polezhayeva. 2015	UIT						
[22]	Schubnell. 2020	PIT			90			Working pressure: 6 bars

References

1. Statnikov, E.S. *Application of Operational Ultrasonic Impact Treatment (UIT) Technologies in Production of Welded Joints*; IIW Doc. XIII-1667-97; International Institute of Welding: Paris, France, 1997. Available online: http://www.appliedultrasonics.com/pdf/APPLICATIONS_OF_OPERATIONAL_ULTRASONIC_IMPACT_TREATMENT_(UIT)_TECHNOLOGIES_IN_PRODUCTION_OF_WELDED_JOINTS.html (accessed on 26 December 2021).
2. Roy, S.; Fisher, J.W.; Yen, B.T. Fatigue resistance of welded details enhanced by ultrasonic impact treatment (UIT). *Int. J. Fatigue* **2003**, *25*, 1239–1247. [CrossRef]
3. Huo, L.; Wang, D.; Zhang, Y. Investigation of the fatigue behaviour of the welded joints treated by TIG dressing and ultrasonic peening under variable-amplitude load. *Int. J. Fatigue* **2005**, *27*, 95–101. [CrossRef]
4. Tominaga, T.; Matsuoka, K.; Sato, Y.; Suzuki, T. Fatigue improvement of weld repaired crane runway girder by ultrasonic impact treatment. *Weld. World* **2008**, *52*, 50–62. [CrossRef]
5. Weich, I.; Ummenhofer, T.; Nitschke-Pagel, T.; Dilger, K.; Eslami, H. Fatigue behaviour of welded high-strength steels after high frequency mechanical post-weld treatments. *Weld. World* **2009**, *53*, 322–332. [CrossRef]
6. Kudryavtsev, Y.; Kleiman, J. *Increasing Fatigue Strength of Welded Joints by Ultrasonic Impact Treatment*; IIW Document XIII- 2338-10; International Institute of Welding: Paris, France, 2010. Available online: http://sintes.ca/documents/IIWDocumentXIII-2338-10.2010UIT-UP.pdf (accessed on 26 December 2021).
7. Maddox, S.J.; Dore, M.J.; Smith, S.D. A case study of the use of ultrasonic peening for upgrading a welded steel structure. *Weld. World* **2011**, *55*, 56–67. [CrossRef]
8. Yıldırım, H.C.; Leitner, M.; Marquis, G.B.; Stoschka, M.; Barsoum, Z. Application studies for fatigue strength improvement of welded structures by high-frequency mechanical impact (HFMI) treatment. *Eng. Struct.* **2016**, *106*, 422–435. [CrossRef]
9. Marquis, G.; Barsoum, Z. IIW Recommendation on High Frequency Mechanical Impact (HFMI) Treatment for Improving the Fatigue Strength of Welded Joints. In *IIW Recommendations for the HFMI Treatment*; Springer: Singapore, 2016; pp. 1–34. Available online: https://link.springer.com/chapter/10.1007/978-981-10-2504-4_1 (accessed on 26 December 2021).
10. Yildirim, H.C. Recent results on fatigue strength improvement of high-strength steel welded joints. *Int. J. Fatigue* **2017**, *101*, 408–420. [CrossRef]
11. Leitner, M.; Barsoum, Z. Effect of increased yield strength, R-ratio, and plate thickness on the fatigue resistance of high-frequency mechanical impact (HFMI)-treated steel joints. *Weld. World* **2020**, *64*, 1245–1259. [CrossRef]
12. Mori, T.; Shimanuki, H.; Tanaka, M. Influence of steel static strength on fatigue strength of web-gusset welded joints with UIT. *J. JSCE* **2014**, *70*, 210–220. (In Japanese) [CrossRef]

13. Lihavainen, V.M.; Marquis, G. Fatigue life estimation of ultrasonic impact treated welds using a local strain approach. *Steel Res. Int.* **2008**, *77*, 896–900. [CrossRef]
14. Quilliec, L.G.; Lieurade, H.P.; Bousseau, M.; Drissi-Habti, M.; Inglebert, G.; Macquet, P.; Jubin, L. Mechanical and modelling of high-frequency mechanical impact and its effect on fatigue. *Weld. World* **2013**, *57*, 97–111. [CrossRef]
15. Schubnell, J.; Hardenacke, V.; Farajian, M. Strain-based critical plane approach to predict the fatigue life of high frequency mechanical impact (HFMI)-treated welded joints depending on the material condition. *Weld. World* **2017**, *61*, 1199–1210. [CrossRef]
16. Schubnell, J.; Pontner, P.; Wimpory, R.C.; Farajian, M.; Schulze, V. The influence of work hardening and residual stresses on the fatigue behavior of high frequency mechanical impact treated surface layers. *Int. J. Fatigue* **2020**, *134*, 105450. [CrossRef]
17. Yıldırım, H.C.; Marquis, G.B. A round robin study of high frequency mechanical impact (HFMI)-treated welded joints subjected to variable amplitude loading. *Weld. World* **2013**, *57*, 437–447.
18. Mikkola, E.; Doré, M.; Marquis, G.B.; Khurshid., M. Fatigue assessment of high-frequency mechanical impact (HFMI)-treated welded joints subjected to high mean stresses and spectrum loading. *Fatigue Fract. Eng. Mater. Struct.* **2015**, *38*, 1167–1180. [CrossRef]
19. Leitner, M.; Stoschka, M.; Ottersböck, M. Fatigue assessment of welded and high frequency mechanical impact (HFMI) treated joints by master notch stress approach. *Int. J. Fatigue* **2017**, *101*, 232–243. [CrossRef]
20. Leitner, M.; Stoschka, M.; Barsoum, Z.; Farajian, M. Validation of the fatigue strength assessment of HFMI-treated steel joints under variable amplitude loading. *Weld. World* **2020**, *64*, 1681–1689. [CrossRef]
21. Tai, M.; Miki, C. Improvement effects of fatigue strength by burr grinding and hammer peening under variable amplitude loading. *Weld. World* **2012**, *56*, 109–117. [CrossRef]
22. Schubnell, J.; Carl, E.; Farajian, M.; Gkatzogianis, P.; Knodel, P.; Ummenhofer, T.; Wimpory, R.; Eslami, H. Residual stress relaxation in HFMI-treated fillet welds after single overload peaks. *Weld. World* **2020**, *64*, 1107–1117. [CrossRef]
23. Yonezawa, T.; Shimanuki, H.; Mori, T. Influence of cyclic loading on the relaxation behaviour of compressive residual stress induced by UIT. *Weld. World* **2020**, *64*, 171–178. [CrossRef]
24. Mikkola, E.; Remes, H.; Marquis, G. A finite element study on residual stress stability and fatigue damage in high-frequency mechanical impact (HFMI)-treated welded joint. *Int. J. Fatigue* **2017**, *94*, 16–29. [CrossRef]
25. Nazzal, S.S.; Mikkola, E.; Yıldırım, H.C. Fatigue damage of welded high-strength steel details impoved by post-weld treatment subjected to critical cyclic loading conditions. *Eng. Struct.* **2021**, *237*, 111928. [CrossRef]
26. Ruiz, H.; Osawa, N.; Rashed, S. Study on the stability of compressive residual stress induced by high-frequency mechanical impact under cyclic loadings with spike loads. *Weld. World* **2020**, *64*, 1855–1865. [CrossRef]
27. Yıldırım, H.C.; Remes, H.; Nussbaumer, A. Fatigue properties of as-welded and post-weld-treated high-strength steel joints: The influence of constant and variable amplitude loads. *Int. J. Fatigue* **2020**, *138*, 105687. [CrossRef]
28. Mikkola, E.; Marquis, G.; Lehto, P.; Remes, H.; Hänninen, H. Material characterization of high-frequency mechanical impact (HFMI)-treated high-strength steel. *Mater. Des.* **2016**, *89*, 205–214. [CrossRef]
29. SONATS. Residual Stress Measurement by X-ray Diffraction. Available online: https://sonats-et.com/en/residual-stress/x-ray-diffraction-services/ (accessed on 26 September 2021).
30. Pirlinga, T. Precise analysis of near surface neutron strain imaging measurements. *Procedia Eng.* **2011**, *10*, 2147–2152. [CrossRef]
31. Kuhlmann, U.; Durr, A.; Bergmann, J.; Thumser, R. *Fatigue Strength Improvement for Welded High Strength Steel Connections Due to the Application of Post-Weld Treatment Methods*; FOSTA, Forschung fur die Praxis P 620: Dusseldorf, German, 2006. (In Germany)
32. Kuhlmann, U.; Gunther, H. *Experimentelle Untersuchungen zur Ermüdungssteigernden Wirkung des PIT-Verfahrens*; Universität Stuttgart Institut für Konstruktion und Entwurf: Stuttgart, Germany, 2009. (In Germany)
33. Tehrani Yekta, R. Acceptance Criteria for Ultrasonic Impact Treatment of Highway Steel Bridges. Ph.D. Thesis, University of Waterloo, Waterloo, AB, Canada, 2012.
34. Suzuki, T.; Okawa, T.; Shimanuki, H.; Nose, T.; Suzuki, H.; Moriai, A. Effect of ultrasonic impact treatment (UIT) on fatigue strength of welded joints. *Adv. Mat. Res.* **2014**, *996*, 736–742. [CrossRef]
35. Ranjan, R.; Ghahremani, K.; Walbridge, S.; Ince, A. Testing and fracture mechanic analysis of strength effects on the fatigue behaviour of HFMI-treated welds. *Weld. World* **2016**, *60*, 987–999. [CrossRef]
36. Leitner, M.; Mossler, W.; Putz, W.; Stoschka, M. Effect of post-weld heat treatment on the fatigue strength of HFMI-treated mild steel joints. *Weld. World* **2015**, *59*, 861–873. [CrossRef]
37. Polezhayeva, H.; Howarth, D.; Kumar, M.; Ahmad, B.; Fitzpatrick, M.E. The Effect of compressive fatigue loads on fatigue strength of non-load carrying specimens subjected to ultrasonic impact treatment. *Weld. World* **2015**, *59*, 713–721. [CrossRef]
38. Anami, K.; Miki, C.; Tani, H.; Yamamoto, H. Improving fatigue strength of welded joints by hammer peening and Tig-dressing. *J. JSCE* **2000**, *17*, 67–78. [CrossRef]
39. Fisher, W.J.; Sullivan, M.D.; Pense, W.A. *Improving Fatigue Strength and Repairing Fatigue Damage*; Fritz Laboratory Reports, Paper 2067; Lehigh University: Bethlehem, PA, USA, 1974.
40. Voce, E. The relationship between stress and strain for homogeneous deformation. *J. Met.* **1948**, *74*, 537–562.
41. Garcia, M. Multiaxial Fatigue Analysis of High-Strength Steel Welded Joints Using Generalized Local Approaches. Ph.D. Thesis, EPFL, Lausanne, Switzerland, 2020.
42. Smith, M. *Abaqus/Standard User's Manual, Version 6.9*; Dassault Systèmes Simulia Corp.: Johnston, RI, USA, 2009.

43. Smith., K.N.; Watson, P.; Topper, T.H. A stress-strain function for the fatigue of metals (stress-strain function for metal fatigue including mean stress effect). *J. Mater.* **1970**, *5*, 767–778.
44. Kujawski, D. A deviation version of the SWT parameter. *Int. J. Fatigue* **2014**, *67*, 95–102. [CrossRef]

metals

Article

Design Implications and Opportunities of Considering Fatigue Strength, Manufacturing Variations and Predictive LCC in Welds

Mathilda Karlsson Hagnell [1,*,†], Mansoor Khurshid [2,3,†], Malin Åkermo [3] and Zuheir Barsoum [3,*,†]

1   The Centre for ECO2 Vehicle Design, KTH Royal Institute of Technology, Teknikringen 8, 10044 Stockholm, Sweden
2   Cargotec Sweden AB Bromma Conquip, Kronborgsgränd 23, Box 1133, 16422 Kista, Sweden; mansoor.khurshid@bromma.com
3   Lightweight Structures, Engineering Mechanics, KTH Royal Institute of Technology, Teknikringen 8, 10044 Stockholm, Sweden; akermo@kth.se
*   Correspondence: mathk@kth.se (M.K.H.); zuheir@kth.se (Z.B.)
†   These authors contributed equally to this work.

Abstract: Fatigue strength dictates life and cost of welded structures and is often a direct result of initial manufacturing variations and defects. This paper addresses this coupling through proposing and applying the methodology of predictive life-cycle costing (PLCC) to evaluate a welded structure exhibiting manufacturing-induced variations in penetration depth. It is found that if a full-width crack is a fact, a 50% thicker design can result in life-cycle cost reductions of 60% due to reduced repair costs. The paper demonstrates the importance of incorporating manufacturing variations in an early design stage to ensure an overall minimized life-cycle cost.

Keywords: manufacturing variations; life-cycle costing; fatigue assessment; welding; welding defects

Citation: Hagnell, M.K.; Khurshid, M.; Åkermo, M.; Barsoum, Z. Design Implications and Opportunities of Considering Fatigue Strength, Manufacturing Variations and Predictive LCC in Welds. *Metals* **2021**, *11*, 1527. https://doi.org/10.3390/met11101527

Academic Editor: Pierpaolo Carlone

Received: 27 August 2021
Accepted: 21 September 2021
Published: 26 September 2021

Publisher's Note: MDPI stays neutral with regard to jurisdictional claims in published maps and institutional affiliations.

1. Introduction

Service interruption due to premature fatigue failure in welded joints can result in significant operational and monetary losses. As premature fatigue failures are often a result of initial manufacturing defects and their variation, as shown by [1], both must be accounted for in the design process. In addition, a structure must also be produced and operated at a competitive cost level, which means life-cycle costing must also be part of the design equation in order to ensure an optimal design.

In a welded joint, fatigue failure is often a result of cracks initiated at the weld toe or the weld root, respectively [2–4]. When a welded joint fails from the root, the fatigue resistance has been found to be affected by several geometrical variables such as weld throat size, plate thickness and depth of weld penetration [5,6]. All these geometrical variables are in turn influenced by manufacturing and variations, and can therefore introduce defects and variations that ultimately govern the final fatigue performance of any particular weld [1]. Strategies to ensure weld qualities are numerous, ranging from weld procedure recommendations [2] to the introduction of post-weld treatments such as HFMI, TIG and burr-grinding [7–9] or post-weld thermal treatments [10,11]. However, all these mentioned quality-ensuring strategies comes at an increased manufacturing cost. A cost, which ultimately has to be compared to that of increased fatigue lifetime, and hopefully reduced operational costs.

Apart from technical capacity, the fatigue design of a structure also dimensions its full life-cycle cost (LCC), and life-cycle energy impact. For example, a design with an improved fatigue behaviour is likely to become more costly to produce as it may include more expensive materials; and/or results in a more involved production process. However, the same design would become less costly throughout its operational use due to its longer technical life and potentially lower maintenance and repair need. Consequently, to fully control and balance both fatigue and life-cycle cost, fatigue assessment and life-cycle cost should ideally be done simultaneously in an early conceptual design stage.

Advances in cost modelling [12–14], have shown that there exists an increased interest in introducing cost assessments in an early design stage. However, life-cycle analysis (LCA), and a traditional LCC, is ultimately still performed in a later design stage as it requires extensive material flow knowledge and large databases. Answering to this, recent research has come to highlight the design benefits of adapting a simplified, more generalized, early-stage LCA. In the study by [15], the method of life cycle energy optimization is proposed and shown useful to assess the importance of recycling of lightweight composite materials. In [16], the authors present a material-selection approach that incorporates both structural design and LCA that spans material systems. In the study by [17], LCA-optimization of a concrete precast bridge provides essential design guidance in relation to the found energy-intensive life-cycle stages of manufacture, use and maintenance. Moreover, for steel bridge construction, Ref. [18] finds that simple production cost optimization leads to increased life-cycle cost. This underlines the importance of designing for full life-cycle cost as opposed to that of only production cost. In addition, LCC is often incorporated in work available on maintenance and repair logistics and optimization, such as work by [19–22]. However, when it comes to LCC coupled to manufacturing variations and fatigue assessment of welded structures, the scientific literature of the field becomes limited. Hence, more research is needed.

This motivates the current study, where the gap is addressed through proposing a predictive LCC (PLCC) scheme, that predicts production, use and end-of-life (EoL) costs of welded structures as a function of governing geometry, complexity and required production and recycling flows. Here, predictive means that the methodology is aimed for use in an early conceptual design stage, in which it can identify general trends that supports holistic design decision while demanding a low amount of user input. Involved production costs are estimated through implementing a previously developed predictive technical cost model by [12,23].

The PLCC scheme is implemented and demonstrated in a parametric case study in which the thicknesses of the main load-carrying members of a representative welded box structure are varied. The different plate thicknesses give rise to different fatigue strength properties and ultimately also different life-cycle costs. Two different fatigue scenarios are considered, each representative of a specific manufacturing variation with respect to lack of penetration depth (LOP). LOP is chosen as a representative manufacturing defect due to its importance to ensure full fatigue life, as shown by [24]. To further assess the production cost impact of penetration depth, an additional assessment of the production cost as a result of penetration depth of the structure is presented. Finally, a discussion is presented on general aspects on managing variations in welding, their impacts on fatigue life as well as important aspects on overarching design related to fatigue assessment methods. Overall, the study spans disciplines on cost and fatigue assessments and addresses important aspects on the multi-disciplinary problem of designing cost-efficient, and robust, welded structures.

2. Scope

The methodology and case-study presented in this paper connects conceptual design of welded fatigue-dimensioned structures to lifecycle costing (LCC). As the focus is early conceptual design, the definition of predictive lifecycle costing (PLCC) is introduced. The methodology suitability is evaluated for application in a variation-driven conceptual design stage. Consequently, discussions on the impact of variation-driven issues with regards to fatigue and predicted life cycle cost are also given as part of the results.

Given the focus on conceptual design, or early stage development, a number of assumptions and limitations need to be posed in order to complement the inherent lack of design knowledge present in such early design stages. Assumptions and limitations include:

- A traditional lifecycle assessment include present distribution channels. However, for the scope of this study, distribution is deemed too uncertain to assess at an early design stage.
- Input cost data such as material cost per kg and equipment investments are retrieved from comparative, published, data to avoid assessing the result of specific supply-chains and internal business partnerships.
- Investment costs are sized for full use in specific production, this means shared equipment systems are not included.
- R&D and overhead costs are excluded from the modelling scope as they are highly tied to specific organizations and often difficult to credit any particular component.
- Inspection and control are not treated in this work, but are instead part of future work.

3. Methodology and Framework

To couple PLCC and conceptual design of fatigue-designed welded joints, it is important to understand, and model, the existing connections and variables between the two. Examples of important variables that connect fatigue performance and individual lifecycle phases are illustrated in Figure 1. The individual coupling variables can either align or converge on life-cycle cost and fatigue behaviour with regards to their optimal value. For example, a low-cost material type minimizes material cost, but likely has an adverse effect on fatigue performance. This in turn affects not only material cost, but also consecutive use phase cost. To further increase the design complexity, the full design space involves a multitude of variables, all with different impact on fatigue behaviour and life-cycle phases. Note that the illustration in Figure 1 lists a full set of connections and variables and depending on specific study, some may not be meaningful in individual cases. For example, for a full structure or a mixed-material system; end-of-life costs are a function of involved geometries as end-of-life disassembly grows in complexity the more complex a geometry or connective welds are. In contrast, end-of-life costs of a single material slab require no disassembly, thus rendering the variable connection unimportant.

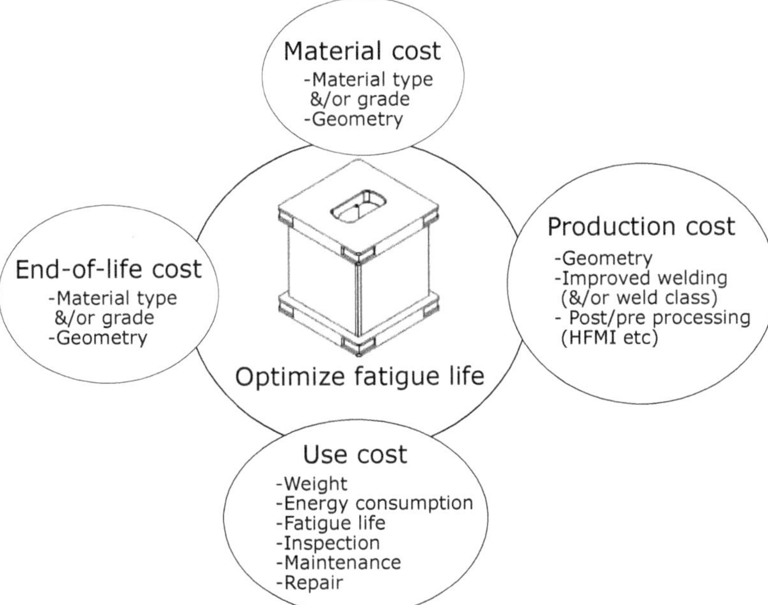

Figure 1. The coupling between LCC and fatigue-dimensioned welded components can be expressed through variables that connects the two for considered life-cycle phases.

4. Predictive Lifecycle Costing (PLCC)

The current study addresses the impact of four life-cycle phases, presented in Figure 2. These phases are raw material acquisition, component production, component use and finally component end-of-life. The life-cycle is considered to be an ideal closed material loop [25], which means that a certain percentage of the entering material is assumed to be recycled to new steel material of the same grade and re-used in a product within the same life-cycle flow. This closed-loop assumption is justified by the fact that steel produced and recycled through an electric arc furnace (EAF) process can be fueled by 100% scrap material [26]. This means that in theory, all scrap material produced during the steel's life-cycle can be recycled. However, as the quality of EAF produced steel is affected if fed by a scrap steel mix containing to many impurities [27], a retrieval rate of 80% is assumed. The material and production costs of a studied component is estimated using a previously developed predictive technical cost model [12,23], while the use phase and end-of-life phase costs are evaluated separately.

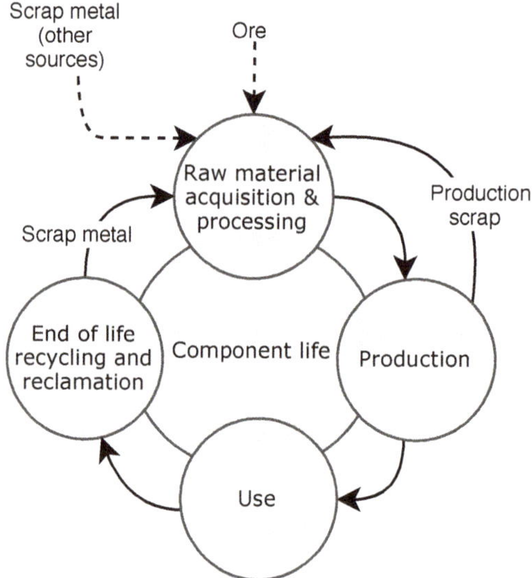

Figure 2. Simplified lifecycle applied within the scope of this paper. Note that a closed loop material flow is assumed, which means dashed material flows (scrap metal originating from other sources and ore) are only drawn for illustrative purposes, and not considered in the following case-study.

5. Predictive Technical Cost Model

To estimate material and production costs of studied component, a previously developed predictive technical cost model [12,23,28] is applied. The model is developed as a stand-alone package in Python [29] and estimates cost through connecting specified production flow to component geometry and complexity [12,23], see Figure 3. The cost model is modular and the production flow is defined by the user through configuring and combining necessary process steps involved in sought production method. The full cost is calculated as the sum of material costs and costs involved within each production process step. Cost categories considered within the scope of the model are defined in Figure 4. Indirect costs such as R&D development and overhead costs are not included within the scope of this model as their size are highly individual and often not known in an early design stage. For the scope of this paper, some extensions to the model library have been made in order to include that of necessary metallic production, including welding processes.

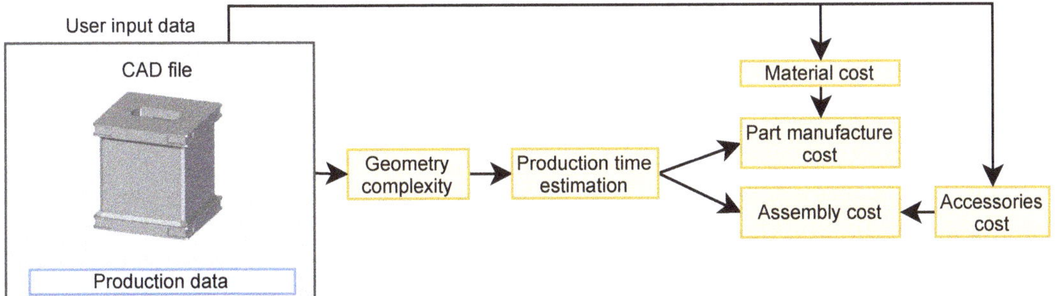

Figure 3. Information flow in applied predictive technical cost model, adapted from previous work [12].

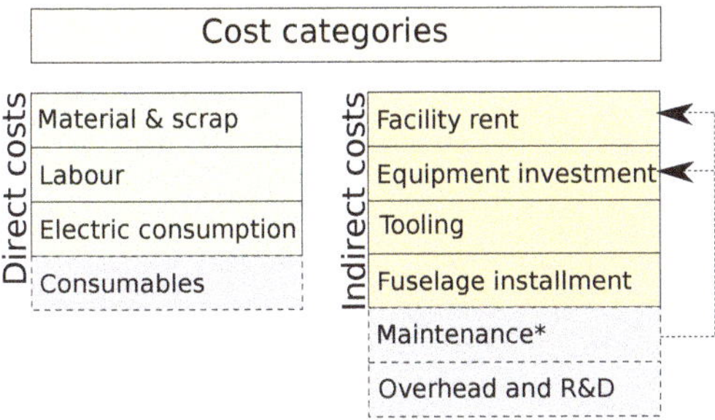

Figure 4. Considered direct and indirect cost categories (drawn in bold line) include material and scrap, labour and power consumption costs. Neglected costs (drawn in dashed line) are overarching costs, such as overhead and R&D, as well as pure consumables (for example basic safety equipment such as gloves, etc). Maintenance costs are not explicitly calculated, but are an implicit part of set equipment utilization and facility costs [28].

5.1. Geometry Complexity

Component complexity is directly correlated to manufacturing and its cost, as the more complex, the more difficult and costly it becomes to manufacture. In the python package, the geometry complexity of the structure to be assessed is calculated and expressed through a complexity factor, C. As a CAD-geometry generally is described using point-clouds and bounding faces, the complexity factor is determined for each such individual bounding face. Factors considered for each bounding face include the face angle, the face curvature radius and the overall face curvature degree. The face angle is the largest internal or neighbour angular transition to the face centre normal. The face curvature radius is the corresponding radius $1/\kappa$ of either the current face, or that neighbour face depending on which produces the highest angular transition. The overall face curvature degree is either single or double, representing a surface bending in single axis versus two axis directions. For more details, see [23].

5.2. Production Time Estimation

Process time is estimated as a function of production process. In general, processing times of machining or additive manufacturing and assembly steps are estimated as a function of the processing rate, r, component complexity, and governing characteristic size, L according to [23] as

$$t = \frac{L}{rC} \tag{1}$$

5.2.1. Machining

Machining of metallic parts are performed using a 5-axis CNC-machine. Machining time is a function of cut length, L_c and table feed rate, v_f, according to [30] as

$$t = \frac{L_c}{v_f} \tag{2}$$

where the table feed rate range from 100 to 250 m/min depending on steel grade [31]. Given the circular cut tool radius of r, each table feed motion removes up to πr^2 m^2 of material per r m feed. An approximation of the removal rate of a face cut-out is therefore

$$Q_A = \frac{A_c}{r v_f}, \tag{3}$$

where A_c is the cut tool area. Using a cut tool of ϕ 15 mm and a table feed rate of 250 m/min, the resulting face removal rate is approximately 0.1 m^2/s. This means the approximate machining time of a face cut-out can be calculate similarly to Equation (2), as $t = A_c/Q_A$.

5.2.2. Welding

For welding methods using solid wires such as metal arc welding (MAG) (ISO 4063-135) and flux cored arc welding (FCAW) (ISO 4063-136), it is proposed that the time to weld a joint between two parts can be expressed as

$$t_{prep} + t_{tack} + t_{pass} + (no_{passes} - 1)t_{repos} + t_{refill}n^\circ_{wires} + t_{post} \tag{4}$$

Variables included are the weld preparation time, t_{prep}, the time to tack weld the plates to achieve sufficient stability for upcoming weld process, t_{tack}, the number of weld passes, no_{passes}, the actual weld pass time, t_{pass}, the time that the welder needs to reposition the wire electrode in between passes, t_{repos}, the solid wire refill time, $t_{refill\ wire}no_{wires}$ and finally the weld post-processing time, t_{post}.

The weld pass time is a function of bead mass, m, and deposition rate, r_w, according to [32] as

$$t_{pass} = \frac{m}{r_w} \tag{5}$$

The bead mass depends on the weld cross-sectional area, $A_{bead\ weld}$, and is here approximated as half of an ellipse and is a function of throat thickness, a and penetration depth i, see Figure 5.

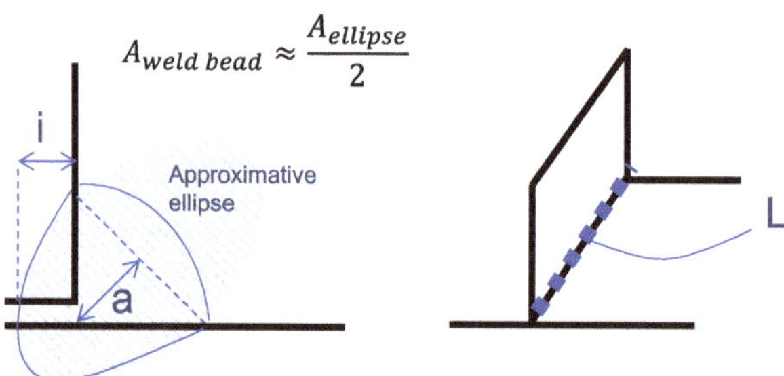

Figure 5. The weld bead mass is calculated from an assumed elliptical cross-sectional area, which is a function of throat thickness, a and penetration depth i.

The deposition rate is a function of the electrode area, A, weld metal density, $\rho_{weld\ metal}$, wire feed speed, v_{feed}, and deposition efficiency, η, and can be expressed as [32]

$$r_w = A\rho_{weld\ metal}v_{feed}\eta \tag{6}$$

The wire feed speed, v_{feed}, can be preliminary determined as a function of weld method and weld complexity [33], if not directly stated by a comparable specified weld process specification (WPS). The deposition efficiency, η, is also a function of welding method and is 0.95 and 0.85 for MAG and FCAW 0.85, respectively [32].

The reposition time, t_{repos}, need to be at least that of the pass cooling time, $\Delta_{t8/5}$, which can be calculated as a function of the preheat temperature, T_0, and the gross heat input per unit length of weld kJ/mm, q_w according to [34] as

$$\Delta_{t8/5} = (6700 - 5T_0)q_w\left(\frac{1}{500 - T_0} - \frac{1}{800 - T_0}\right) \tag{7}$$

5.2.3. Weld Preparation

Weld preparation can consist of different processes such as gouging [32], manual abrasive cleaning or the application of a chemical cleaner. In general, the time to perform these type of processes is proposed to be expressed as a function of the weld length, L, the weld leg hypotenuse, b, the process speed, v, and the weld complexity factor, C_w, according to

$$t_{prep} = bLvC_w \tag{8}$$

In an early conceptual stage, the complexity factor, C_w, can be assumed equal to that of a difficulty factor as suggested by [33].

5.3. Material Cost

The material cost is calculated as a function of the scrap rate, r_{scrap}, cost per kg material, $C_{kg\ cost}$, and the weight of the component, w, according to [23] as

$$C_{mtrl\ cost} = wC_{kg\ cost}(1 + r_{scrap}) \tag{9}$$

The cost per kg material is usually negotiated with respect to purchased quantity and supplier relationship status; however, as the cost model is to be applied in an early stage conceptual phase, guiding price indexes such as that of steel price index figures [35] and other material data bases [36] are used throughout the model.

5.4. Part Manufacture Cost

Direct costs such as labour and power consumption are a function of individual cost driver, i.e. hourly cost and electricity cost respectively, and process time t. Indirect costs such as facility rent, equipment investment and tooling are calculated as a function of manufacturing volume, n, or parts per year. Necessary number of facility and production lines, n_p, are therefore calculated as a function of process time, annual work time, t_{tot} and the annual production volume according to [23] as

$$n_p = \left\lceil \frac{tn}{t_{tot}} \right\rceil \tag{10}$$

Given Equation (10), the investment cost of a certain manufacturing method is calculated as $n_p C_I/\delta$, where C_I is the investment cost of a piece of equipment and δ its depreciation rate. To account for installation and drive-in costs, an extra 20% is added to the investment cost of acquired machinery.

Machining

The investment cost of a CNC-machine can vary from 50 k€ for a 2-axis-machine to beyond 500 k€ for a multi-spindle, multi-axis machine [37]. A 5-axis CNC-machine for 300 k€ is considered to fulfil most needs and feed speeds for a medium to high production setup [23].

5.5. Assembly Cost

The predictive technical cost model covers several different assembly processes, both manual and automatic. The assembly method of importance for a later case study is that of metallic welding.

Metal Arc Welding and Flux Cored Arc Welding

The investment cost of a welding machine varies from 700 € for a small 140 A manual welder to 17 k€ for an industrial-graded 600A synergic welder [38]. In the predictive technical cost model, welding equipment is selected based on sufficient power output for manual or synergic, see Figure 6. A manual welding machine demands a more skilled operator, but is generally less costly than a synergic, or pulse-controlled, machine.

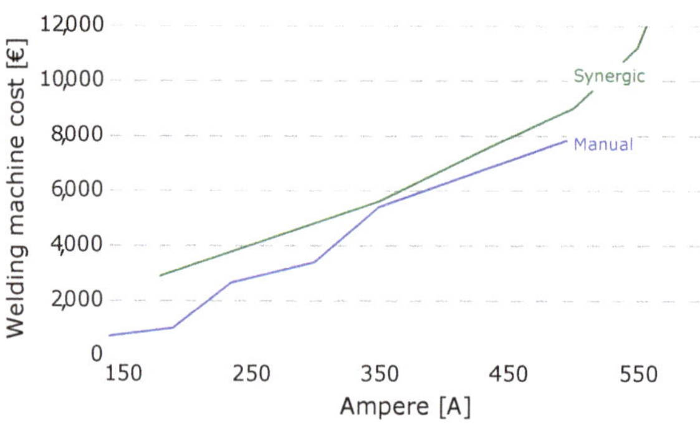

Figure 6. Investment costs of different welding machines [38,39] as a function of highest Ampere output.

6. Fatigue Assessment

Apart from manufacturing variations, fatigue assessment in itself give rise to hidden design variations as different fatigue assessment strategies yield different fatigue strength. In fact, a design can become either under-performing or overly conservative if the chosen fatigue life assessment strategy is not sufficient for the particular design and loading scenario.

Currently, there are mainly four established methods for fatigue life assessment of welded structures [2–4]; nominal stress approach, structural/geometrical "hot-spot" stress approach, effective notch stress approach and linear elastic fracture mechanical crack growth approach. Fatigue resistance of complex welded components based on stress analysis performed with FEA can be assessed in many ways with varying degrees of time consumption and accuracy. A large model will increase both the model preparation and the computational time. Large and complex FEA models may include several critical locations and complex boundary conditions, see example in Figure 7 where the stress value is continually changing at different locations. Nominal stress values are in some of the critical sections difficult or impossible to define. Even if a nominal stress can be defined, one must select from a catalogue of details, the geometry most closely resembling the actual welded detail. In many cases, the actual weld has little similarity to one of the geometries shown in the standards [40–42]. A schematic overview over complexity and work effort for different design methods are presented in Figure 8.

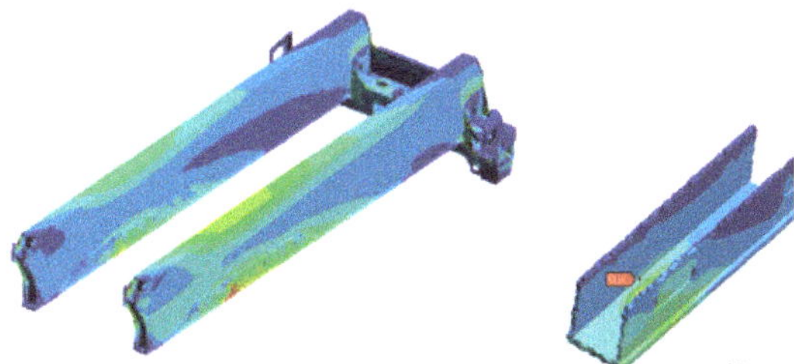

Figure 7. Stresses in Telescopic beam unit of a spreader, red colour corresponds to high and blue to low stresses.

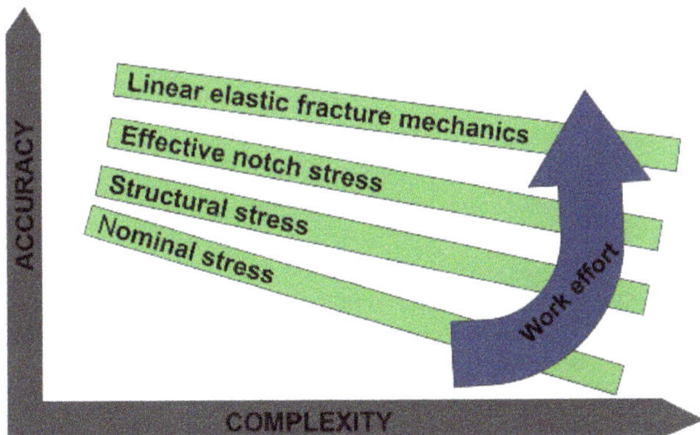

Figure 8. Schematic overview accuracy, complexity and work effort associated with the different fatigue assessment methods for welded structures.

Various studies have been conducted in order to investigate the accuracy and to map the source of variation for these different fatigue assessment methods. Ref. [43] carried out fatigue assessment using conventional methods and compared the results with fatigue testing of welded A-stay beam structure in an articulated hauler. The failure observed was weld root failure, and it was observed that the estimated fatigue life showed large scatter.

In a recent study, Ref. [44] carried out a round robin fatigue strength assessment of the welded box structure, identical to the structure investigated in the current study. The aim of the study was to identify variation and sources of variation in welding production, map scatter in fatigue life estimation and define and develop concepts to reduce these in all steps of product development. The estimated fatigue lives were also compared with fatigue testing, where the objective was weld root failure where different amounts of weld root penetration were studied. Differences were identified between both methods and participants using the same code/recommendations. It was concluded that for the applied cases, the nominal stress method overestimated the fatigue life and the effective notch method is conservative in comparison to the life of tested components.

Delkhosh et al. [45] studied the fatigue strength of the component in this study using Linear Elastic Fracture Mechanics (LEFM) approach. A parametric study was conducted to study the effect of various weld parameters on the fatigue strength, such as lack of weld metal penetration, load position, and plate thicknesses. The LEFM approach could capture the crack propagation from the weld root reasonably well and estimate the fatigue life. It was observed that compared to fatigue life estimations by nominal stress method or effective notch stress method, the LEFM approach could estimate the residual life more accurately, see Figure 9.

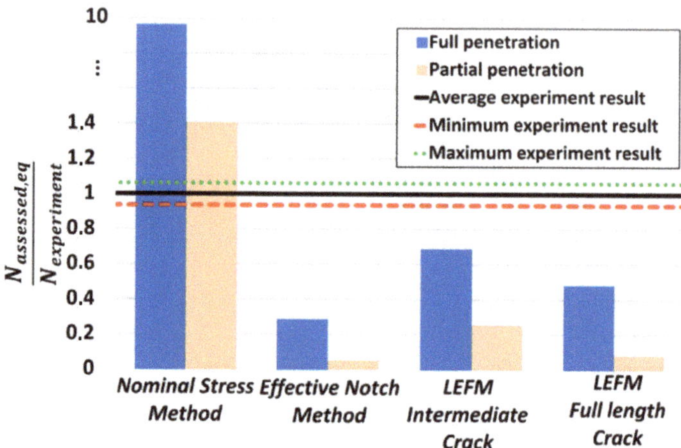

Figure 9. Results range for the different methods and level of the fatigue life assessed, from [45].

Residual stresses in welded joints can have beneficial or detrimental effect on the fatigue strength [2–4]. Residual stresses in the box welded structure considered in this study were evaluated experimentally and numerically by [46]. The effect of residual stress state on the fatigue strength of box welded structures has been discussed by Delkhosh et al. (2020). It was concluded that residual stresses did not significantly affected the fatigue strength of the box welded structure.

7. Case Study: PLCC of a Welded Box Structure

The investigated structure is a representative welded box structure, see Figure 10. The structure is an important member in a spreader. The box structure consists of two flange plates and four web plates [45]. The plates are manufactured from high-strength-structural steel (HSS) where the flange plates are manufactured from S700QL, a quenched and tempered grade, while the web plates are manufactured from S600MC, a hot-rolled structural steel made for cold forming. The structure is assembled through four longitudinal welds and two circumferential bevel welds, see Figure 10. From the perspective of fatigue, the circumferential bevel welds are the critical welds.

Parameters investigated in the PLCC case study include varied flange plate thickness, t_f, varied web plate thickness, t_w and off-center hole position. Design cases and parameter configurations of current case study are given in Table 1. Note that the shift of the hole position, 30 cm towards the critical web plate, effect the symmetry of the loading. This causes the loading mode to become eccentric and thus effects the fatigue life of the box.

Table 1. Design cases and parameter values [45].

Specimen Group	Case ID	Flange Thickness t_f [mm]	Web Thickness t_w [mm]	Loading Mode
Reference case	MTA	30	10	Centric
Varied t_f	MTB	40	10	Centric
	MTC	60	10	Centric
Varied t_w	MTD	30	8	Centric
Varied hole position	MTE	30	10	Eccentric
	MTF	40	10	Eccentric

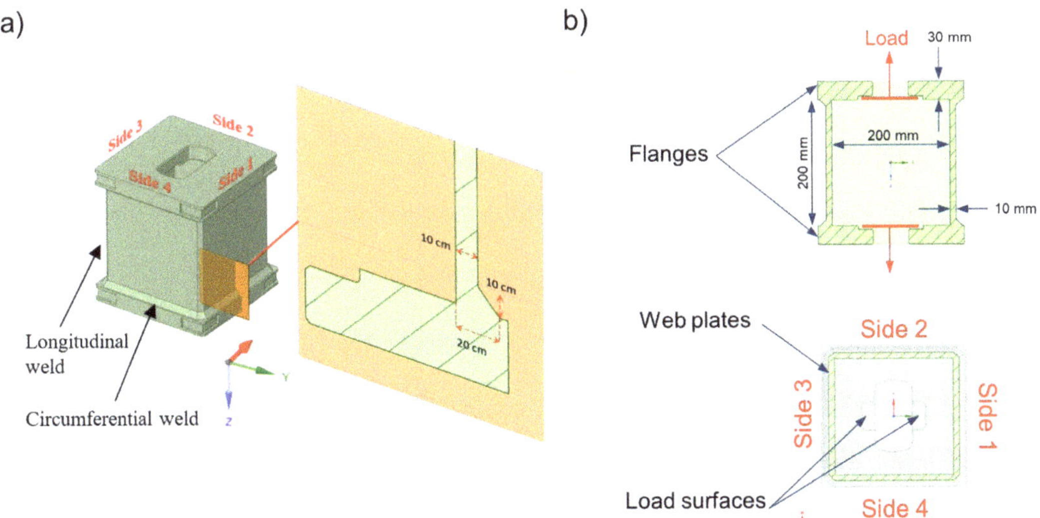

Figure 10. The welded box structure in (**a**) and applied load and dimension details in (**b**) [45].

7.1. Fatigue Behavior and Estimated Life

Each weld box design case is evaluated with respect to four fatigue scenarios of different severity, see Table 2. In each fatigue scenario, it is assumed that the welded box fails as a result of LOP. Note that full weld penetration in this case still assumes a slight lack of penetration of 0.5 mm.

Table 2. Fatigue failure scenarios [45].

Scenario ID	Description	LOP [mm]
S1	Full-length crack, partial penetration	4
S2	Intermediate crack (width = 40 mm), partial penetration	4
S3	Full-length crack, full penetration	0.5
S4	Intermediate crack (width = 40 mm), full penetration	0.5

The predicted fatigue life for all weld box setups and fatigue failure scenarios are given in Table 3.

Table 3. Estimated fatigue life (cycles) for each investigated design case according to [45].

Fatigue Scenario	MTA	MTB	MTC	MTD	MTE	MTF
S1	102,841	287,933	1,103,500	67,768	33,366	96,467
S2	589,198	1,538,700	5,486,300	458,650	182,450	517,390
S3	311,108	925,034	Infinite	275,370	102,480	304,010
S4	838,979	2,265,700	Infinite	664,850	255,420	731,070

7.2. Material Cost

Material acquisition cost is calculated according to Equation (9) using material cost parameters in Table 4. Costs per kg material are here retrieved from steel price index figures [35] and other material data bases [36]. Production scrap rates are estimated from the ratio between cut-out size (including trimmings) and original feeding material size [47].

Table 4. Material cost data [35].

Steel Type	Density [kg/m^3]	Cost [€/kg]	Scrap Rate [%]
Cold rolled, quenched and tempered, HSS plate	7850	0.7	26
Hot rolled HSS plate	7850	0.65	6

Apart from used raw material, the cost of the weld filler material is also a factor. Filler material cost used is defined in Table 5.

Table 5. Weld filler material data.

Filler Material	[€/kg]
ER 70S-6	5.3

7.3. Production Cost

The production cost is a function of the manufacture of the plates and their assembly. The manufacture involves CNC-machining of HSS plate perimeters and cut-outs. The assembly involves mounting and weld tacking followed by several weld passes. It is assumed that the production is performed at an hourly labour cost of 4–5 € [48], and electricity fuselage annual cost of 3000 € for a 600 kW fuse [49]. To investigate the cost implication of introducing simple weld preparation and post-processing, manual cleaning is considered in both cases for the most involved production flow process. The process flow is specified in Figure 11. For more welding details, refer to [45,46,50].

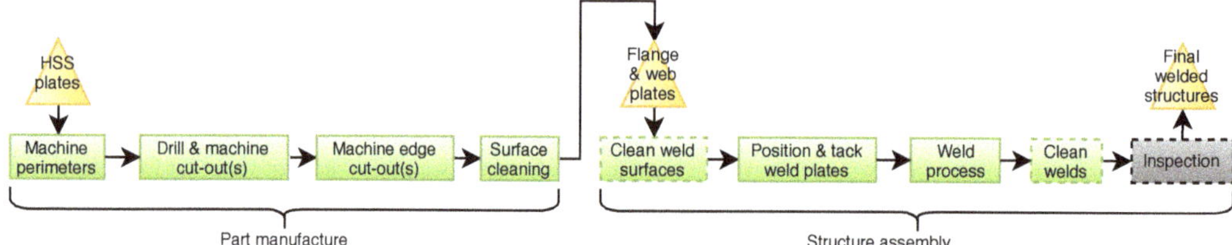

Figure 11. Welded box production flow. As the cost-impact of introducing weld preparation and weld post-processing is investigated, these processes are marked with a dashed line. Inspection cost is not the topic of this paper and is therefore excluded and marked in grey.

7.4. Use Phase Cost

The use phase cost is a direct function of fatigue design together with application and governing duty cycle, or the proportion of time during which a piece of equipment is in active use. For a crane hoist, that means the proportion of time during which the hoist is lifting or lowering a load. Assuming a severe service [51], the crane hoist is estimated to operate in a harbour setting where it lifts containers weighing on average 25 tonnes. The crane hoist is estimated to lift 12 such containers per hour and is operated for 10 h a day, 250 days a year and has a technical lifetime of 20 years. The hoist crane is lifted using a 200 kW hoisting motor for all three cases. Continuous electricity cost is a function of the European weighted average electricity cost of 0.125 € per kwh [52] and required fuselage cost [23].

Apart from costs given by the explicit duty cycle, other use phase costs include inspections and maintenance. The American Occupational Safety and Health Administration [53] dictates that frequent and periodic inspections are to be carried out. Frequent, daily, inspections can be performed by the crane operator before use. In this paper, the costs of frequent inspections are implicitly recorded through assuming that visual inspections are performed by the operator throughout the 10 h operating time. Periodic inspections are to be performed by experts, at a 1–12 month interval. In this paper periodic inspections by experts are considered to be performed annually. The annual inspection is expected to result in half a day shut-down and a cost resulting from the combination of downtime cost and expert inspection cost. The downtime cost corresponds to the explicit income loss of planned downtime. The income loss is considered to be the terminal handling charge (THC) loss, which range from 88–210 € for a 20 ft container [54]. In the performed PLCC, a THC-charge of 200 € as reported by Rotterdam, Netherlands [55] is used. The expert inspection cost fee is assumed equal to that of a flying dispatch cost [56], at an hourly rate of 85 € .

If fatigue failure or failure initiation is discovered during the technical lifetime of the crane hoist, repair and resulting downtime costs are also part of the life-cycle cost, see Figure 12. Similarly to that of annual inspection, it is assumed that a discovered failure requires half a day shut-down for repair work. Apart from the downtime and flying dispatch cost, the repair cost also includes the direct welding cost as calculated from the predictive technical cost model described in Section 5.2.2. The cumulative fatigue damage level, C, is evaluated over the estimated technical lifetime, and it is assumed that fatigue failure occurs if Miner's rule predicts a cumulative damage level of 70% according to

$$C = \frac{n_i S_i}{N_i S_i} > 0.7 \rightarrow fatigue\ failure \tag{11}$$

as a function of the comparative stress level, S_i, the current number of cycles, n_i and the total number of cycles to failure, N_i, from Table 3. It is assumed that one fatigue cycle corresponds to the lifting of one container. The comparative stress level is given by geometry, gravity, flange plate mass, m_i, and representative flange plate area A_i, according to $\tan 45 g m_i / A_i$. If fatigue failure occurs, it is assumed that repair work resets the crane hoist to a fully operational, undamaged state ($n_i = 0$, $C = 0$).

7.5. End-of-Life Cost

To allow for some alternative waste flows, a retrieval rate of 80% is accounted for at the final end-of-life stage. Production scrap is not credited to the full cycle, and is therefore simply treated as a cost. It is assumed that the steel of a retrieved component is valued at a UK market price of 0.26 € per kg [57].

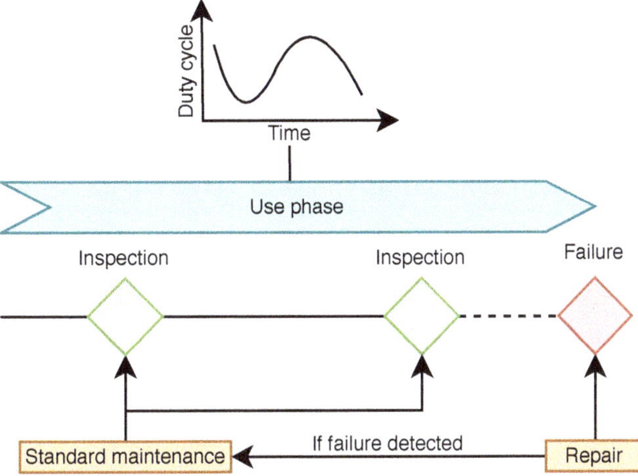

Figure 12. Important use phase costs include actual application use and resulting inspection, maintenance, repair and downtime.

8. Results

The results include sections on all investigated design cases. In addition, as the fatigue scenarios explored in the parametric PLCC concern the impact of weld penetration depth; a complementing pure production cost assessment is presented for three different penetration depths, *i*. These penetration depths are $i = 6, 8, 10$ mm, where 10 mm penetration depth corresponds to ideal full penetration. This assessment directly couples penetration depth to production cost and therefore shows how manufacturing efforts that ensure weld quality also drive production costs.

8.1. PLCC as a Function of Flange Plate Thickness, Web Plate Thickness and Hole Position

The predicted life-cycle cost grow for each year, as presented in Figures 13–15 for each design case and fatigue scenario. The different design cases and fatigue scenarios show different step-wise incremental life-cycle cost increases over the years. These steps often correspond to repair and downtime costs as a result of fatigue damage, as shown by the R&I (research and inspection) label in the figures to the right in Figures 13–15. For the posed technical lifetime of 20 years, the initial production cost becomes negligible in all design cases. Different flange plate thicknesses (MTA-MTC) and shifted hole position (MTE-MTF) have the highest impact on the overall lifecycle cost, see Figures 13 and 15, respectively. A lowered web thickness (MTD), is shown in Figure 14, to have little effect on the overall lifecycle cost for all fatigue scenarios.

The flange plate thickness is in Figure 13 shown to have a high impact on the lifecycle cost, and most dominantly so in the most severe fatigue scenario, S1. At full technical life, increasing the flange thickness 10 mm (MTB) reduces the overall lifecycle cost by 40%, while increasing the flange thickness 30 mm (MTC) reduces the lifecycle cost by 60%. In the same fatigue scenario, it can be noted that repairs for the cases with increased flange thicknesses, MTB and MTC, are needed each third and ninth year, respectively. The baseline on the other hand, needs continuous annual repairs throughout its lifetime for the same fatigue scenario. For the other three fatigue scenarios, S2–S4, increased flange plate thicknesses also result in reduced lifecycle costs and less frequent repairs.

Thinner web plates are shown to have low impact on the life-cycle cost of the full box, see Figure 14. For the fatigue scenarios S1 and S3, we find that all considered cases need annual repairs thus making the design cases equal in terms of life-cycle cost. A small impact can be noticed in the less severe fatigue scenarios, S2 and S4, where the life-cycle cost at full technical life is a few percentage higher for the case with thinner webs. In each fatigue scenario, the thinner webs case, MTD, need to be repaired with one respectively two years shorter intervals compared to the baseline box.

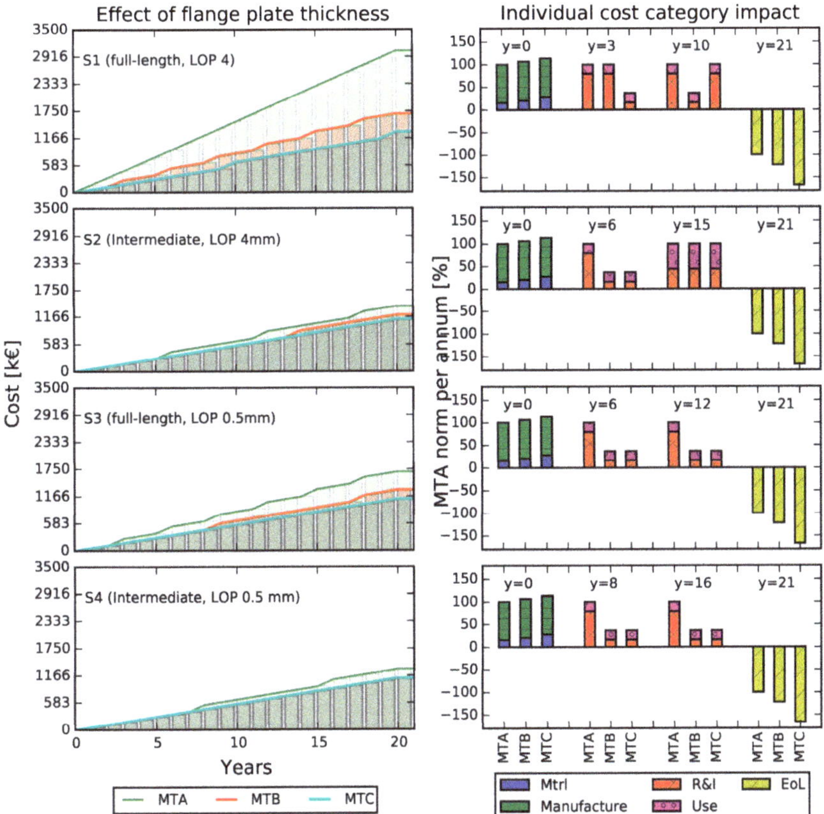

Figure 13. Annual accumulated lifecycle cost for each fatigue scenario and design cases MTA-MTC, in which flange thickness is varied according to $t_f = 30, 40, 60$ mm, respectively. (**left**) accumulated costs per year. (**right**) the divisions between cost contributions for specified year, normalized with respect to the baseline case (MTA).

The position of the hole on the flange plates is shown in Figure 15 to have high impact on the fatigue scenarios S2–S4. For the most severe fatigue scenario, S1, we find that all considered cases need annual repairs. The highest impact of eccentricity is shown in fatigue scenario S3, where the life-cycle cost becomes close to 70% higher as opposed to the baseline box. In all fatigue scenarios it is shown that increasing the flange thickness by 10 mm, case MTF, compensates for the hole eccentricity. Thus, in all fatigue scenarios case MTF returns a similar life-cycle cost as that of the baseline box.

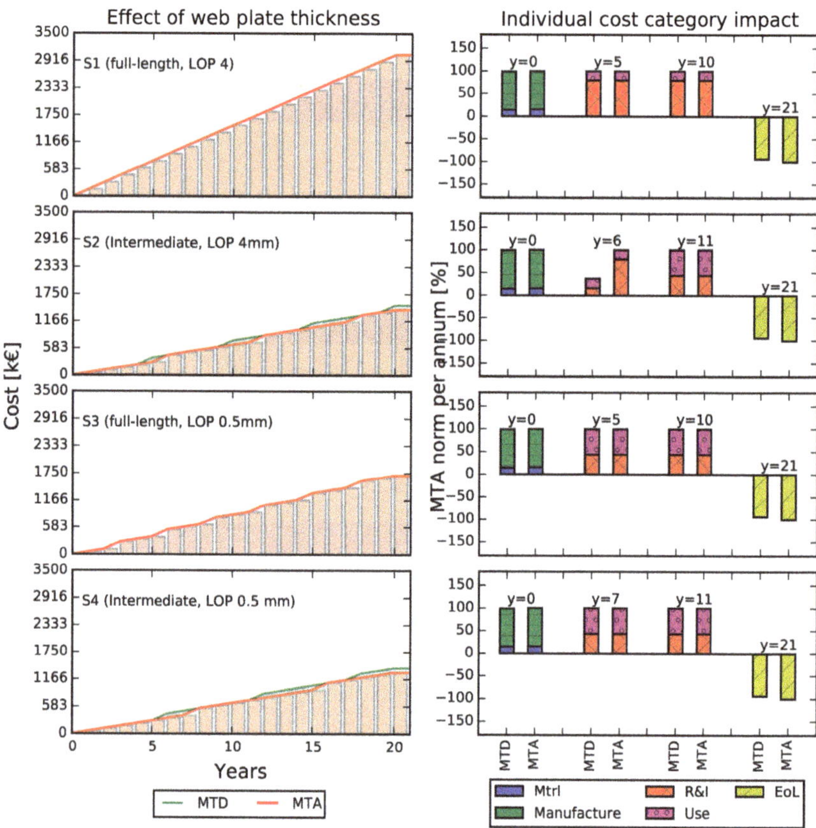

Figure 14. Annual accumulated life-cycle cost for each fatigue scenario and design cases MTA and MTD, in which the web plate thickness is varied according to $t_w = 8, 10$ mm, respectively. (**left**) accumulated costs per year. (**right**) the divisions between cost contributions for specified year, normalized with respect to the baseline case (MTA).

Production Cost

The production cost of the welded box as a function of annual manufacturing volume for each flange plate thickness case is given in Figure 16. The production cost reduces with increasing annual manufacturing volume. This is a result of how the cost for lower annual manufacturing volume is dominated by manufacturing and indirect costs such as investments, see Figure 17. For larger annual manufacturing volumes, the production cost stabilize at a lowest cost per part, which corresponds to a stable division between cost categories, governed by direct costs such as material and operator costs. When the lowest stable cost has been reached, a flange plate thickness increase of 10 mm increases the production cost with 24% while a flange plate thickness increase of 20 mm increases the production cost with 43%.

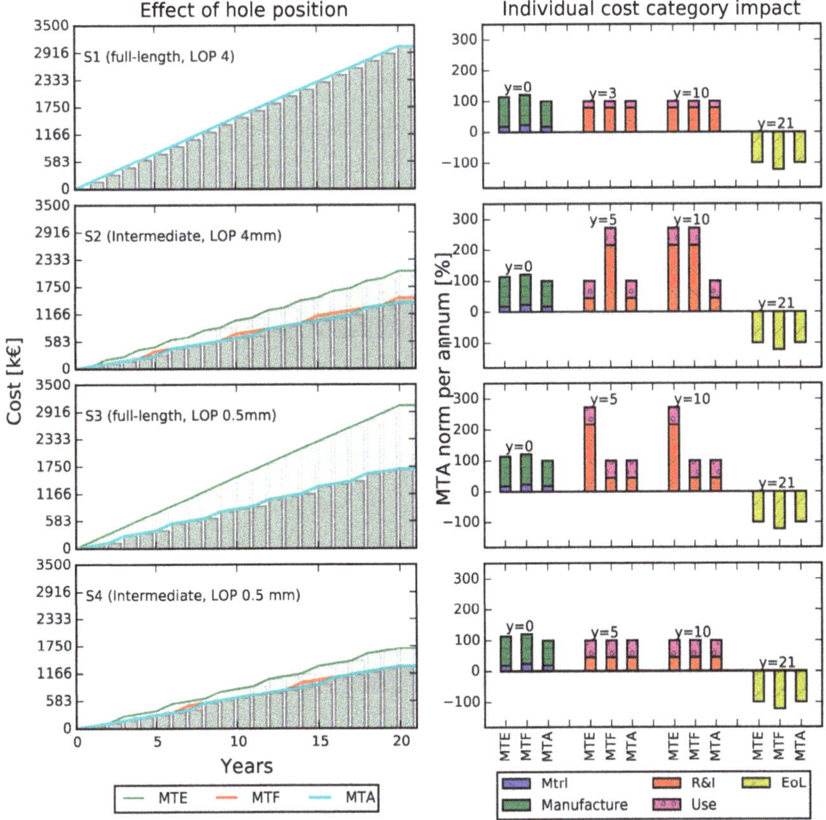

Figure 15. Annual accumulated life-cycle cost for each fatigue scenario and design cases MTE, MTF and MTA, where each represents a different hole position (eccentric vs. centric). (**left**) accumulated cost for per year. (**right**) the divisions between cost contributions for specified year, normalized with respect to the baseline case (MTA).

How the cost shifts importance between indirect costs (investments and facility costs) to direct costs (operator, material and direct electricity costs) is similarly shown for the process of assembly, i.e., welding, of the box structure, shown in Figure 18.

8.2. The Impact of Penetration Depth on Production Cost

The production costs as a function of annual manufacturing volume of the welded box structure for different penetration depths are given in Figure 19. It is clear that only accounting for different penetration depths, or in fact different bead mass, has little impact on the overall production cost. However, to ensure full penetration depth, the welding process need to be more involved. If considering increased effort through applying pre- and post-weld cleaning and assuming previous hourly labour rates, the cost increases between 16–30% depending on the rate of cleaning, see Figure 19.

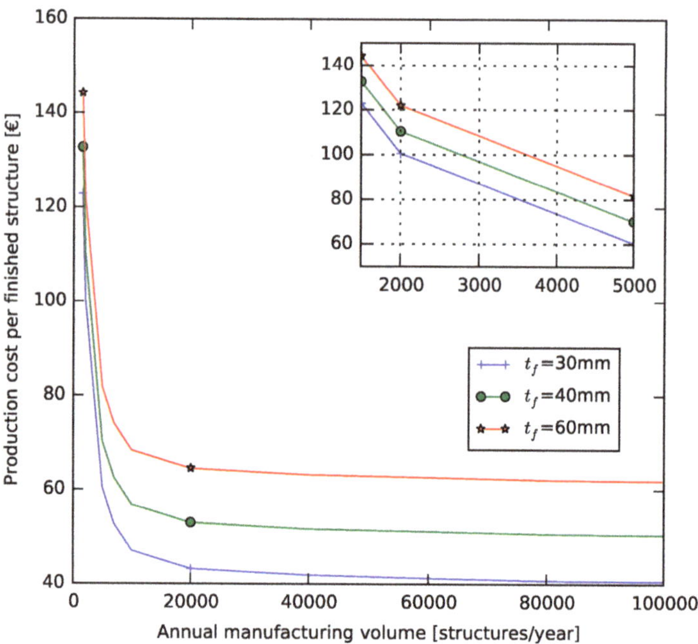

Figure 16. Production cost as a function of annual manufacturing volume for cases $t_f = 30, 40, 60$ mm.

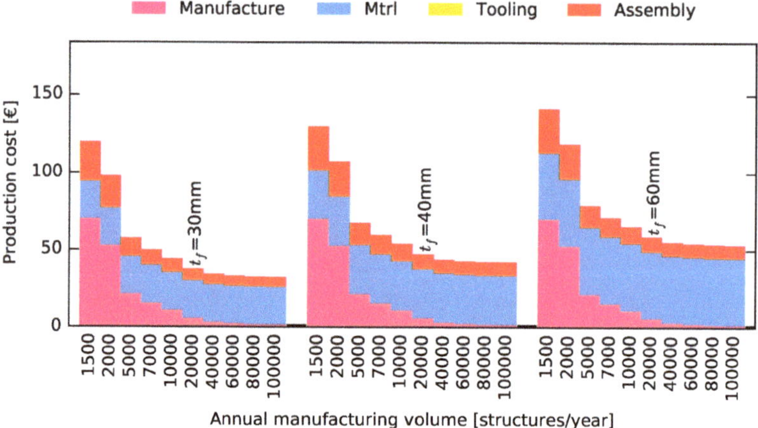

Figure 17. Cost division between manufacture, material and assembly (i.e., welding) costs for different annual manufacturing volumes for the case of $t_f = 30, 40, 60$ mm.

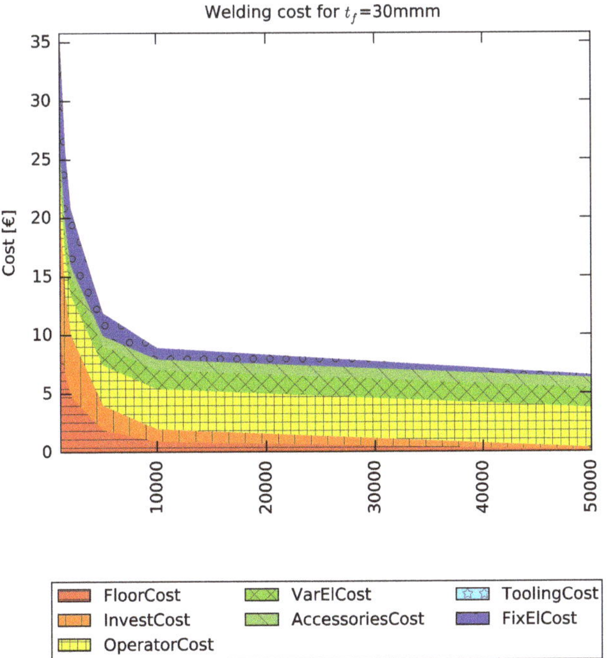

Figure 18. Cost division between cost categories for the welded box assembly, i.e., welding, for the case of $t_f = 30$ mm.

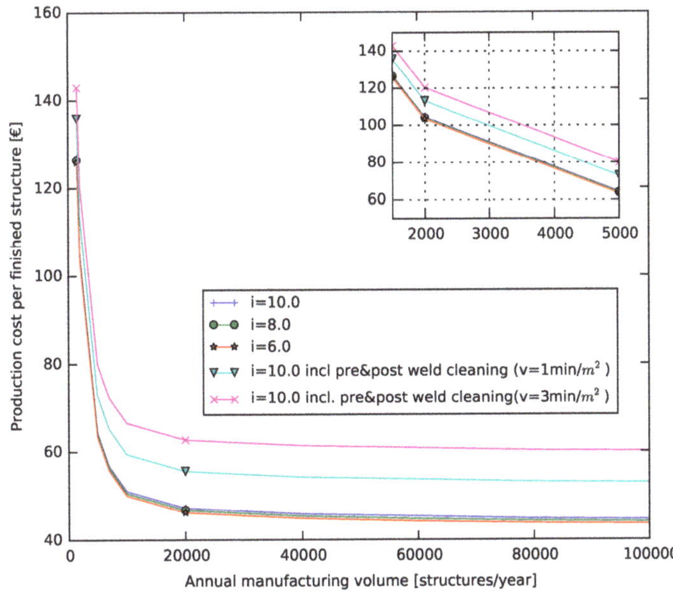

Figure 19. Production cost as a function of annual manufacturing volume for different penetration depths together with pre- and post-weld cleaning $i = 6, 8, 10$ mm.

9. Discussion

The presented PLCC scheme and case study have shown the importance of proper design and fatigue assessment in an early conceptual design stage in order to ascertain an optimized design and control life-cycle costs. Furthermore, the fact that maintenance and repair costs in the most severe fatigue scenarios tend to drive the life-cycle costs highlights the importance to properly control and inspect manufacturing variations present in welding.

The life-cycle assessment has demonstrated that the most important life-phase is the operational use and that the costs to minimize are operation, maintenance and repair. Ultimately, the size of the use phase costs in comparison to the production costs means most fatigue-enhancing efforts, such as increasing the thickness of the load-carrying member (t_f), change of steel steel grade used or introducing improved weld classes, are justified even when resulting in significantly increased production costs. When increasing the flange thickness for example, a 20 mm flange plate thickness increase comes at a cost of 43% production cost increase (or about 18 €), which results in a 60% overall life-cycle cost reduction (or about 1.75 M€). Efforts addressed at managing manufacturing variations, such as flange hole position eccentricity, can similarly also have beneficial impact on the overall lifecycle cost of a design. For example, the impact of the flange hole position eccentricity is in design case MTF shown to be fully compensated by a flange plate thickness increase of 10 mm.

However, it is important to express that minimized production costs are needed to ensure a competitive supply chain and final component. This means all design improvements that maximise the structure fatigue life of a welded structure need to be evaluated for their efficiency before implemented. For example, although ensuring consistent penetration depth is key to improved fatigue behaviour, different strategies to achieve this are more or less costly. For example, the simple introduction of pre- and post-cleaning has been shown to have a large impact on the final, stabilized, production cost, resulting in production cost increases of 16-30% for large annual manufacturing volumes. Thus, to avoid implementing a fatigue improvement strategy that is less cost-efficient than another, predictive technical cost modelling and cost assessment becomes valuable tools.

The welding extension added to the used predictive technical cost model has been verified to that of actual weld specification documents, and the predicted weld time has been shown to comply well with that of recorded figures. This means the estimated production costs are accurately sized. However, it is important to note that the final production costs are a function of defined process scheme and indeed, country of manufacture. Particularly operator and electricity costs vary greatly between nations.

Finally it is important to emphasize that the incorporation of life-cycle costing in an early design stage is not only beneficial from a design perspective, but will also be a key driving factor towards enabling future circular economy, and indeed a fully sustainable production system.

View of Variation

The variation in production can be viewed differently depending on the different stages of the product development stage and role dependent; different roles within the organization needs different information, hence context dependent. In the manufacturing flow, the variation could occur between factories, technology suppliers, batches and within one part. Variation could also be observed within one single weld; along the weld, within the cross section and weld toe geometry. Hence, variation will occur in all product development phases which will have a significant impact on the product quality. Therefore, it is important to identify the sources of variations in the different product development phases in order to minimize their cause. Figure 20 illustrates examples of different sources of variation that could occur in the welding production.

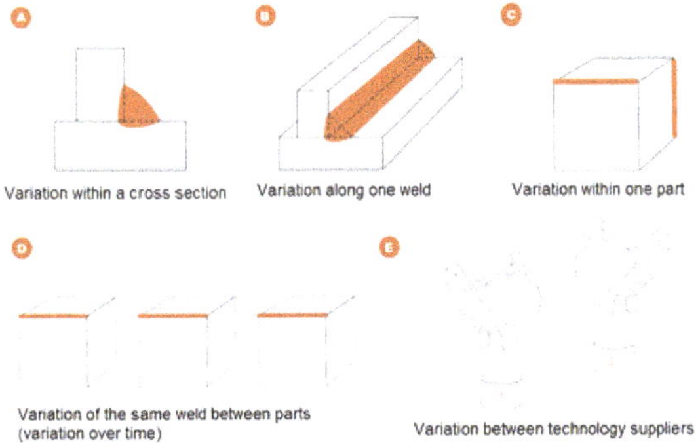

Figure 20. Different sources of variation (**A–E**).

10. Conclusions

Presented PLCC method and case study have shown the importance of designing a welded structure with respect to structural demands and variation in manufacture and corresponding predicted life-cycle cost. In addition, used predictive technical cost model has been demonstrated valuable for investigating the cost-efficiency of fatigue-enhancing strategies of the same welded structure. Some important conclusions are:

- Welding and production costs are negligible in relation to re-occurring repair costs for all considered design cases and fatigue scenarios.
- Repair and maintenance costs outweighs that of operational costs for the more severe fatigue scenarios considered.
- Increased flange plate thicknesses is the most effective means to reduce overall life-cycle costs, as it can increase the intervals between repairs significantly. Ultimately, increased flange plate thicknesses can result in lifecycle cost reductions of up to 60% (or about 1.75 M€) and fully compensate for hole position variations considered.

Author Contributions: Conceptualization, M.K.H. and M.K.; methodology, M.K.H.; software, M.K.H.; validation, M.K.H., M.K. and Z.B.; formal analysis, M.K.H.; investigation, M.K.H.; resources, M.K.H.; data curation, M.K.H.; writing—original draft preparation, M.K.H. and Z.B.; writing—review and editing, Z.B., M.K. and M.Å.; visualization, M.K.H. and M.K.; supervision, M.Å. and Z.B.; project administration, M.K.H and Z.B; funding acquisition, M.K.H, Z.B. and M.K. All authors have read and agreed to the published version of the manuscript.

Funding: This work was supported by the Centre for ECO2 Vehicle Design, funded by the Swedish Innovation Agency Vinnova (Grant Number 2016-05195) and the project VariLight, funded by the Swedish Innovation Agency Vinnova through Grant Number 2016-03363.

Institutional Review Board Statement: Not applicable.

Informed Consent Statement: Not applicable.

Data Availability Statement: The data presented in this study are available within the article itself, drawing on provided reference list.

Conflicts of Interest: The authors declare no conflict of interest.

References

1. Öberg, A.E.; Åstrand, E. Improved productivity by reduced variation in gas metal arc welding (GMAW). *Int. J. Adv. Manuf. Technol.* **2017**, *92*, 1027–1038. [CrossRef]
2. Hobbacher, A. *Recommendations for Fatigue Design of Welded Joints and Components*, 2nd ed.; IIW Collection; Springer: Berlin/Heidelberg, Germany, 2016. [CrossRef]
3. Radaj, C.D.; Fricke, W. *Fatigue Assessment of Welded Joints by Local Approaches*, 2nd ed.; Woodhead Publishing Limited: Sawston, UK, 2006.
4. Fricke, W. IIW guideline for the assessment of weld root fatigue. *Weld. World* **2020**, *57*, 753–791. [CrossRef]

5. Frank, K.H.; Fisher, J.W. Fatigue strength of fillet welded cruciform joints. *J. Struct. Div.* **1979**, *105*, 1727–1740. [CrossRef]
6. Guha, B. A new fracture mechanics method to predict the fatigue life of welded cruciform joints. *Eng. Fract. Mech.* **1995**, *52*, 215–229. [CrossRef]
7. Marquis, G.B. ; Barsoum, Z. *IIW Recommendations for the HFMI Treatment For Improving the Fatigue Strength of Welded Joints*; Springer: Singapore, 2016; ISBN 978-981-10-2503-7.
8. Aldén, R.; Barsoum, Z.; Vouristo; T.; Al-Emrani, M. Robustness and effect of weld quality on the fatigue life improvement of welded joints using HFMI techniques. *Weld World* **2020**, *64*, 1947–1956. [CrossRef]
9. Yekta, R.T; Ghahremani,K.; Walbridge, S. Effect of quality control parameter variations on the fatigue performance of ultrasonic impact treated welds. *Int. J. Fatigue* **2013**, *55*, 245-256. [CrossRef]
10. Silva,F.J.G.; Pinho, A.P; Pereira, A.B.; Paiva, O.C. Evaluation of Welded Joints in P91 Steel under Different Heat-Treatment Conditions. *Metals* **2020**, *10*, 99. [CrossRef]
11. Sousa, V.F.C.; Silva,F.J.G.; Pinho, A.P; Pereira, A.B.; Paiva, O.C. Enhancing Heat Treatment Conditions of Joints in Grade P91 Steel: Looking for More Sustainable Solutions. *Metals* **2021**, *11*, 495. [CrossRef]
12. Hagnell, M.K.; Åkermo, M. Cost efficiency, integration and assembly of a generic composite aeronautical wing box. *Compos. Struct.* **2016**, *152*, 1014–1023. [CrossRef]
13. Rybicka, J.; Purse, T.; Parlour, B. A Generic Cost Estimating Approach for a Composite Manufacturing Process Assessment. *Adv. Manuf. Technol. XXXIV* **2021**. [CrossRef]
14. Tierney, C.; Higgins, C.; Quinn, D.; Backer, J.D.; Allen, C.; Örtnäs, A.; Persson, J.; McClelland, J.; Higgins, P.; Murphy, A. A scalable cost modelling architecture for evaluating the production cost-effectiveness of novel joining techniques for aircraft structures. *Procedia Manuf.* **2021**, *54*, 7–12. [CrossRef]
15. Bouchouireb, H.; O'Reilly, C.J.; Göransson, P.; Schöggl, J.P.; Baumgartner, R.J.; Potting, J. Towards Holistic Energy-Efficient Vehicle Product System Design: The Case for a Penalized Continuous End-of-Life Model in the Life Cycle Energy Optimisation Methodology. *Proc. Des. Soc. Int. Conf. Eng. Des.* **2019**, *1*, 2901–2910. [CrossRef]
16. Poulikidou, S.; Schneider, C.; Björklund, A.; Kazemahvazi, S.; Wennhage, P.; Zenkert, D. A material selection approach to evaluate material substitution for minimizing the life cycle environmental impact of vehicles. *Mater. Des.* **2015**, *83*, 704–712. [CrossRef]
17. Penadés-Plà, V.; García-Segura, T.; Martí, J.V.; Yepes, V. An Optimization-LCA of a Prestressed Concrete Precast Bridge. *Sustainability* **2018**, *10*, 685. [CrossRef]
18. Lee, K.M.; Cho, H.N.; Choi, Y.M. Life-cycle cost-effective optimum design of steel bridges. *J. Constr. Steel Res.* **2004**, *60*, 1585–1613. [CrossRef]
19. Zou, G.; Banisoleiman, K.; González, A. Probabilistic maintenance optimization for fatigue-critical components with constraint in repair access and logistics. In Proceedings of the 14th International Conference on Probabilistic Safety Assessment and Management (PSAM 14), Los Angeles, CA, USA, 16–21 September 2018.
20. Kim, S.; Frangopol, D.M.; Soliman, M. Generalized Probabilistic Framework for Optimum Inspection and Maintenance Planning. *J. Struct. Eng.* **2013**, *139*, 435–447. [CrossRef]
21. Kim, S.; Ge, B.; Frangopol, D.M. Effective optimum maintenance planning with updating based on inspection information for fatigue-sensitive structures. *Probabilistic Eng. Mech.* **2019**, *58*, 103003. [CrossRef]
22. Turan, O.; Ölçer, A.İ.; Lazakis, I.; Rigo, P.; Caprace, J.D. Maintenance/repair and production-oriented life cycle cost/earning model for ship structural optimisation during conceptual design stage. *Ships Offshore Struct.* **2009**, *4*, 107–125. [CrossRef]
23. Hagnell, M.K.; Åkermo, M. A composite cost model for the aeronautical industry: Methodology and case study. *Compos. Part B Eng.* **2015**, *79*, 254–261. [CrossRef]
24. Öberg, A.E. VariLight Reduced VARIation in the manufacturing processes enabling LIGHTweight welded structures, public report. *Fordonsstrategisk Forsk. Och Innov.* **2019**.
25. Broadbent, C. Steel's recyclability: demonstrating the benefits of recycling steel to achieve a circular economy. *Int. J. Life Cycle Assess* **2016**, *21*, 1658–1665. [CrossRef]
26. Worrel, E.; Reuter, M.A. (Eds.) *Handbook of Recycling*; Elsevier: Amsterdam, The Netherlands, 2014.
27. Harvey, L.D.D. Iron and steel recycling: Review, conceptual model, irreducible mining requirements, and energy implications. *Renew. Sustain. Energy Rev.* **2021**, *138*, 110553. [CrossRef]
28. Hagnell, M.K. Technical Cost Modelling and Efficient Design of Lightweight Composites in Structural Applications. Ph.D. Thesis, KTH Royal Institute of Technology, Stockholm, Sweden, 2019.
29. Foundation, P.S. Python Language Reference, Version 2.7. Available online: http://www.python.org (accessed on 31 March 2021).
30. Coromant, S. Milling Formulas and Definitions. Available online: https://www.sandvik.coromant.com/en-gb/knowledge/machining-formulas-definitions/pages/milling.aspx (accessed on 31 March 2021).
31. SSAB. Machining Recommendations for Strenx®. Available online: https://www.ssab.com/support/processing#downloads (accessed on 24 September 2021).
32. Weman, K. *Welding Processes Handbook*, 2nd ed.; Woodhead Publishing: Sawston, UK, 2012.
33. Jarmai, K.; Farkas, J. Cost calculation and optimisation of welded steel structures. *J. Constr. Steel Res.* **1999**, *50*, 115–135. [CrossRef]
34. Zhu, J.; Khurshid, M.; Barsoum, Z. Accuracy of computational welding mechanics methods for estimation of angular distortion and residual stresses. *Weld. World* **2019**, *63*, 1391–1405. [CrossRef]

35. MEPS International Ltd. World Steel Prices. Available online: https://worldsteelprices.com/; https://www.meps.co.uk/gb/en/products/europe-steel-prices (accessed on 31 March 2021).
36. London Metal Exchange (LME). Metal Prices. Available online: https://www.lme.com/ (accessed on 31 March 2021).
37. Maes, J. Make It from Metal. Available online: https://makeitfrommetal.com/how-much-does-a-cnc-machine-cost/ (accessed on 31 March 2021).
38. EWM AG. Welding Machines. Available online: https://www.ewm-sales.com/ (accessed on 31 March 2021).
39. The Indiana Oxygen Company. Welding Supplies from Ioc. Available online: https://www.weldingsuppliesfromioc.com/ (accessed on 31 March 2021).
40. BS7608. *Guide to Fatigue Design and Assessment of Steel Products, 2014–2015*, 2nd ed.; BSI Standards: London, UK, 2015.
41. EN1993-1-9. *Eurocode 3: Design of STEEL structures—Part 1–9: Fatigue, 2003*; BSI Standards: London, UK, 2003.
42. EN13001. *Cranes—General design—Part 1: General Principles and Requirements, 2015*; BSI Standards: London, UK, 2015.
43. Pettersson, G.; Barsoum, Z. Finite element analysis and fatigue design of a welded construction machinery component using different approaches. *Eng. Fail. Anal.* **2012**, *26*, 274–284. [CrossRef]
44. Hultgren, G.; Khurshid, M.; Haglund, P.; Barsoum, Z. Mapping of scatter in fatigue life assessment of welded structures—A round robin study. *Weld. World* **2021**, *65*, 1841–1855. [CrossRef]
45. Delkhosh, E.; Khurshid, M.; Barsoum, I.; Barsoum, Z. Fracture mechanics and fatigue life assessment of box-shaped welded structures: FEM analysis and parametric design. *Weld. World* **2020**, *64*, 1535–1551. [CrossRef]
46. Zhu, J.; Khurshid, M.; Barsoum, Z. Assessment of computational weld mechanics concepts for estimation of residual stresses in welded box structures. *Procedia Struct. Integr.* **2019**, *17*, 704–711. [CrossRef]
47. Kaltenbach. The Scrap Calculation in a CNC Plate Processing Line. Available online: https://www.kaltenbach.com/en/media/talking-steel/the-scrap-calculation-in-a-cnc-plate-processing-line/ (accessed on 31 March 2021).
48. Eurostat. Hourly Labour Costs. Available online: https://ec.europa.eu/eurostat/statistics-explained/index.php?title=Hourly_labour_costs (accessed on 20 September 2021).
49. The World Bank. Available online: https://www.worldbank.org/ (accessed on 31 March 2021).
50. Haglund, M.K.P.; Barsoum, Z. Mapping of scatter in fatigue life assessment of welded structures—A Round Robin Study. In Proceedings of the IIW Annual Assembly and International Conference, Bratislava, Slovakia, 7–12 July 2019.
51. Acculift. Demystifying The Duty Cycle of a Hoist. Available online: https://acculift.com/demystifying-the-duty-cycle-of-a-hoist/ (accessed on 20 September 2021).
52. Eurostat. Electricity Price Statistics. Available online: https://ec.europa.eu/eurostat/statistics-explained/index.php?title=Electricity_price_statistics (accessed on 20 September 2021).
53. Occupational Safety and Health Administration (OSHA). 1910.179—Overhead and Gantry Cranes. Available online: https://www.osha.gov/laws-regs/regulations/standardnumber/1910/1910.179 (accessed on 31 March 2021).
54. European Commission Competition. Terminal Handling Charges during and after the Liner Conference Era. Available online: https://ec.europa.eu/competition/sectors/transport/reports/terminal_handling_charges.pdf (accessed on 31 March 2021).
55. Container News. ZIM Updates THC/THD Rates. Available online: https://container-news.com/zim-updates-thc-thd-rates/ (accessed on 31 March 2021).
56. Mazzella Companies. Lifting & Rigging Learning Center. Available online: https://www.mazzellacompanies.com/Resources/Blog/how-much-does-an-overhead-crane-inspection-cost (accessed on 20 September 2021).
57. Free Encyclopedia for UK Steel Construction Information. Recycling and Reuse. Available online: https://www.steelconstruction.info/Recycling_and_reuse (accessed on 3 March 2021).

Article

Influences of Residual Stress, Surface Roughness and Peak-Load on Micro-Cracking: Sensitivity Analysis

Jairan Nafar Dastgerdi [1,2,*], **Fariborz Sheibanian** [1], **Heikki Remes** [2] and **Hossein Hosseini Toudeshky** [1]

[1]   Department of Aerospace Engineering, Amirkabir University of Technology, 424 Hafez Avenue, Tehran 15875-4413, Iran; f.sheibanian@aut.ac.ir (F.S.); hosseini@aut.ac.ir (H.H.T.)

[2]   Department of Mechanical Engineering, School of Engineering, Aalto University, P.O. Box 14300, Aalto, FIN-00076 Espoo, Finland; heikki.remes@aalto.fi

*   Correspondence: jairan.nafardastgerdi@aalto.fi; Tel.: +98-21-6454-5636

Abstract: This paper provides further understanding of the peak load effect on micro-crack formation and residual stress relaxation. Comprehensive numerical simulations using the finite element method are applied to simultaneously take into account the effect of the surface roughness and residual stresses on the crack formation in sandblasted S690 high-strength steel surface under peak load conditions. A ductile fracture criterion is introduced for the prediction of damage initiation and evolution. This study specifically investigates the influences of compressive peak load, effective parameters on fracture locus, surface roughness, and residual stress on damage mechanism and formed crack size. The results indicate that under peak load conditions, surface roughness has a far more important influence on micro-crack formation than residual stress. Moreover, it is shown that the effect of peak load range on damage formation and crack size is significantly higher than the influence of residual stress. It is found that the crack size develops exponentially with increasing peak load magnitudes.

Keywords: surface roughness; residual stresses; peak load; finite element method; micro-crack formation

Citation: Nafar Dastgerdi, J.; Sheibanian, F.; Remes, H.; Hosseini Toudeshky, H. Influences of Residual Stress, Surface Roughness and Peak-Load on Micro-Cracking: Sensitivity Analysis. *Metals* **2021**, *11*, 320. https://doi.org/10.3390/met11020320

Academic Editor: Janice Barton

Received: 11 January 2021
Accepted: 8 February 2021
Published: 12 February 2021

1. Introduction

The effect of surface treatments and different processing methods on the performance of high-strength steel under high peak stresses, either as single events or as a part of service loading, is the main concern of strength design and life estimation. The main parameters affecting and describing surface integrity are surface roughness, residual stress, and the material properties in the surface layer. These parameters can vary separately as a result of the manufacturing procedures and machining conditions [1–3]. Therefore, the influences of these affecting parameters on the performance and failure damage mechanism of high-strength steels in real engineering applications should be investigated.

It has long been identified that fatigue cracks generally nucleate from the free surface and the local microscopic stress and strain concentration at the surface defects are remarkable factors for crack nucleation and propagation [4–8]. Surface roughness is a significant index for characterizing surface micro-topography, and it has a critical effect on fatigue life [9,10]; thus, the influence of surface roughness on fatigue performance has been a comprehensive research area for several years, and numerous studies have investigated the effects and function mechanism of surface roughness caused by different surface processing methods on the fatigue behavior of diverse materials [11–15]. Despite significant advances in understanding the effect of surface roughness on the fatigue performance of various kinds of materials, the role of surface roughness in crack nucleation and the damage mechanism under peak load, e.g., as an initial step before fatigue loading, is poorly understood. Moreover, the influences of residual stress and material properties on the surface layer are still obscure, considering the simultaneous impact of surface roughness on crack formation and failure mechanism.

However, some researchers have experimentally studied the effect of surface integrity on the fatigue strength of high-strength steels [16–19]. They have provided a comprehensive overview of the influence of these parameters on fatigue strength, although there is still an insufficient understanding of the effects of surface roughness, hardness, and residual stress on failure mechanism, crack initiation, and propagation under loading for high-strength steel materials. Therefore, predicting the structure strength and mechanical performance under loading with new manufacturing processes and machining parameters is not possible except by performing new time-consuming and expensive tests. A better model, encompassing the detailed characteristics of the surfaces with all affecting parameters, is expected to capture the failure mechanism over a broad range of conditions. Thus, it is imperative to study the combined effect of surface integrity parameters on the damage mechanisms and micro-cracking of engineering structures and components under peak loads, which is the focus of this investigation.

Ductile fracture has been extensively investigated for the purpose of modeling and assessing the failure mechanisms of materials and structures, especially in the context of metals and alloys used in different engineering practices using various proposed ductile fracture criteria [20–23]. These criteria have been extended based on various assumptions, hypotheses, and experimental observations related to ductile fracture. The phenomenological ductile fracture criterion has been used to develop models for the prediction of fracture processes such as crack nucleation, propagation and failure mechanisms [24–26]. To implement the ductile fracture criterion into surface integrity analysis, a new finite element (FE) modeling approach was recently developed by the authors. The developed approach fully captures the complexity of the surface roughness using actual two-dimensional surface topography and the effect of residual stress by introducing global layer-wise modeling, with a constant temperature in each layer following the measured residual distribution [27].

In this study, the micro-mechanism motivated phenomenological damage model was applied to predict ductile fracture initiation in the context of stress triaxiality and equivalent plastic strain [28]. The ductile fracture criterion has been calibrated for the prediction of fracture locus with an inverse numerical-experimental approach in order to determine the material constants in the criterion [27]. Although the approach is successfully calibrated with experiments, it has not yet been applied for the systematic analysis of surface integrity effects. Since surface integrity includes several influencing factors, in addition to experimental observation, numerical simulation is necessary to reveal the main affecting factors. For instance, compressive loads tend to relax compressive residual stresses in proportion to their magnitude. Relaxation due to a single peak load is usually evaluated as being of greater significance than gradual cyclic relaxation [29,30]. As residual stress relaxation pertains to the correlation of the local stresses and the local yield strength [31], surface roughness as an affecting parameter on the local stress concentration influences relaxation behavior. It is unclear under what conditions residual stress relaxation arises, or what the impact of residual stress distribution on crack formation and damage mechanism may be. It is well known that residual stress relaxation is a complex phenomenon [32]. Thus, the integration of residual stress in predictive modeling computation, without consideration of relaxation during operation, results in an imprecise prediction for the trustworthiness and reliability of the components and structures. It is worthwhile mentioning that this study is the second part of the authors' recent work [27], which provides a further understanding of the influence of compressive peak load on the micro-crack formation, crack size, and residual stress state of sandblasted high-strength steels in real engineering applications when peak load is applied to the surface before or during fatigue loading. The influence of peak load on residual stress distribution was studied, and further analysis was carried out to investigate the effects of peak load, affecting the parameters of fracture locus, and residual stress, affecting damage mechanism and crack size.

2. Materials and Methods

2.1. Characteristics of Sandblasted High-Strength Steel Surface

This study focuses on 15-mm-thickness sandblasted 690 high-strength steel plate. In this case, previous experimental investigations provide good descriptions of roughness measurement, residual stress, and material properties [27]. The surface roughness measurements were carried out according to SFS-EN ISO 4288 [33], and the size of the surface profiles was defined for the rolled plate. The surface contour is depicted in Figure 1. The arithmetical average value, R_a, and the average of the five largest peak-to-peak values, R_z, were identified for this surface contour. These values were 128 µm and 235 µm, respectively. The elastic properties of the material are described using a Young's modulus of E = 210 GPa, and a Poisson ratio of $\nu = 0.3$. The chemical composition of the studied material is presented in Table 1. Due to the complexity of the surface topology, high local stresses are expected to arise at the surface. To evaluate the effect of local plasticity on the near surface stress fields, the von Mises yield criterion is used in the simulations, assuming associated plastic flow and isotropic hardening.

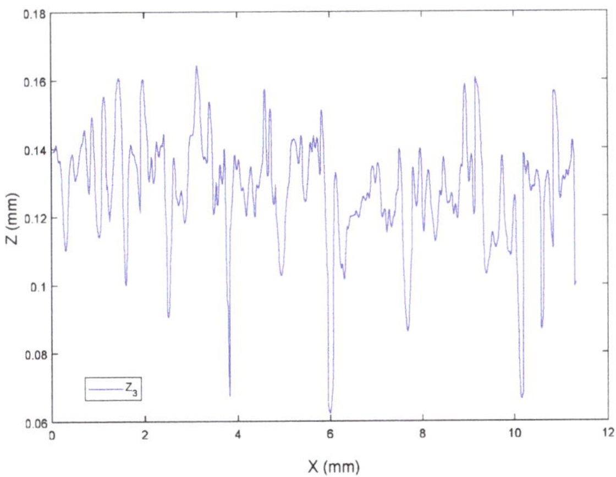

Figure 1. Profile of the surface roughness [27].

Table 1. Chemical composition (%). Reprinted with permission from ref. [18]. Copyright 2021 Springer Nature.

Material	C	Si	Mn	P	S	Al	Nb	V	Ti	Cu	Cr	Ni	Mo	B	N
S690	0.16	0.21	1.39	0.011	0.001	0.047	0.015	0.018	0.007	0.01	0.25	0.06	0.502	0.001	0.002

In this study, the through-thickness residual stress distribution with the maximum stress value (−80 MPa) measured at the surface [27] is employed. This distribution is obtained from the strain-gage hole drilling measurements for sandblasted steel samples [32]. Furthermore, to investigate the compressive residual stress effect on micro-crack formation and crack size, the residual stress value at the surface is increased from −80 MPa to −320 MPa, as reported experimentally for the grinding manufacturing process of 690 high-strength steel plate [1].

2.2. Ductile Fracture Criterion

In this simulation, the micro-mechanism-inspired phenomenological damage model described in [28], which uses a fracture strain that is dependent on stress triaxiality, is employed for the evolution of damage. The damage model and calibration approach is briefly presented in this section in order to describe the physical basis for and the parameters that affect the numerical simulations carried out in this paper. The calibration approach employed, which is a new and efficient method, was explained in details in Ref. [27] by the authors.

The criterion is constructed according to the damage accumulation induced by the nucleation, growth and shear coalescence of voids [28], which can be written as follows:

$$\left(\frac{2\tau_{\max}}{\overline{\sigma}}\right)^{C_1} \times \left(\frac{\langle 1+3\eta\rangle}{2}\right)^{C_2} \times \overline{\varepsilon_f} = C_3 \qquad \langle x\rangle = \begin{cases} x, & \text{when } x \geq 0 \\ 0, & \text{when } x < 0 \end{cases} \tag{1}$$

The model is founded based on the microscopic analysis of ductile fracture, where void nucleation is explained as a function of the equivalent plastic strain, the void growth is demonstrated as a function of stress triaxiality, as $1+3\eta$, and the coalescence of voids is represented by the normalized maximal shear stress, defined as $\tau_{\max}/\overline{\sigma}$. The influence of nucleation, growth, and coalescence of voids is controlled by the two calibration exponents: C_1 and C_2. The material constant C_3 is equal to the equivalent plastic strain required to fracture in uniaxial tension. This model is able to explain these various phenomena, and the shape of the fracture locus can be easily controlled using the three material constants. Thus, it is preferable to use this criterion. The aforementioned equation gives the fracture strain for the full range of stress triaxiality.

Ideally, the input fracture criterion $\overline{\varepsilon_f}$ in Equation (1) should be specified based on tests covering the full range of stress triaxiality. For instance, in the study in which the fracture model was introduced [28], the material constants C_1, C_2 and C_3 were determined by fitting a curve to the experimental data. The same approach is applied in this study, but only uniaxial tension test data, together with single peak tensile load, are used. The input fracture criterion in the uniaxial tension range is calibrated based on the equivalent plastic strain computed at fracture initiation, failure strain $\overline{\varepsilon_f}$. The $\overline{\varepsilon_f}$ at uniaxial tension $\eta = 1/3$ gives the value for material constant C_3. This is calculated using an approach that first computes the true stress-strain curve of S690 high-strength steel under uniaxial loading. Then, the defined true stress-strain curve is implemented in the FE simulation using a UMAT subroutine to compute the equivalent von Mises plastic strain at fracture initiation. The algorithm of this approach is explained in detail in Reference [27]. In this study, the numerical simulation is carried out with Abaqus version 6.14 (Dassault Systèmes Simulia Corporation, Johnston, RI, USA).

To provide a full-range true stress-strain ($\sigma_t - \varepsilon_t$) curve for S690 high-strength steel under uniaxial loading, the instantaneous area method is applied to analyze the data provided using a digital imaging correlation technique. This technique has been employed to measure deformation fields of steel coupons during the whole range of deformation. Using this approach, the constitutive model for S690 high-strength steel materials is introduced, and the true strain is computed as follows [34]:

$$\sigma_t(\varepsilon)\begin{cases} \sigma_t = E\varepsilon_t & \text{for } \varepsilon_t \leq \varepsilon_y \\[4pt] \sigma_t = S_y \times \left[1 + 3\times 10^{-3}\times\left(\frac{\varepsilon_t}{\varepsilon_y}-1\right)\right] & \text{for } \varepsilon_y \leq \varepsilon_t \leq 6\varepsilon_y \\[4pt] \sigma_t = S_y \times \left\{1.015 + 0.1\times\left[1 - 0.01\times\left(0.6\left(\frac{\varepsilon_t}{\varepsilon_y}-6\right)-10\right)^2\right]\right\} & \text{for } 6\varepsilon_y \leq \varepsilon_t \leq 15\varepsilon_y. \\[4pt] \sigma_t = S_y \times \left\{1.094 + 0.1\times\left(\frac{\varepsilon_t/\varepsilon_y-15}{100}\right)^{0.45}\right\} & \text{for } 15\varepsilon_y \leq \varepsilon_t \leq 130\varepsilon_y. \\[4pt] \sigma_t = S_y \times \left\{1.2 - 0.09\times\left[\left(\frac{\varepsilon_t}{\varepsilon_y}-130\right)^{1.1}/340\right]\right\} & \text{for } 130\varepsilon_y \leq \varepsilon_t \leq 1.2. \end{cases} \tag{2}$$

where $E = 210$ MPa and $S_y = 770$ MPa is the yield strength of the material, and ε_y, ε_t are the strain at yielding and fracture, respectively. In this study, the ductile fracture criterion is employed for damage initiation, and a fracture energy base criterion is applied for damage evolution.

Using the proposed approach, the value of the material constant C_3 is 1.2, which gives a $\overline{\varepsilon_f}$ at uniaxial tension of $\eta = 1/3$. Then, various values for C_1, C_2 were considered on the basis of the common range ($1 < C_1 < 8$ and $0 < C_2 < 1$) for these parameters [28]. Employing an inverse numerical–experimental approach proposed by the authors, C_1,

C_2 can be calibrated in such a way that FE simulation for pre-notched specimens with ductile damage can predict the micro-cracks observed from SEM images in the proximity of the notch under peak load conditions [27]. The material constant C_1 modulates the effect of the normalized maximal shear stress on the shear coalescence of voids during plastic deformation. As C_1 increases, the influence of the maximal shear stress on ductile fracture increases, and accordingly, the fracture strain at $\eta = 0$ decreases. The sensitivity analyses for investigating the effects of the material constants C_1 on the ductile fracture criterion and micro-crack formation reveal that damage-induced micro-crack formation is insensitive to the values of C_1 under compressive peak load conditions [27]. A common mean value, $C_1 = 4$, is observed to give a reliable estimation for surface integrity analysis.

For negative stress triaxialities, the fracture is governed by the shear mode. Power exponent C_2 is added to the void growth function in order to represent void coalescence. Therefore, if parameter C_2 increases, void growth in compression will be greatly suppressed, and the progress of void growth under shear stress will be slower, with void coalescence being distinctly compressed. Since the surface has negative residual stress and micro-defects and is subjected to the compressive overload, the damage occurs mainly on the compression side of the fracture locus, and thus, C_2 is the most influential parameter on micro-crack formation, as depicted in Figure 2. This figure shows the calibrated fracture locus for various values of C_2. Employing an inverse numerical–experimental approach, the correct value for the constant C_2 is selected based on the micro-crack length obtained from microscopic analysis of the surface. When the micro-crack length is known, the calibration of the constants C_2 is easy, since the length of the formed micro-cracks is sensitive to the constant C_2.

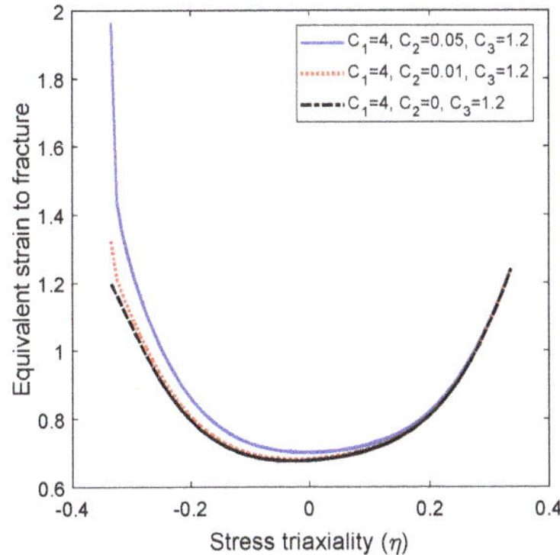

Figure 2. Fracture locus for high-strength steel 690.

This new special approach is employed in this study, and the fracture locus constructed using this calibration approach with the material constants $C_1 = 4$, $C_2 = 0.01$, $C_3 = 1.2$ is further used in the FE simulation, as shown in Figure 2. Moreover, a sensitivity analysis is carried out on the effect of C_2 as the most influential parameter with respect to micro-crack formation and crack length under compressive overload conditions using different fracture locus.

2.3. Numerical Simulation

The FE model was built using the two-dimensional real surface contour of the specimen measured by profilometry to capture the complexity of the surface roughness as depicted in Figure 3. The final dimensions of the model are considered to be 31.1 mm × 11.28 mm × 15 mm. The size of the FE model was determined based on the length of the measured profile, thickness (15 mm), and the width of the studied sample, in such a way to be large enough as a representative model for the real engineering situation.

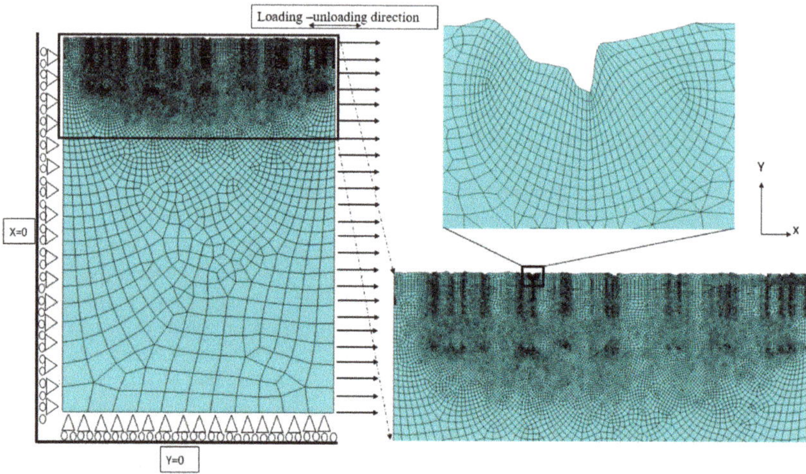

Figure 3. Two-dimensional real surface topography model and local mesh [27].

Since the surface roughness consists of different micro-notches with sharp tips in some areas, a free mesh cannot provide accurate results. It is important to use the well-defined mesh especially close to the notch tip. Thus, the mesh element is refined in the micro-defect root to be as small as possible while still being valid for continuum mechanics. The minimum element size is defined as being three times the average grain size of the material, which is equal to roughly 10 µm for the studied steel. With this approach, the material model is valid for a group of grains instead of for an individual grain.

The residual stress effect is considered by proposing layer-wise global modeling with a constant temperature in each layer following the measured residual distribution obtained in the experiments. In Figure 4a, the measurement-based residual stress distribution is shown as a solid smooth line, while the stepwise continuous lines demonstrate the discontinuous residual stress distribution employed in the layer-wise FE model. Figure 4b shows the applicability of the proposed modeling approach in defining layers with constant temperatures using the estimated through-thickness residual stress distribution.

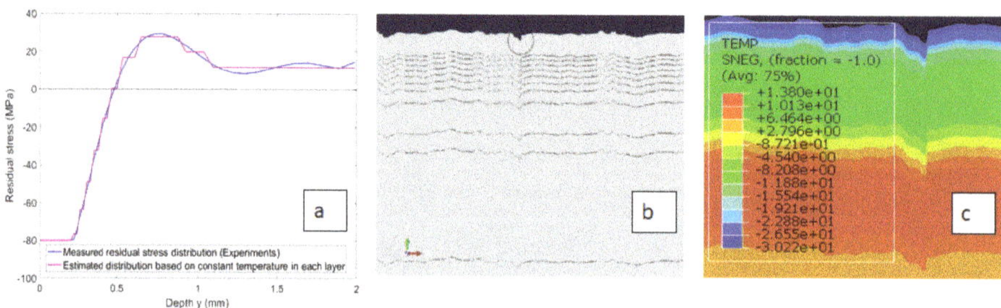

Figure 4. (**a**) Continuously approximated residual stress distribution (solid line) and discontinuously approximated residual stress distribution with constant values in each layer (stepwise continuous line) [27], (**b**) layer-wise global modeling, (**c**) the applied temperature field using the proposed approach.

In this method, several layers are defined in the FE model, and the residual stress distributions are presented in the FE analysis as temperature fields [27]. The temperature fields are determined using field distribution, field magnitude, and the thermal expansion coefficient. A typical value of $\alpha = 1.2 \times 10^{-5}$ (1/°C) was used in this study for the thermal expansion coefficient for steel. The temperature field, applied using the proposed global layer-wise approach, is depicted in Figure 4c for the measured residual stress distribution. In this approach, the temperature field is applied discontinuously, with a constant temperature in each layer following the initial measured distribution. The temperature field is defined as a pre-step, and is applied to the initial undeformed mesh.

The loading condition for the FE simulation is considered to be a single compressive peak load, as shown in Figure 5. The specimen is subjected to compressive loading in order to relax or redistribute the residual stress. The compressive peak load is around the yield stress $(-1.1S_y)$, similar to a situation in which a ship is launched or finds itself in severe weather conditions. Then, one tensile load cycle is considered for this material, representing the typical fatigue loading in service.

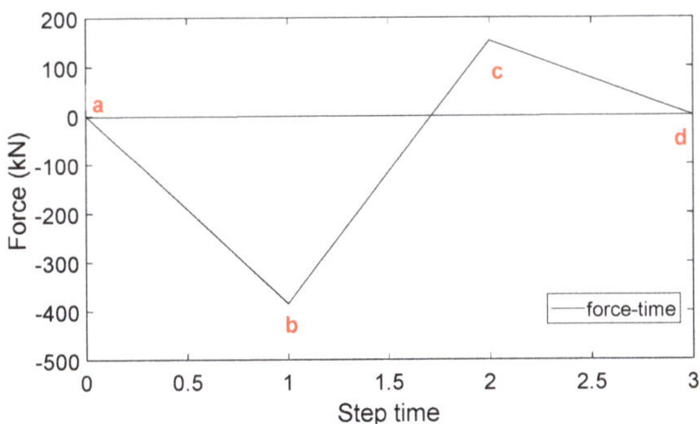

Figure 5. Loading condition in one cycle [27].

3. Results and Discussion

A sensitivity analysis was carried out for surface roughness, residual stress, and peak load effects. Firstly, the residual stress effect on the damage initiation, evolution, and crack formation was investigated. Moreover, the peak load effect on residual stress distribution and relaxation behavior was studied. Secondly, the influences of peak load and the factors affecting fracture locus were examined.

3.1. Influence of Residual Stress on Micro-Crack Formation

Figure 6 shows the damage initiation, evolution, and crack formation based on element deletion for the most critical micro-notch caused by the surface roughness using ductile fracture criteria without the residual stress effect. In Figure 7, the residual stress of -80 MPa is also considered in the FEM, using layer-wise global modeling. The result in terms of damage, i.e., formed crack size, is very similar to the analysis cases performed both with and without residual stresses.

After the peak load, the sample is nickel plated, with the nickel layer having a thickness of 20–30 µm, to preserve the surface geometry. Then, sample preparation and polishing are carried out, and the top and the bottom side of the specimen are studied using scanning electron microscopy (SEM). The experimental observations from SEM images after peak load demonstrate the maximum damage size and micro-crack length to be an average of 75 µm; Figure 8a,b. The longest micro-crack on the top surface of the sample is shown with the arrow.

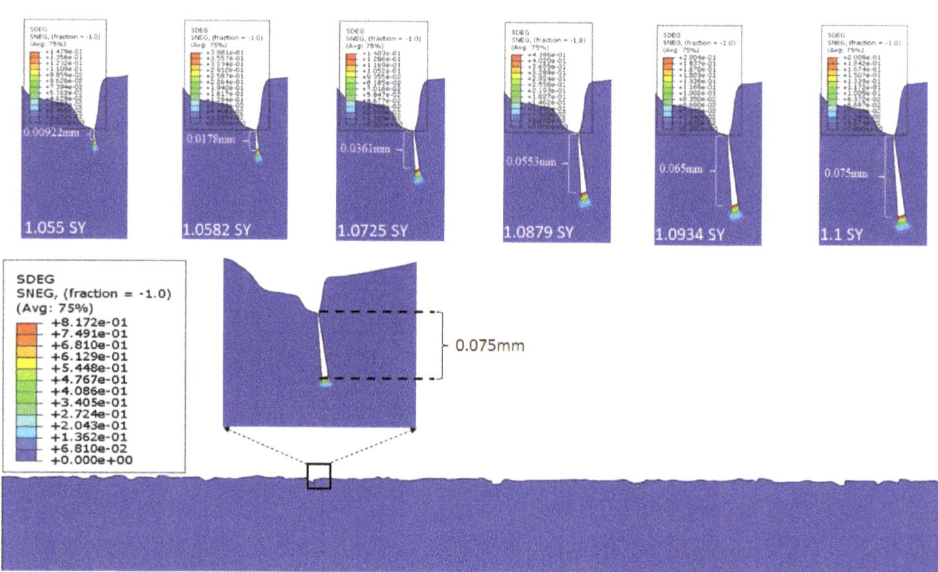

Figure 6. Damage initiation, evolution and crack formation for the most critical micro-notch at the surface without residual effect.

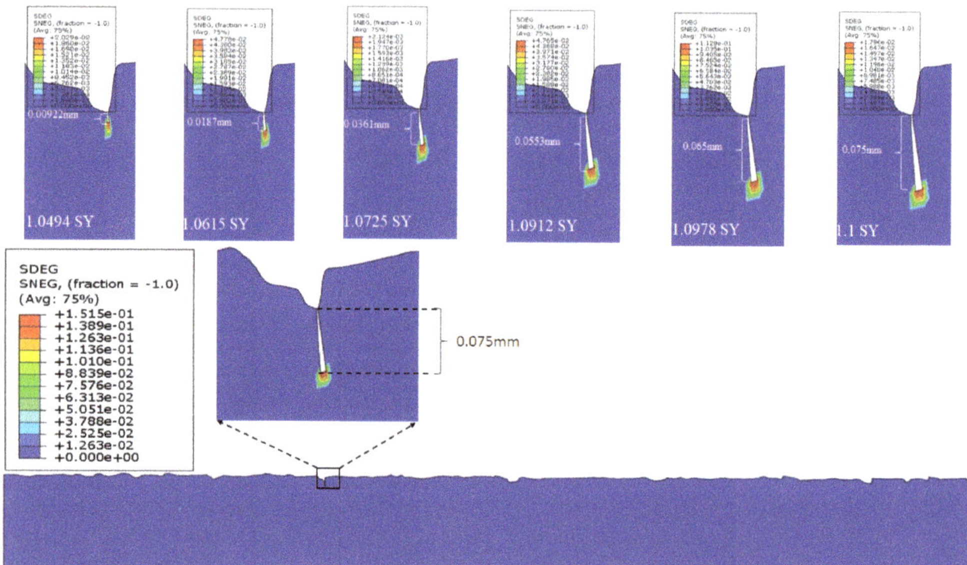

Figure 7. Crack formation for the most critical micro-notch at the surface with residual effect, at −80 MPa.

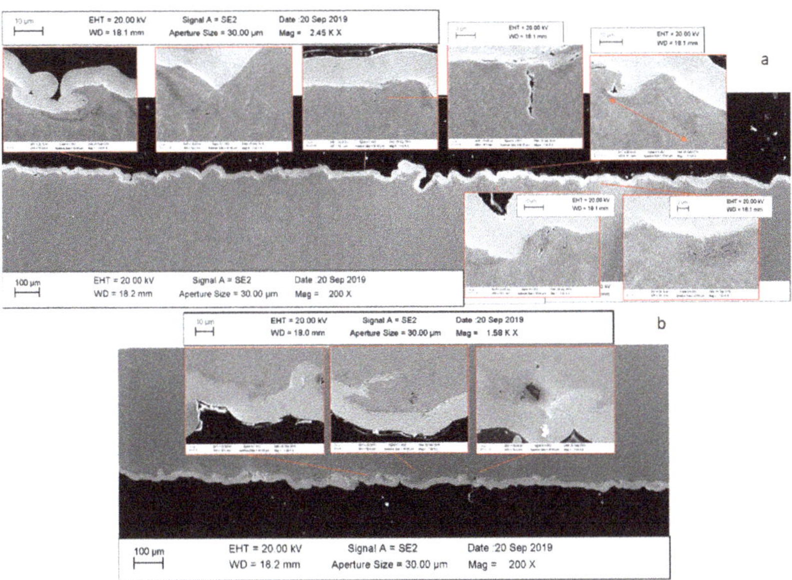

Figure 8. The SEM images from (**a**) the top and (**b**) the bottom side of the sample after peak load [27].

The obtained results, as presented in Figures 6 and 7, raise an important question and demonstrate that further understanding and analysis are required in order to determine the effect of residual stress on damage mechanisms. For this purpose, the first element precisely below the critical notch was observed during damage with and without the residual stress effect. Figure 9a shows the stress values versus time step for this element with and without the residual stress effect. The loading time step is considered for the compression peak load from point a to point b, as depicted in Figure 5. In the case in which the residual stress effect is considered, there is a pre-step that we use to define the residual stress. In this figure, the time step from −1 to 0 represents this pre-step. As shown in this figure, the presence of residual stress as defined based on the temperature field in the pre-step causes the element to experience a stress value of −328 MPa at the beginning of the loading step, while in the other case, without the residual stress effect, the stress values in the element start from zero at the beginning of the loading step. It can be seen, at a time step of around 0.3, that the element in both the cases with and without residual stress experiences the same stress values. Figure 9b shows the equivalent plastic strain versus time step for the first element precisely below the critical notch with and without the residual stress effect. If the equivalent plastic strains, which define the failure and removal of the element, are also similar, then we cannot expect differences in the damage initiation and crack formation for these two cases.

This phenomenon might be related to the relaxation of residual stress occurring under this loading condition. The initial residual stress field inherent to or produced during the manufacturing process may change and may not be constant during the operation life of the finished component under residual stress. These residual stresses may be reduced and redistributed, and this diminution is referred to as relaxation. If the sum of the applied and residual stresses locally exceeds the yield stress of a material, residual stress relaxation occurs. Thus, residual stresses do not remain constant, but are relaxed or altered and redistributed during service. Generally, a large relaxation would be expected in the high-stress region, similar to what occurred in this study with an external load value of −1.1 S_y. Figure 10a depicts the stress distribution at different time steps of the loading. It can be seen that the multi-layer residual stress distribution disappears at step −0.58 S_y of the loading, or around the middle of the loading step from point a to point b, as depicted in Figure 10b.

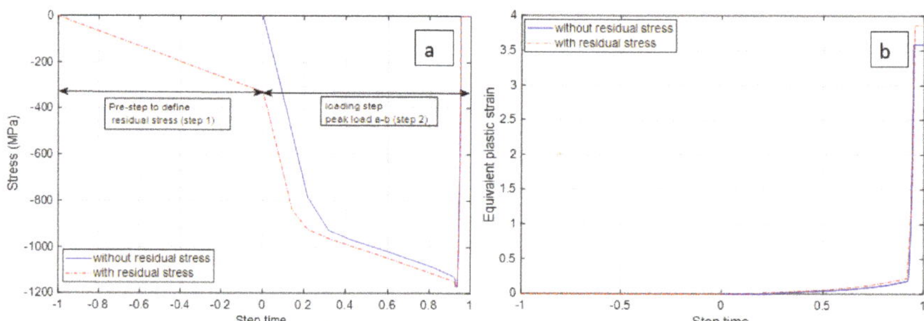

Figure 9. (**a**) Stress values versus time step and (**b**) strain versus time step for the first element precisely below the critical notch, without and with a residual stress of −80 MPa.

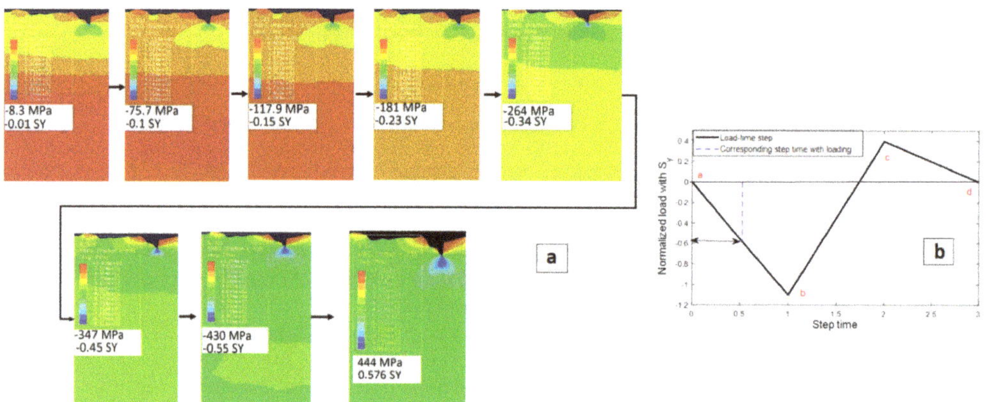

Figure 10. (**a**) stress distribution at different time steps of the loading with stress residual effect −80 MPa, (**b**) effective time for residual stress before being relaxed.

For further analysis, the residual stress value is increased from −80 MPa to −320 MPa in order to more precisely investigate the residual stress effect on the damage process. With the increased value of residual stress, it can be seen that there are some differences between the cases with and without residual stress with respect to the length of the crack, as can be seen from a comparison between Figures 6 and 11; however, the crack size does not significantly increase, i.e., the increase is from 75 μm to 84.7 μm. For higher residual stress (−320 MPa), the notch stress value already reaches the yield stress of the material during the pre-step, and the local stress value at the beginning of the loading time step is (for the first element precisely below the critical notch) −947 MPa; see Figure 12.

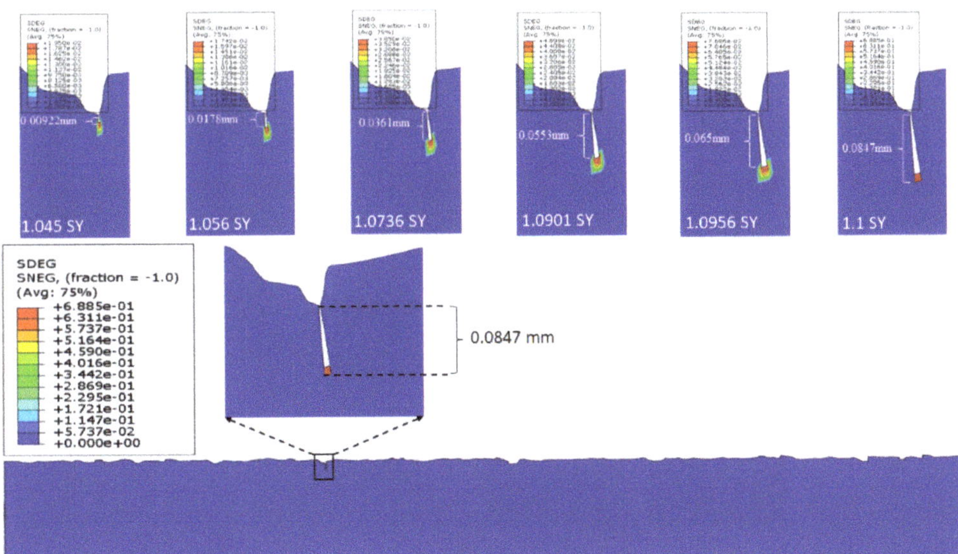

Figure 11. Crack formation for the most critical micro-notch at the surface with a residual stress effect of −320 MPa.

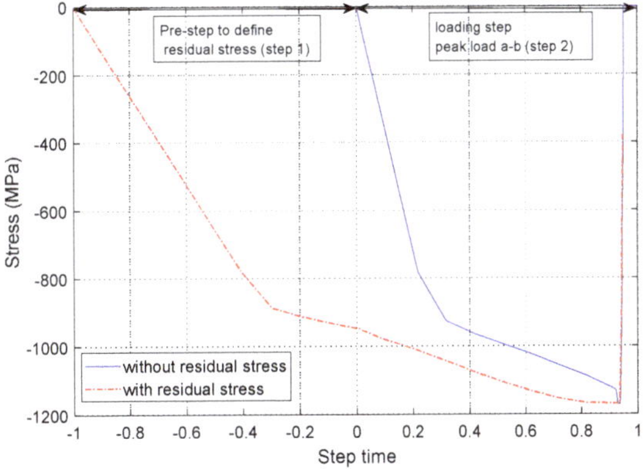

Figure 12. Stress values versus time step for the first element precisely below the critical notch without and with residual stress of −320 MPa.

The stress distribution at different time steps of loading for the case of a residual stress of −320 MPa is shown in Figure 13. It can be observed that the global multi-layer residual stress distribution remains effective at step 9 of loading, corresponding to $-0.45\,\sigma_y$ of loading; however, it relaxes after some further time steps, corresponding to $-1\,\sigma_y$ of loading. It can be concluded that the residual stress relaxation rate, in this case, is lower in comparison with the other case with a lower residual stress value (−80 MPa), while the elements do not undergo plastic deformation in the pre-step with the residual stress effect at the beginning of the loading step. These results also clarify the reason for further crack formation in the case with a higher residual stress value, compared to the case with a lower value.

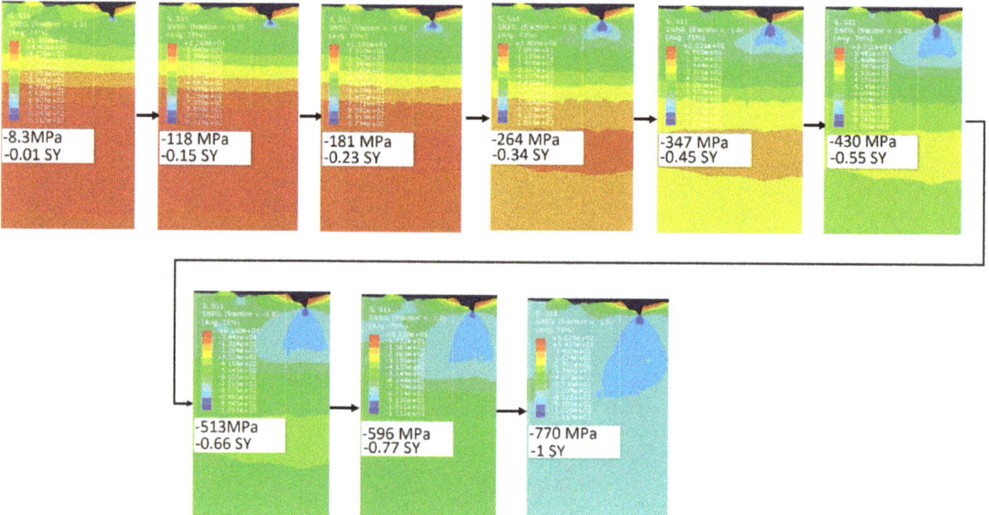

Figure 13. Stress distribution at different time steps of the loading with stress residual effect −320 MPa.

Figure 14 shows the crack formation (crack length) as a function of normalized stress value with the nominal yield strength of the material with different residual stress values, i.e., −80 MPa and −320 MPa. The slope of crack length versus load is quite similar for both cases; however, this slope is higher for the residual stress value of −320 MPa at the final step of loading.

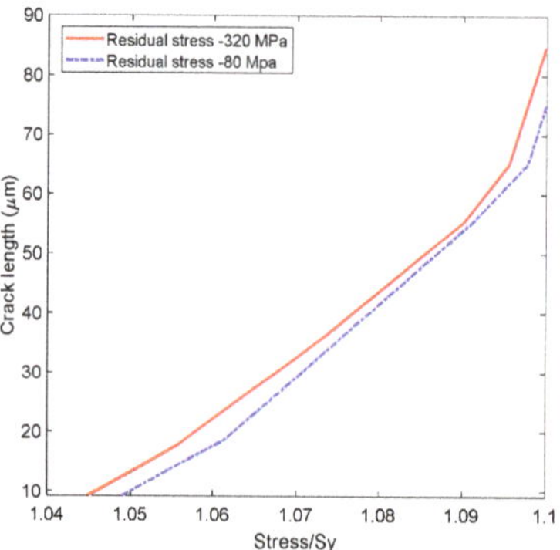

Figure 14. Crack length versus normalized stress with the nominal yield strength of the material with different values of residual stress (−80 MPa and −320 MPa).

3.2. Influence of Peak Load on Relaxation of Residual Stress

To investigate the influence of peak load values on residual stress relaxation and distribution, further analyses were carried out in cases where there was no overload, while considering the effect of residual stress. To this end, the load history, in this case, was a-c-d, as depicted in Figure 5. Figure 15 shows the stress distribution at different time steps of the loading. It was found that the residual stress remained effective at the end of the loading step; however, it was redistributed during the loading time step. Moreover, the predicted residual stress distribution at the beginning of loading and the stress distribution at the end of the loading step along the path below the critical notch are depicted in Figure 16. The predicted stress exhibits higher values closer to the notch tip due to the stress concentration factor and the high localized plastic deformation; however, the residual stress remains effective at the end of the loading step without residual stress relaxation. It can be concluded that in the presence of residual stress, the overload value affects the residual stress relaxation rate and distribution.

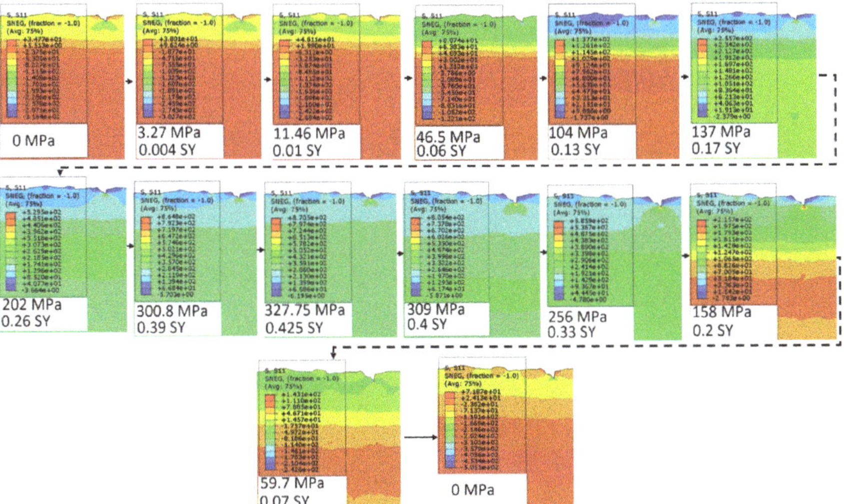

Figure 15. The stress distribution at different time steps of the loading without peak load and with the effect of residual stress.

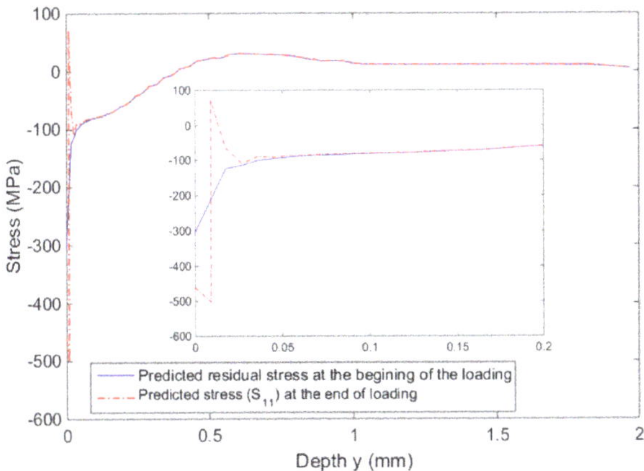

Figure 16. The predicted residual stress distribution at the beginning of the loading and the stress distribution at the end of loading without peak load and with the effect of residual.

Further analyses are required to validate these findings; thus, the relaxation of residual stress for different overload values was considered with residual stress values of −80 MPa and −320 MPa being induced in the material during the manufacturing process. Figure 17a,b shows the stress distribution at different time steps of loading for a peak stress magnitude of −1.15 S_y with residual stresses of −80 MPa and −320 MPa, respectively. For the higher value of residual stress (−320 MPa), it can be observed that the residual stress relaxation rate decreased significantly in comparison to the other case with a residual stress effect of −80 MPa. Figure 17c shows the duration of the residual stress effect before relaxation at a peak stress magnitude of −1.15 S_y with residual stress −80 MPa and

−320 MPa. It can be seen for a residual stress of −80 MPa that residual stress relaxation occurs at the time step corresponding to −0.55 S_y, while for a residual stress of −320 MPa, the residual stress relaxation occurs later, at the time step corresponding to −0.95 S_y.

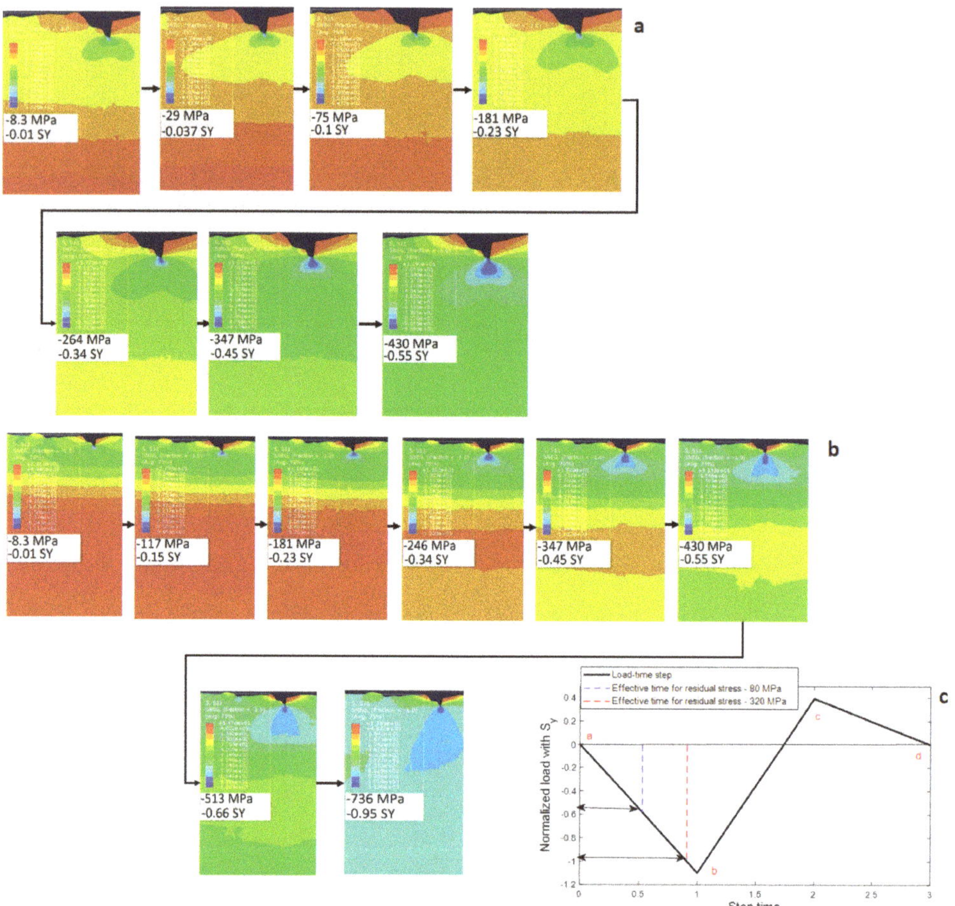

Figure 17. The stress distribution at different time steps of loading for an overload magnitude 1.15 S_y before relaxation of the residual stress: (**a**) with residual stress −80 MPa, and (**b**) with residual stress −320 MPa. (**c**) The effective time for different residual stress values before relaxation.

3.3. Influence of Fracture Locus and Load Level on Micro Crack Formation

The role of material constants (C_1, C_2, C_3) on the ductile fracture criterion has been studied recently [27]. As discussed, C_2 is the most significant material constant in the ductile fracture criterion. Parameter C_2 determines the equivalent plastic strain to fracture at negative stress triaxiality in the fracture locus, as depicted in Figure 2. In this section, the effect of C_2 and overload on crack formation are investigated. The other material constants, C_1 and C_3, were assumd to be 4 and 1.2, respectively.

Figure 18a,b show the crack size at the most critical notch against different C_2 values and their corresponding fracture locus with the overload values −1.1 S_y and −1.15 S_y in cases of two different residual stress values: −80 MPa and −320 MPa, respectively. This figure shows the effect of overload, the C_2 parameter (fracture locus), and residual stress effects on the damage mechanism and crack size, simultaneously. It can be seen that the influence of overload on damage formation and crack size is greater than the residual stress effect, and there are significant differences in the crack size between the overload values $-1.1\,S_y$ and $-1.15S_y$ at a constant magnitude of residual stress (-80 MPa), as shown in Figure 18a. However, there is no significant variance in the crack size between the different magnitudes of residual stress, -80 MPa and -320 Mpa, with a constant overload value, as depicted in Figure 18b. The relaxation of residual stress is the main reason for this behavior.

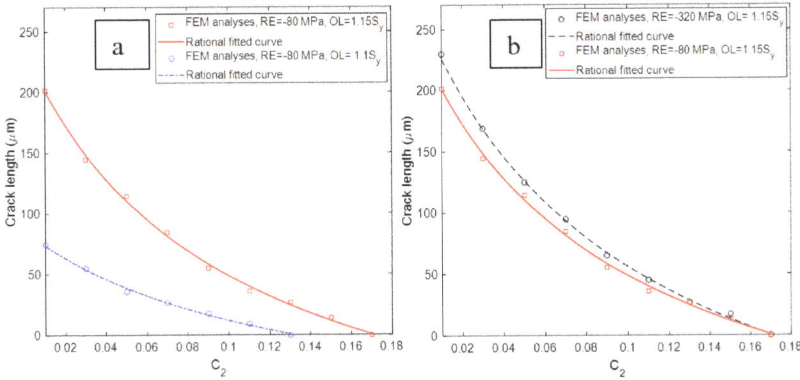

Figure 18. Influence of (**a**) overload and (**b**) residual stress on the damage and crack size with the C_2 parameter effect.

To further study the effect of the overload, five different magnitudes of peak stress, ranging from $-0.95\,S_y$ to $-1.15\,S_y$, were considered, and the crack sizes for $C_2 = 0.01$ and $C_2 = 0.03$ with a residual stress of -80 MPa are presented in Figure 19. It can be seen that there is no damage for overload values lower than $-1\,S_y$, and then, the crack sizes increase exponentially with increasing overload values.

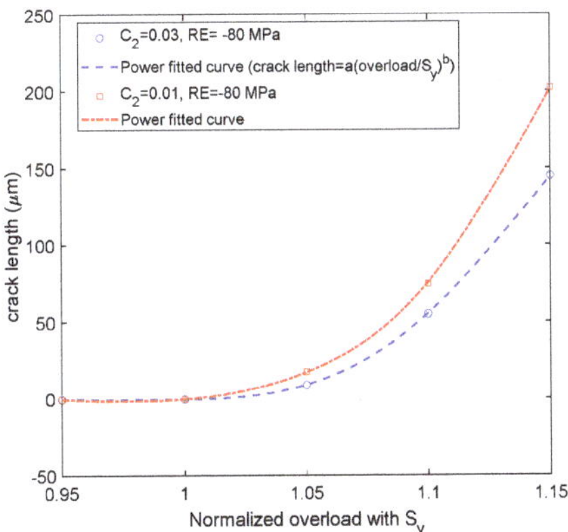

Figure 19. Crack length versus normalized overload values (overload/S_y) with different values of the C_2 parameter: 0.01 and 0.03.

To clarify the effect of surface integrity when the surface is subjected to peak load as a part of loading before other fatigue loading scenarios, the behavior of different elements located close to and far from the crack (see Figure 20a) were studied during the loading time step (see Figure 20b) for three different magnitudes of peak stress, as depicted in Figure 21a–c. It can be seen that the element located close to the crack (element number

2) experienced high stress at time step 3, at the c point of the force-time profile; however, the force magnitude subjected to the element at this stage is low. Thus, this element and the other elements located in the region close to the crack tip are prone to damage even with low magnitudes of loading in other load scenarios after the material has been subjected to peak load. To define the size of the affected region close to the tip of the crack, the stress magnitudes at different locations in the path below the crack tip are depicted in Figure 22a,b for three various overload magnitudes and two different fracture loci, with $C_2 = 0.01$ and $C_2 = 0.03$. This figure shows that the size of the affected region close to the crack tip is more extensive for higher magnitudes of peak stress, especially in cases where the fracture locus has a lower equivalent plastic on at the compression side when $C_2 = 0.01$.

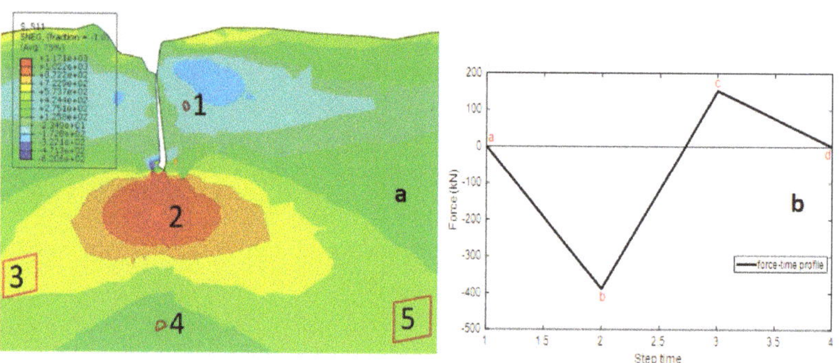

Figure 20. (**a**) Locations of five different elements close to and far from the crack, (**b**) loading time profile.

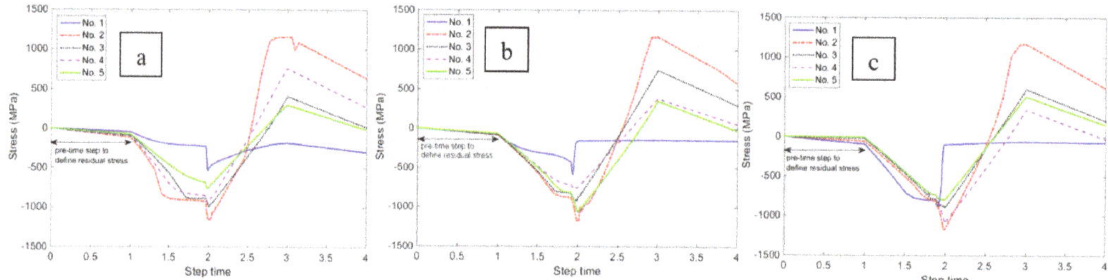

Figure 21. Behavior of different elements located close to and far from the crack for three magnitudes of peak stress: (**a**) $-1.05\,s_y$, (**b**) $-1.1\,s_y$, (**c**) $-1.15\,s_y$.

This study additionally establishes new research related to the effect of surface integrity when the peak load is applied to the surface before or during fatigue loading. The existing analytical approach for cut-plate edges, as well as welded joints, uses the original geometry as an initial point for fatigue assessment; meanwhile, the possible influence of peak load on the geometrical parameters, residual stress state, and microcrack formation has been neglected [35–37]. The findings of this study highlight this fact, indicating that considering the original geometry without investigating the possibility of microcrack formation when peak load is a part of the loading scenario can lead to significant inaccuracy in fatigue life prediction. The degree of inaccuracy depends on the peak load magnitude. Since it was found in this study that the crack size increases exponentially with increasing magnitudes of peak load, this neglect could lead to catastrophic problems, especially when engineering components and structures are exposed to high peak stresses and overloads during normal operation and under severe conditions. It should be noted that this study is focused on the numerical investigation of surface integrity, and its influence on microcrack formation of high-strength sandblasted steel under peak load conditions. In future work, further analysis considering the experimental aspects will be carried out in real engineering situations in order to highlight the importance and effectiveness of the proposed numerical approach.

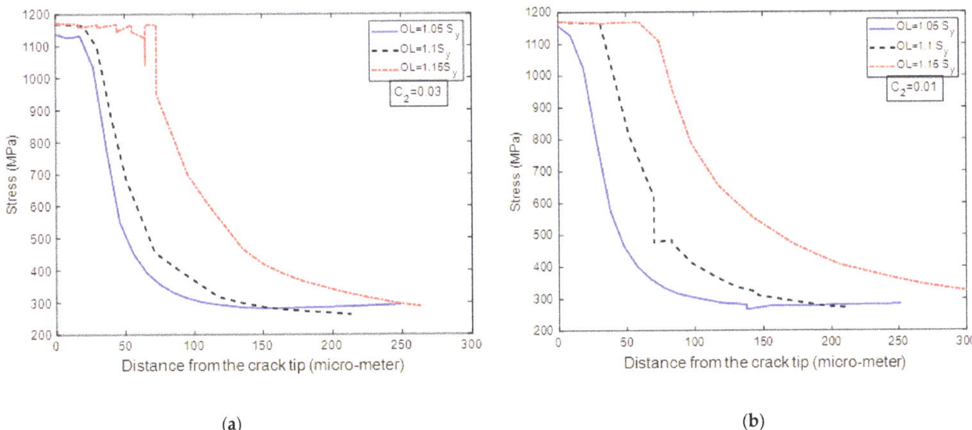

Figure 22. Stress distribution along the path below the tip of the crack for (**a**) $C_2 = 0.03$ and (**b**) $C_2 = 0.01$, with a residual stress of -80 MPa.

4. Conclusions

In this study, a new FEM approach and simulations were employed to characterize the peak load effect on microcrack formation and residual stress state. The numerical simulations simultaneously accounted for the influences of surface roughness and residual stress on the performance of high-strength steel under peak load conditions with ductile fracture criterion. By using this approach, it is possible to monitor and evaluate residual stress relaxation in real time. The main findings of this study can be concluded as follows

- With increasing values of residual stress, the residual stress relaxation rate decreases in comparison to the other studied case with lower magnitudes of residual stress. The residual stress relaxation rate is an affecting parameter with respect to the damage mechanism, and crack size increased with lower residual relaxation rates; however, this effect was not very significant.
- Under peak load conditions, surface roughness has a far more important influence on microcrack formation than residual stresses.
- The influence of compressive overload on damage formation and crack size is greater than the residual stress effect, and there are significant differences in the crack size between various overload values with constant residual stress magnitude. There was no significant variance in crack size between different residual stress magnitudes with a constant overload value. Meanwhile it was found that crack size increases exponentially with increasing magnitudes of peak load in cases with a constant magnitude of residual stress.
- Material areas located in regions close to the crack tip are prone to damage even with low loading magnitudes in other load scenarios after subjecting the material to peak load. The size of the affected region close to the crack tip is more extensive with higher magnitudes of peak stress, especially in cases where the fracture locus has a lower equivalent plastic strain on the compression side.

Author Contributions: Conceptualization, J.N.D. and H.R.; methodology, J.N.D.; software, F.S.; validation, J.N.D. and F.S.; Formal analysis, J.N.D.; resources, H.R.; data curation, F.S.; writing—

original draft preparation, J.N.D.; writing—review and editing, J.N.D., H.R., H.H.T.; visualization, H.H.T. All authors have read and agreed to the published version of the manuscript.

Funding: This research received no external funding.

Institutional Review Board Statement: Not applicable.

Informed Consent Statement: Not applicable.

Data Availability Statement: Not applicable.

Acknowledgments: The present research was supported by project RAMSSES that has received funding under the European Union's Horizon 2020 research and innovation program under the grant agreement No 723246. The information contained herein reflects the views only of the authors, and the European Union cannot be held responsible for any use which may be made of the information contained herein. All financial support is gratefully appreciated.

Conflicts of Interest: The authors declare no conflict of interest.

References

1. Diekhoff, P.; Hensel, J.; Nitschke, T.; Dilger, K. Investigation on fatigue strength of cut edges produced by various cutting methods for high-strength steels. *Weld. World* **2020**, *64*, 545–561. [CrossRef]
2. Sperle, J.O. *Influence of Parent Metal Strength on the Fatigue Strength of Parent Material with Machined and Thermally Cut Edges. IIW Document XIII-2174-07*; International Institute of Welding: Paris, France, 2007.
3. Alhussein, A.; Capelle, J.; Gilgert, J.; Tidu, A.; Hariri, S.; Azari, Z. Static, dynamic and fatigue characteristics of the pipeline API 5L X52 steel after sandblasting. *Eng. Fail. Anal.* **2013**, *27*, 1–15. [CrossRef]
4. McKelvey, S.A.; Fatemi, A. Surface finish effect on fatigue behavior of forged steel. *Int. J. Fatigue* **2012**, *36*, 130–145. [CrossRef]
5. Hussain, K.; Wilkinson, D.; Embury, J. Effect of surface finish on high temperature fatigue of a nickel based super alloy. *Int. J. Fatigue* **2009**, *31*, 743–750. [CrossRef]
6. Murakami, Y.; Endo, M. Effect of defects, inclusions and inhomogenities on fatigue strength. *Int. J. Fatigue* **1994**, *16*, 163–182. [CrossRef]
7. Sasahara, H. The effect on fatigue life of residual stress and surface hardness resulting from different cutting conditions of 0.45%C steel. *Int. J. Mach. Tools Manuf.* **2005**, *25*, 131–136. [CrossRef]
8. Małecka, J.; Rozumek, D. Metallographic and mechanical research of the O-Ti2AlNb alloy. *Meterials* **2020**, *13*, 3006. [CrossRef] [PubMed]
9. Ryu, H.R.; Kim, W.S.; Ha, H.I.; Kang, S.W.; Kim, M.H. Effect of toe grinding on fatigue strength of ship structure. *J. Ship Prod.* **2008**, *24*, 152–160. [CrossRef]
10. Xun, L.I.; Guan, C.; Zhao, P. Influences of milling and grinding on machined surface roughness and fatigue behavior of GH4169 superalloy workpieces. *Chin. J. Aeronaut.* **2018**, *31*, 1399–1405.
11. Zhang, M.; Wang, W.Q.; Wang, P.F.; Liu, Y.; Li, J. The fatigue behavior and mechanism of FV520B-I with large surface roughness in a very high cycle regime. *Eng. Fail. Anal.* **2016**, *66*, 432–444. [CrossRef]
12. Lopes, H.P.; Elias, C.N.; Vieira, M.V.; Vieira, V.T.; de Souza, L.C.; Dos Santos, A.L. Influence of surface roughness on the fatigue life of nickel-titanium rotary endodontic instruments. *J. Endod.* **2016**, *42*, 965–968. [CrossRef]
13. Pegues, J.; Roach, M.; Scott Williamson, R.; Shamsaei, N. Surface roughness effects on the fatigue strength of additively manufactured Ti-6Al-4V. *Int. J. Fatigue* **2018**, *116*, 543–552. [CrossRef]
14. Novovic, D.; Dewes, R.C.; Aspinwall, D.K.; Voice, W.; Bowen, P. The effect of machined topography and integrity on fatigue life. *Int. J. Mach. Tools Manuf.* **2004**, *44*, 125–134. [CrossRef]
15. Haghshenas, A.; Khonsari, M.M. Damage accumulation and crack initiation detection based on the evolution of surface roughness parameters. *Int. J. Fatigue* **2018**, *107*, 130–144. [CrossRef]
16. Gao, Y.; Li, X.; Yang, Q.; Yao, M. Influence of surface integrity on fatigue strength of 40CrNi2Si2MoVA steel. *Mater. Lett.* **2007**, *61*, 466–469. [CrossRef]
17. Remes, H.; Korhonen, E.; Lehto, P.; Romanoff, J.; Ehlers, S.; Niemela, A.; Hiltunen, P.; Kotakanen, T. Influence of surface integrity on fatigue strength of high-strengh steel. *J. Constr. Steel Res.* **2013**, *89*, 21–29. [CrossRef]
18. Lillemae-Avi, I.; Liinalampis Lehtimalil, E.; Remes, H.; Lehto, P.; Romanoff, J.; Ehlers, S.; Niemela, A. Fatigue strength of high strength steel after shipyard production process of plasma cutting, grinding and sandblasting. *Weld. World* **2018**, *62*, 1273–1287. [CrossRef]
19. Rozumek, D.; Lewandowski, J.; Lesiuk, G.; Correia, J. The influence of heat treatment on the behavior of fatigue crack growth in welded joints made of S355 under bending loading. *Int. J. Fatigue* **2020**, *131*, 105328. [CrossRef]
20. Bao, Y.B.; Wierzbicki, T. On fracture locus in the equivalent strain and stress triaxiality space. *Int. J. Mech. Sci.* **2004**, *46*, 81–98. [CrossRef]
21. Papasidero, J.; Doquet, V.; Mohr, D. Ductile fracture of aluminum 2024-T351 under proportional and non-proportional multi-axial loading: Bao–Wierzbicki results revisited. *Int. J. Solids Struct.* **2015**, *69–70*, 459–474. [CrossRef]

22. Lou, Y.; Huh, H. Prediction of ductile fracture for advanced high strength steel with a new criterion: Experiments and simulation. *J. Mater. Process. Tech.* **2013**, *213*, 1284–1302. [CrossRef]
23. Kõrgesaar, M.; Romanoff, J.; Remes, H.; Palokangas, P. Experimental and numerical penetration response of laser-welded stiffened panels. *Int. J. Impact. Eng.* **2018**, *114*, 78–92. [CrossRef]
24. Hu, Q.; Li, X.; Han, X.; Chen, J. A new shear and tension based ductile fracture criterion: Modeling and validation. *Eur. J. Mech. A-Solids* **2017**, *66*, 370–386. [CrossRef]
25. Mohr, D.; Marcadet, S.J. Micromechanically-motivated phenomenological Hosford–Coulomb model for predicting ductile fracture initiation at low stress triaxialities. *Int. J. Solids Struct.* **2015**, *67–68*, 55. [CrossRef]
26. Brünig, M.; Albrecht, D.; Gerke, S. Modeling of ductile damage and fracture behavior based on different micromechanisms. *Int. J. Damage Mech.* **2011**, *20*, 558–577. [CrossRef]
27. Nafar Dastgerdi, J.; Sheibanian, F.; Remes, H.; Lehto, P.; Hosseini Toudeshky, H. Numerical modeling approach for considering effects of surface integrity on micro-crack formation. *J. Constr. Steel Res.* **2020**, *175*, 106387. [CrossRef]
28. Lou, Y.; Huh, H.; Lim, S.; Pack, K. New ductile fracture criterion for prediction of fracture forming limit diagrams of sheet metals. *Int. J. Solids Struct.* **2012**, *49*, 3605–3615. [CrossRef]
29. McClung, R.C. A literature survey on the stability and significance of residual stresses during fatigue. *Fatigue Fract. Eng. Mater. Struct.* **2007**, *30*, 173–205. [CrossRef]
30. Dalaei, K.; Karlsson, B.; Svensson, L.E. Stability of shot peening induced residual stresses and their influence on fatigue lifetime. *Mater. Sci. Eng.* **2011**, *A528*, 1008–1015. [CrossRef]
31. Farajian-Sohi, M.; Nitschke-Pagel, T.; Dilger, K. Residual stress relaxation of quasi-statically and cyclically-loaded steel welds. *Weld. World* **2010**, *54*, 49–60. [CrossRef]
32. Laamouri, A.; Sidhom, H.; Braham, C. Evaluation of residual stress relaxation and its effect on fatigue strength of AISI 316L stainless steel ground surfaces: Experimental and numerical approaches. *Int. J. Fatigue* **2013**, *48*, 109–121. [CrossRef]
33. SFS-EN ISO 4288. *Geometrical Product Specifications (GPS)—Surface Texture: Profile Method—Rules and Procedures for the Assessment of Surface Texture*; ISO: Genève, Switzerland, 1996.
34. Hoa, H.C.; Chunga, K.F.; Liua, X.; Xiaoa, M.; Nethercot, D.A. Modelling tensile tests on high strength S690 steel materials undergoing large deformations. *Eng. Struct.* **2019**, *192*, 305–322. [CrossRef]
35. Radaj, D.; Sonsino, C.M.; Fricke, W. Recent developments in local concepts of fatigue 26 assessment of welded joints. *Int. J. Fatigue* **2009**, *31*, 2–11. [CrossRef]
36. Arola, D.; Ramulu, M. An examination of the effects from surface texture on the strength of fiber reinforced plastics. *J. Compos. Mater.* **1999**, *33*, 102–123. [CrossRef]
37. Arola, D.; Williams, C.L. Estimating the fatigue stress concentration factor of machined surfaces. *Int. J. Fatigue* **2002**, *24*, 923–930. [CrossRef]

Article

Probabilistic Surface Layer Fatigue Strength Assessment of EN AC-46200 Sand Castings

Sebastian Pomberger [1,*], **Matthias Oberreiter** [1], **Martin Leitner** [1], **Michael Stoschka** [1] **and Jörg Thuswaldner** [2]

[1] Christian Doppler Laboratory for Manufacturing Process based Component Design,
    Chair of Mechanical Engineering, Montanuniversität Leoben, Franz-Josef-Straße 18, 8700 Leoben, Austria;
    matthias.oberreiter@unileoben.ac.at (M.O.); martin.leitner@unileoben.ac.at (M.L.);
    michael.stoschka@unileoben.ac.at (M.S.)
[2] Chair of Mathematics and Statistics, Montanuniversität Leoben, Franz-Josef-Straße 18, 8700 Leoben, Austria;
    joerg.thuswaldner@unileoben.ac.at
*   Correspondence: sebastian.pomberger@unileoben.ac.at; Tel.: +43-3842-402-1470

Received: 20 March 2020; Accepted: 5 May 2020; Published: 9 May 2020

Abstract: The local fatigue strength within the aluminium cast surface layer is affected strongly by surface layer porosity and cast surface texture based notches. This article perpetuates the scientific methodology of a previously published fatigue assessment model of sand cast aluminium surface layers in T6 heat treatment condition. A new sampling position with significantly different surface roughness is investigated and the model exponents a_1 and a_2 are re-parametrised to be suited for a significantly increased range of surface roughness values. Furthermore, the fatigue assessment model of specimens in hot isostatic pressing (HIP) heat treatment condition is studied for all sampling positions. The obtained long life fatigue strength results are approximately 6% to 9% conservative, thus proven valid within an range of 30 µm $\leq Sv \leq$ 260 µm notch valley depth. To enhance engineering feasibility even further, the local concept is extended by a probabilistic approach invoking extreme value statistics. A bivariate distribution enables an advanced probabilistic long life fatigue strength of cast surface textures, based on statistically derived parameters such as extremal valley depth Sv_i and equivalent notch root radius $\bar{\rho}_i$. Summing up, a statistically driven fatigue strength assessment tool of sand cast aluminium surfaces has been developed and features an engineering friendly design method.

Keywords: cast aluminium; fatigue strength assessment; surface layer porosity; areal roughness parameter; hot isostatic pressing; extreme value statistics; probabilistic long life fatigue strength

---

1. Introduction

For fatigue strength assessment of metallic castings in mechanical engineering the designer has to consider a manufacturing process based on local material properties such as shrinkage pores or surface texture based notches. Neglecting the effect of defects on fatigue strength will result in oversizing of mechanical components to maintain globally sufficient component safety. As nowadays lightweight construction demands and sustainable designs are encouraged, aluminium sand cast components are often utilised which enable complex geometries and thus support significant weight savings of up to 50% [1]. However, it is well known that aluminium castings inherit both internal casting defects, particularly shrinkage pores, as well as surface texture related micro and macro notches driven by the surface geometrical structure (SGS), affecting the local fatigue strength. Therefore, applicable assessment methodologies, considering these local influences, are an advantageous tool in fatigue design.

In terms of porosity effect on the cast component fatigue strength various studies have been conducted, thoroughly investigating crack growth behaviour [2–5] as well as statistical description taking into account the effect of pore size, location and shape [6–21]. Therein, fatigue assessment concepts like the $\sqrt{area}$-approach of Murakami [22], or threshold-based concepts like the concept of Kitagawa and Takahashi [23], are frequently applied due to their engineering feasible applicability. Defect size and spatial location within the components are thereby stated to be the driving forces in terms of fatigue strength reduction effects. Other studies [15,16,22] point out that surface defects are more critical than internal ones, which emphasises the crucial influence of the cast surface layer and its defect distribution on the local fatigue resistance.

Aside from internal inhomogeneities, the surface texture, or surface roughness, plays a decisive role on cast fatigue strength. Surface pits, caused by the SGS, basically act as micro and macro notches and increase the stress concentration, thus reducing the local fatigue strength. Therefore, cast mechanical components are often additionally surface finished by machining or by polishing, in order to counteract the detrimental surface roughness effect. Thus, many studies contribute to the surface roughness effect on fatigue strength, investigating machined surfaces obtaining periodic surface textures [24–29]. As in the literature [30,31] various analytical equations are introduced to characterise geometrical notches, these formulations are adapted to assess machined [32–35] or even cast [36–38], and more recently also additively manufactured [39] surfaces.

As stated in [37,38] for cast surfaces, areal roughness parameters should be used in order to characterise the surface texture thoroughly. The areal roughness evaluation methodology published by the authors of [38] enables such a holistic characterisation of cast, as well as of additively manufactured, surface textures. The presented sub-area analysis provides additional information about local roughness parameters, which focusses on more distinctive, and therefore more fatigue crack-initiating surface pits. A modification of Peterson´s stress concentration factor [31] and its application on investigated sand cast aluminium surfaces lead to an engineering feasible areal fatigue assessment approach, as presented in [36]. The modified stress concentration factor $K_{t,mod}$, as introduced in [36], is given in Equation (1) and utilises the local surface pit depth Sv_{local}, the mean value of the crack initiating cast surface pit depth Sv_{rev} and an equivalent notch root radius $\bar{\rho}$. Taking the local notch sensitivity into account, the surface fatigue notch factor $K_{f,s}$ can be subsequently evaluated. Figure 1 depicts the result of an exemplary areal fatigue assessment for a sand cast surface layer, leading to sub-area based $K_{f,s}$ values. The red cross marks the technical crack initiation point of the tested sample, evaluated by means of fracture surface analysis. The presented concept also features the assessment of surface layer porosity, that is, pores directly beneath the surface which are partly broached by the surface texture and therefore interact with the surface roughness based notches.

$$K_{t,mod} = 1 + 2 \left(\frac{\left(\frac{Sv_{local}}{Sv_{rev}} \right)^{a_1} \cdot Sv_{rev}}{\bar{\rho}} \right)^{\frac{1}{a_2}} \tag{1}$$

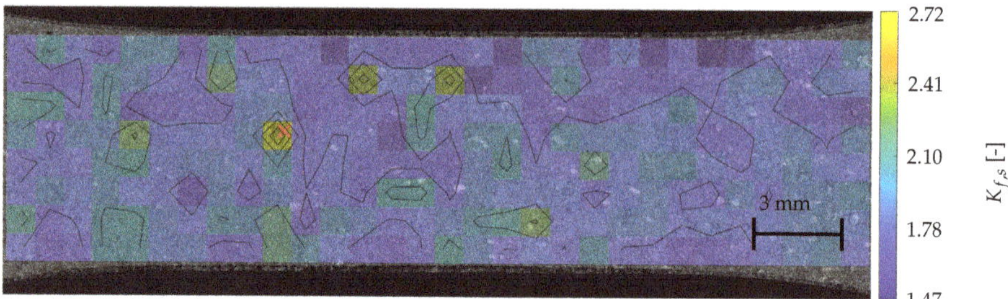

Figure 1. Exemplarily mapping of the surface fatigue notch factor $K_{f,s}$ with 1 mm $\times$ 1 mm sub-areas.

This methodology may be validated even further by additional datasets, which should provide significantly different cast surface textures in order to broaden the applicability of the method towards a wider range of surface roughness values. In addition, as this areal sand cast surface characterisation method is based on local roughness values, the fatigue designer has to have knowledge about these manufacturing process dependent values. However, such as localised information is in general not available for cast surface structures. Therefore, this concept is extended by a probabilistic approach as previously recommended in [36]. Moreover, the effect of sub-area size ought to be investigated to cover miscellaneous evaluation area magnitudes as well.

Therefore, this paper scientifically contributes to the following key parts.

- Extension of the assessment methodology presented in [36] utilising an additional aluminium sand cast surface exhibiting a significantly varying surface roughness structure.
- As-cast surfaces in the T6 heat treatment condition were the main research target in [36]; the applicability of the assessment model to cast specimens with additional hot isostatic pressing (HIP) heat treatment is evaluated.
- Robustness study of the presented method in terms of sub-area magnitude or sample size and their effect on the evaluated statistical distribution.
- Statistical characterisation of the sand cast surface texture and subsequently probabilistic evaluation of the manufacturing process related surface fatigue strength as design recommendations of cast components.

2. Investigated Material

The investigated aluminium alloy's EN numerical designation is EN AC-46200. The gravity sand cast components are crankcases, manufactured by means of the core package system (CPS) casting process [40–43]. For details about the specimen geometry and the nominal chemical composition of the material the authors refer to the work in [36] as reference. As an additional different surface roughness texture is investigated for validation in this study, the specimen series with this new sampling position is denoted by P2, while the original specimen series investigated in [36] are labelled as P1 in the following. Beside the variation of the specimen position the effect of an additional HIP heat treatment (HIP+T6), as also studied in [44], is investigated and compared to T6 heat treatment condition. The HIP+T6 specimens are subsequently labelled as HIP. Due to the HIP process, shrinkage pores within the bulk material shall be closed and its effect on the surface layer will be studied. Typically applied HIP parameters for aluminium alloys [44–48], such as temperature T, pressure p and time t, are given in Table 1.

Table 1. Typical hot isostatic pressing (HIP) parameters for Al alloys [44–48].

T [°C]	p [MPa]	t [h]
510–521	103	2–6

Metallographic analysis revealed that the HIP specimens do not differ in microstructure in terms of secondary dendrite arm spacing (DAS) in comparison to T6 specimens, thus matching the findings in [49,50]. The DAS has been evaluated as described in [51], and the mean values of sampling position P1 and P2 in T6 as well as HIP heat treatment condition are listed in Table 2. For both, T6 P1 (Figure 2b) and HIP P1 (Figure 2a) a mean DAS of about 26 µm was evaluated. For specimens at sampling position P2 a slight DAS gradient is observable, see Figure 2c. This is caused by the increased solidification rate within the surface layer at this sampling position. Near the surface a DAS of about 21 µm was measured, which slowly increased to about 28 µm at a distance of 9 mm measured from the cast surface. This matches the results of Aigner [49], investigating the bulk material of EN AC-46200 at sampling position P2. Although in Figure 2c a specimen with T6 heat treatment is presented, the HIP micostructure in terms of DAS is identical as the process only affects the bulk material porosity. However, as DAS does not affect the fatigue strength in the presence of defects according to [6,13], differences in DAS can be neglected.

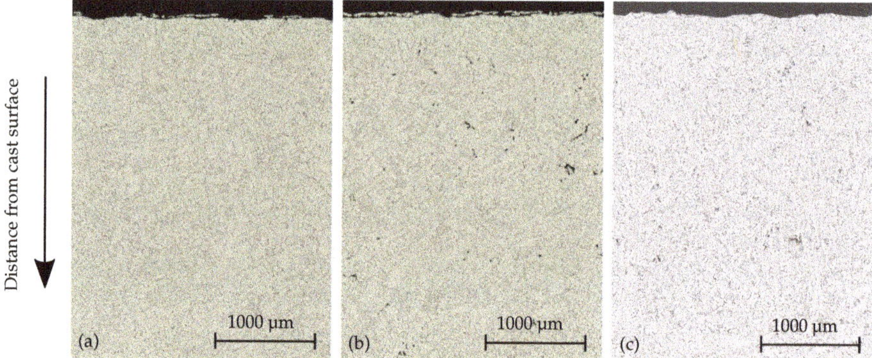

Figure 2. Microstructure of (**a**) HIP P1, (**b**) T6 P1 and (**c**) T6 P2.

Table 2. Evaluated mean dendrite arm spacing (DAS) for T6 and HIP at P1 and P2.

Measurement Position	T6 P1	HIP P1	HIP P2	T6 P2 [49]	HIP P2 [49]
surface DAS [µm]	27.3	26.5	21.4	-	-
bulk DAS [µm]	26.1	25.1	28.8	27.5	27.2

The tested material properties of both HIP and T6 heat treatment are opposed in Table 3. While the ultimate tensile strength R_m as well as Vickers hardness $HV10$, Young's modulus E and yield strength $R_{P0.2}$ only differ slightly, the elongation at rupture A is significantly increased, which matches the findings in [20,44].

Table 3. Tested material properties of EN AC-46200 with T6 and HIP heat treatment.

Alloy	$HV10$ [-]	E [MPa]	R_m [MPa]	$R_{P0.2}$ [MPa]	A [%]
EN AC-46200 T6	123	74,300	300	285	0.51
EN AC-46200 HIP	124	74,600	320	260	1.78

3. Experimental

Within this work, fatigue tests were performed on HIP sand cast surfaces at sampling positions P1 and P2. Although three test series have been experimentally investigated (HIP P1, HIP P1(2), HIP P2), only two of them (HIP P1 and HIP P2) will be presented in detail to enhance clarity. However, fatigue test results such as S/N-parameters are evaluated and tabulated for all three investigated series.

All testing series possess cast surfaces and the fatigue tests were performed identically to the procedure described in [36] utilising a Rumul Cracktronic®. Due to the load stress ratio of R = 0 under bending load the highly tensile-stressed region of the specimen is set to the cast surface layer. The S/N-curves are statistically evaluated following the procedure applied in [36]. Figure 3 depicts the nominal bending S/N-curve of the HIP test series at the new sampling position P2, possessing a significantly reduced surface roughness, as discussed in Section 5. Within all S/N-figures, the stress amplitude σ_a is normalised to the material's near defect-free long life fatigue strength $\sigma_{LLF,0}$. As stated in [36], the value of $\sigma_{LLF,0}$ was experimentally evaluated by means of HIP specimens with machined and subsequently polished surface condition. Thus, the observed fatigue strength is unaffected both from porosity effect and surface roughness effect.

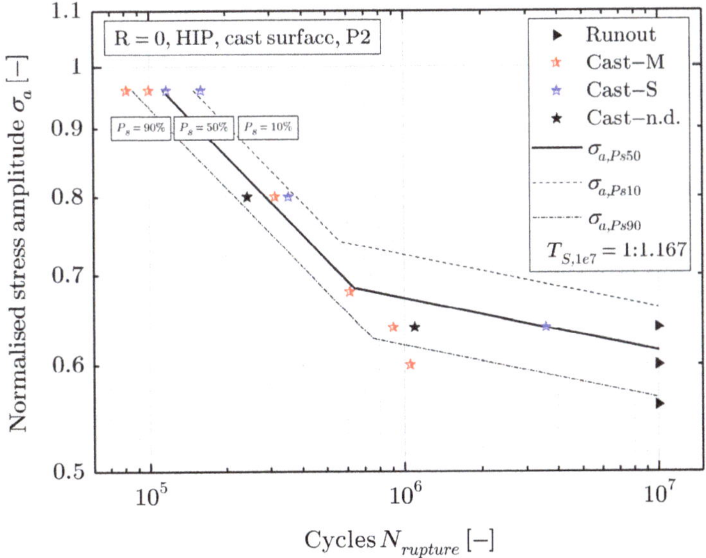

Figure 3. Normalised S/N-curve of the specimen series HIP P2.

The crack initiation cause is marked red if the crack initiated at a combination of a surface pit and a micropore located directly within the surrounding area (Cast-M), and blue if only the cast surface texture (Cast-S), that is, surface roughness, caused technical crack initiation. Figure 4 exemplary depicts the fractographically evaluated crack initiating defects of Cast-S and Cast-M specimens of the HIP P1 and HIP P2 testing series by means of scanning electron microscopy (SEM). Fractographic images of HIP P1 are also representative for the T6 P1 testing series of [36]. For almost all Cast-M specimens shrinkage porosity was observed to participate in crack initiation. Only few cases revealed gas pores, bifilms or intermetallic phases to be critical. In contrast with the locations of the pores of Cast-M P1 specimens, those of Cast-M P2 rarely have been broached, but were found to be located about 10 µm to 30 µm beneath the surface. This may be caused by the elevated solidification rate within the surface layer at this sampling position. For Cast-n.d. specimens, no distinct crack initiation cause could be clearly determined by fracture surface analysis, which is why they are not subsequently taken into account for validation.

The S/N-curves (Figures 3 and 5) are given with their 90% and 10% probability of survival and the stress scatter index $T_{S,1e7}$ is calculated according to [52] by means of Equation (2) at ten million load cycles. The evaluated S/N-curve provides in Table 4 the value of the inverse slopes k_1 and k_2, which is five times k_1 [53], the transition knee point N_T and the normalised long life fatigue strength $\sigma_{a,Ps50}$ as well as the stress scatter index $T_{S,1e7}$.

$$T_{S,1e7} = 1 : \frac{\sigma_a(P_s = 10\%)}{\sigma_a(P_s = 90\%)} \tag{2}$$

Figure 5 depicts the evaluated S/N-curve of the HIP specimen series at sampling position P1, which is similar to the original sampling position presented in [36]. The coloration of the markers is the same as in Figure 3. The accompanying fracture surface analysis revealed great similarity to those samples of the T6 P1 testing series in [36]. As the evaluation of the long life fatigue strength of the HIP P1 specimen series by means of the $\arcsin\sqrt{p}$ method [54] would lead to a smaller scatter within the long life region compared to the finite life region, the normalisation process of S/N-curves was applied as proposed in [52,55], and also accordingly executed in [36]. The evaluated S/N-curve results are listed in Table 4 as well. Additionally, within Figure 5 the evaluated long life fatigue strength $\sigma_{a,Ps50}$ of the T6 testing series with cast surface, as sketched in [36], is highlighted by the purple dash-dotted line for comparison.

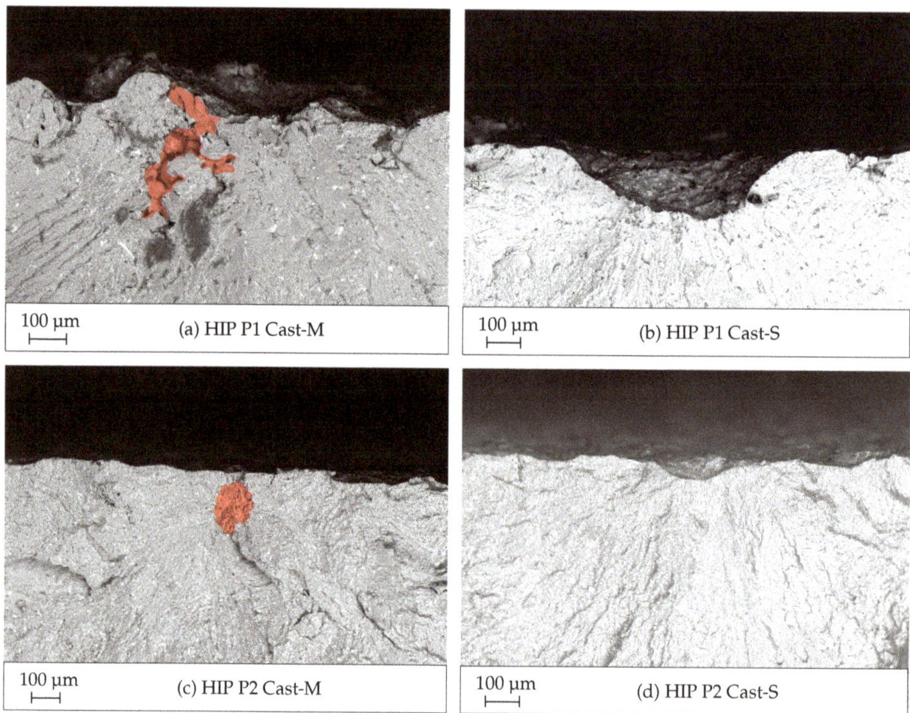

Figure 4. Defect cases: (**a**) HIP P1 cast-M, (**b**) HIP P1 Cast-S, (**c**) HIP P2 Cast-M and (**d**) HIP P2 Cast-S.

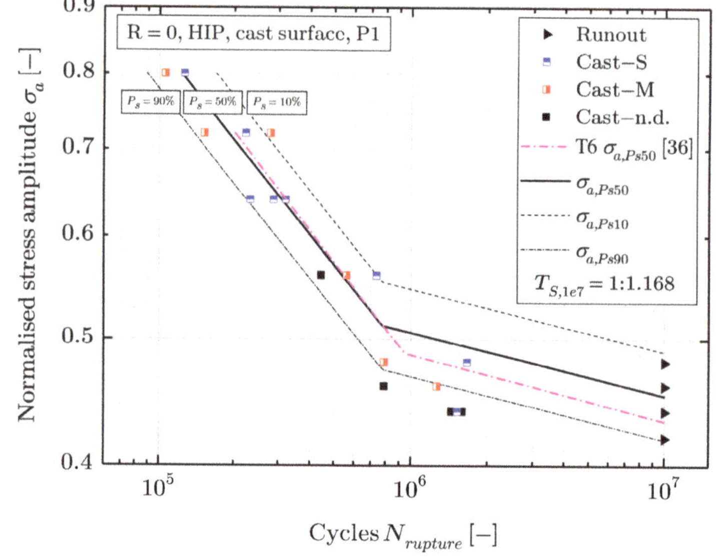

Figure 5. Normalised S/N-curve of the specimen series HIP P1.

At this point, it must be stated that the fact that Cast-M specimen fatigue results are similar to Cast-S specimens does not imply that surface layer porosity can be neglected. Both surface layer porosity and the cast surface texture showed similar fatigue results if the are evaluated independently. Thus, they both affect the fatigue strength in a comparable manner. However, if the cracks initiate combinatorial, as observed in Cast-M specimens, see Figure 4a,c, they lead to similar fatigue test results even if the defects are smaller. Summing up, smaller surface layer inhomogeneities and less detrimental surface texture combinatorial considered may be more crucial than a single, distinct surface pit. Therefore, they may be treated in a combined manner. For more information regarding the combinatorial failure mechanism see the work in [36].

Table 4. Evaluated results of the fatigue tests of the investigated specimen series.

Specimen Series	k_1 [-]	k_2 [-]	$\sigma_{a,Ps50}$ [-]	N_T [-]	$T_{S,1e7}$ [-]
T6 P1 [36]	3.97	19.85	0.435	927.960	1:1.290
HIP P1	4.14	20.68	0.452	779.323	1:1.168
HIP P1(2)	3.29	16.48	0.411	734.094	1:1.150
HIP P2	5.11	25.56	0.614	641.770	1:1.167

HIP Effect on the Cast Surface Layer

While the HIP process leads to significantly enhanced fatigue strength results in terms of bulk material testing [45,50,56–59], this effect was not present for any of the investigated HIP cast surface series. While for one HIP P1 testing series the evaluated long life fatigue strength $\sigma_{a,Ps50}$ was above the T6 value (as presented in Figure 5), the other HIP P1(2) testing series the $\sigma_{a,Ps50}$ was slightly lowered, but both within the 90% and 10% stress scatter band of HIPped samples. In terms of Cast-S specimens, it was found that the evaluated surface roughness values, and therefore the estimated fatigue strength by means of the introduced fatigue assessment model, was comparable to those of the cast T6 P1 series. Therefore, it is reasonable that the Cast-S points of the HIP P1 specimens in Figure 5 fit to the cast T6 P1 S/N-curve. However, regarding the Cast-M specimens, one may expect a significantly higher fatigue strength due to closed shrinkage porosity. Investigations on metallographic T6 specimens revealed a higher porosity within the surface layer, especially up to a certain depth, see Figure 6. Almost without exception these were shrinkage pores evolving during the solidification process. The left sub-figure shows the detected micropores, while on the right diagram the evaluated degree of porosity is plotted. By means of a user defined routine, pores are detected on a metallographic specimen, which have been captured by means of a digital optical microscope. The degree of porosity was then evaluated by counting the black pixels of the picture, which have been determined as pores, in relation to the white pixels within the same horizontal line. Subsequently, mean values of the degree of porosity have been calculated within a vertical range of 250 µm. It is clearly recognisable that the highest degree of porosity occurs in about 1 mm to 2 mm depth measured from the cast surface. To achieve information about the spatial distribution of the micropores, the metallographic specimen was subsequently grinded, thereby removing about 100 µm, and subsequently evaluated again. This methodology hast been carried out several times for four T6 specimens, resulting in 140 metallographic analysis in total. It emerged that the trend of degree of porosity, as depicted in Figure 6, is representative for the T6 P1 specimens series. An increased degree of porosity near the cast surface of AlSi castings was also stated by Leitner et al. [60]. This increased porosity formation may be reasoned by the oxide entrainment mechanism as a result of turbulent mould filling, see [61–63]. However, a comprehensive insight in the degree of porosity can be more properly evaluated by means of XCT-scans [21,60].

The same investigations and evaluation procedure have been conducted for four HIP specimens, again resulting in 140 metallographic analysis slices. A representative result is depicted in Figure 7. It is clearly recognisable, that the HIP process lead to significantly reduced, or even partially completely suppressed porosity within the bulk material. Only within the first mm in depth, porosity was still observable whereat the HIP process did not close these micropores. As this is within the highly stressed region of the specimen, crack initiation is still caused by those surface roughness and microporosity mixed cases (Cast-M), thus resulting in similar long life fatigue strength values as previously discussed.

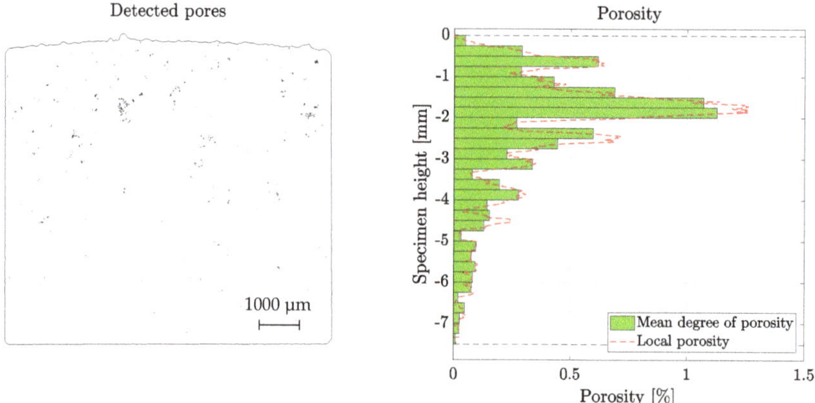

Figure 6. Detected pores and degree of porosity of an exemplary cast surface T6 P1 cross section.

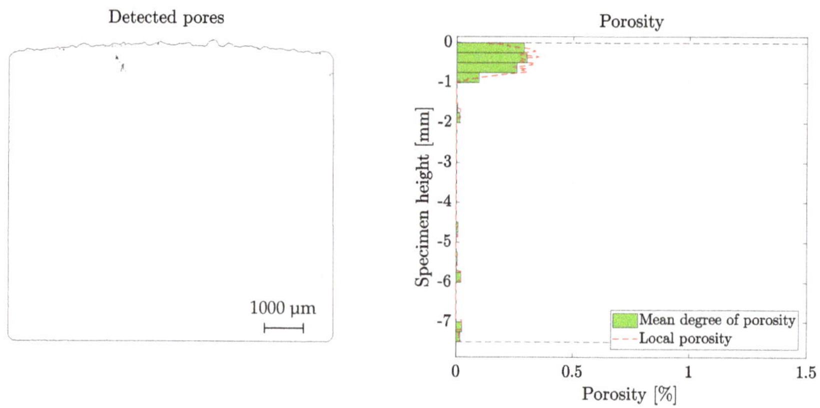

Figure 7. Detected pores and degree of porosity of an exemplarily cast surface HIP P1 cross section.

The comparably high possibility of observing a mixed (Cast-M) defect case becomes visible by a comparison of the cast surfaces in T6 and HIP condition by means of SEM. Figure 8a,b depicts SEM images of the cast T6 surface at sampling position P1. Both sub-figures show that the cast surface is frequently broached by cavities, or shrinkage pores. Those cavities can be found both within surface pits as well as at surface peaks. In Figure 8b, even the dendritic structure of the α-phase is observable. Thus, a relatively high chance of crack initiation occurring at a combination of both surface pits due to the surface roughness and surface layer porosity is present.

In Figure 9, the SEM images of the investigated cast surfaces in HIP condition are illustrated for both sampling position P1 and P2. For cast surface texture comparison purpose, Table 5 lists the mean values Sa_{mean} of the global Sa roughness parameter of all specimens, in respect to the sampling position. Additionally, the 10–90% scatter values are given. The cast surface of P1 (Figure 9a) is basically identical to the cast surface in T6 heat treatment condition presented in Figure 8a. Especially broached pores are again observable in a similar amount. Those cavities and dendritic canals, created by the solidification process, can reach down to about 1 mm in depth in some cases, as those micropores can not be closed by the HIP process. This matches the statement of Atkinson [56] on surface connected porosity. Thus, it can be stated that if, subsequently to the casting process, the machined surface finish is conducted with the aim of removal of surface near porosity, the process has to cover a certain depth. For the exemplified case of Figure 7, removing only 0.5 mm of the cast surface would lead to a broached pore. As broached pores essentially decrease the local fatigue strength as well, such a machining process would not have the intended favourable fatigue effect and lead to similar fatigue strength results as those including cast surface, as presented in [36]. For the cast surface in HIP condition at sampling position P2, see Figure 9b, no broached pores were recognisable on the surface. This matches the results of the HIP P2 fracture surface analysis, where predominantly surface layer pores have been observed which are not broaching the cast surface, but are located about 10 µm to 30 µm beneath. As already mentioned, this might be caused by the significantly increased solidification rate at this sampling position P2 compared to P1.

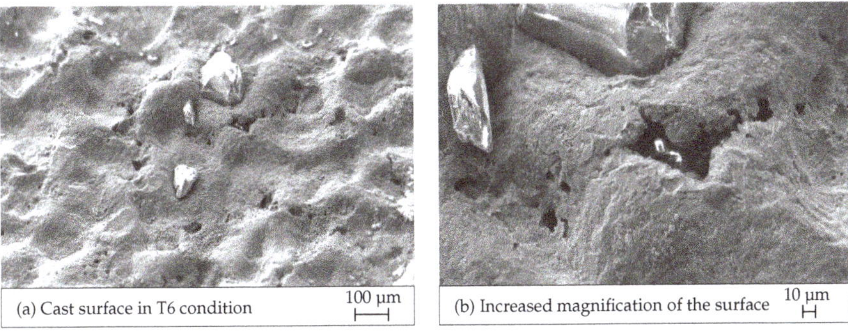

<table>
<tr><td>(a) Cast surface in T6 condition</td><td>100 µm</td><td>(b) Increased magnification of the surface</td><td>10 µm</td></tr>
</table>

Figure 8. SEM image of the specimen series T6 P1 investigated in [36].

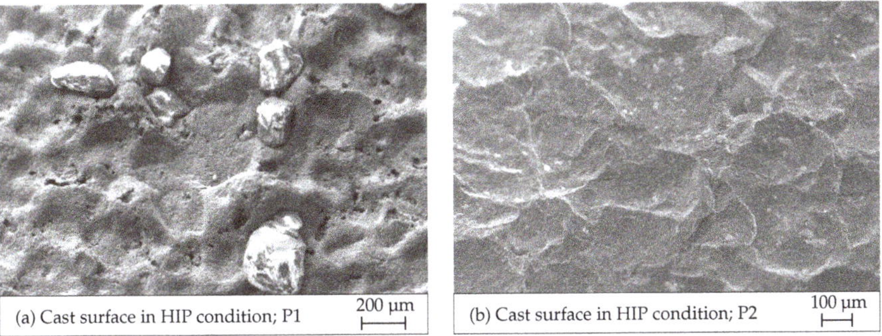

<table>
<tr><td>(a) Cast surface in HIP condition; P1</td><td>200 µm</td><td>(b) Cast surface in HIP condition; P2</td><td>100 µm</td></tr>
</table>

Figure 9. SEM image of the specimen series (**a**) HIP P1 and (**b**) HIP P2.

Table 5. Mean value Sa_{mean} of the global Sa values with its 10–90% scatter.

Roughness Parameter	T6 P1	HIP P1	HIP P2
Sa_{mean} [µm]	17.9 ± 6.4	18.7 ± 8.0	9.3 ± 3.6

4. Fatigue Assessment Model

This section contributes to the alteration of the fatigue assessment model as originally presented in [36] on HIP surfaces as well as on sampling position P2, which possesses a significantly different surface roughness. Thereby, the model's application range in terms of surface roughness parameter values is studied.

4.1. Modification of the Model

First, the local roughness values Sv_{local} and $\bar{\rho}$ at the crack initiation point as well as the modified parameter Sv_{rev} were evaluated. Originally, Sv_{rev} was introduced as the mean value of crack initiating surface pits and it has been substituted as a statistical parameter, based on the most critical surface pits. It represents the value with 50% probability of occurrence of the five biggest Sv sub-area valley depth distribution (Sv_i GEV), which is referenced in detail within Section 5. It was found, that the in [36] presented exponents of $a_1 = 0.6$ and $a_2 = 2$ caused too conservative results for Cast-S specimen failures taken from sampling position P2. Thus, these parameters have been adapted to $a_1 = 0.4$ and $a_2 = 1.8$ instead, to improve the range of applicability of the basic assessment concept.

For the Cast-M specimen failures, the introduced neural network (NN) in [36] has been adapted to only four neurons and four input variables. The pore location and elongation parameters e_{min}, e_{max} and α have been replaced by the statistical roughness value Sv_{rev}. Thus, the overall condition of geometry dependent distinct roughness values is now considered by Sv_{rev}. Therefore, only the defect size $\sqrt{area}$, the local maximum pit height Sv_{local}, the equivalent notch root radius $\bar{\rho}$ and the statistical pit depth with 50% probability of occurrence Sv_{rev} act as input variables. The network was further trained by four specimens of the HIP P2 testing series in addition to the twenty-five T6 P1 specimens from [36]. The coefficient of determination for the training set was $R^2 = 0.988$, resulting in the interaction coefficients ψ, listed in Table 6. Therein, the mean values ψ_{mean} as well as its standard deviation ψ_{std} and the minimum ψ_{min} and maximum ψ_{max} values are listed in detail. The individual interaction coefficient ψ of each Cast-M specimen is subsequently used within Equation (4) for calculation of the mixed fatigue strength reduction factor $K_{f,m}$ by taking the fatigue strength reduction factor $K_{f,p}$ of surface layer microporosity and the surface fatigue notch factor $K_{f,s}$ of surface roughness-based notches as combinatorical defect case into account. $K_{f,p}$ is calculated by Equation (3), as introduced in [36].

Table 6. Evaluated interaction coefficients of the NN training series of T6 P1 [36] and HIP P2.

Training Series	ψ_{mean} [-]	ψ_{std} [-]	ψ_{min} [-]	ψ_{max} [-]	Sample Size [-]
T6 P1 [36]	0.598	0.034	0.547	0.674	25
HIP P2	0.420	0.025	0.384	0.473	4

$$K_{f,p} = \frac{1.6 \cdot HV}{C_1 \cdot \frac{HV + C_2}{(\sqrt{area})^{1/6}}} \tag{3}$$

$$K_{f,m} = (K_{f,s} \cdot K_{f,p})^{\psi} \tag{4}$$

4.2. Validation of the Model

After modification of the concept in order to improve the overall performance, the model was validated by means of the HIP P2 Cast-S data as well as the remaining two Cast-M specimens, which have not yet been taken into account for training of the modified neural network. The fatigue assessment result is depicted in Figure 10. The fatigue strength is normalised by the material's near defect-free long life fatigue strength $\sigma_{LLF,0}$, which was evaluated at a load stress ratio of R = 0 under bending load at ten million load cycles. The black dashed line marks the long life fatigue strength with 50% probability of survival, taken from the fatigue testing, see Figure 3. The fatigue strength values for each specimen, received by the fatigue assessment model, are depicted in blue for Cast-S specimens and in red for Cast-M specimens. Additionally, for each specimen, the experimental fatigue test data point is extrapolated to one million load cycles and additional plotted into the Figures 10 and 11 represented by $\sigma_{a,1e7}$. Overall, the fatigue strength assessment is 7% conservative, regarding the fatigue strength $\sigma_{LLF,*,Ps50}$ with a probability of survival of 50% in respect to the experimental fatigue strength result $\sigma_{a,Ps50}$. The evaluated fatigue strength $\sigma_{LLF,*,Ps10}$ with a probability of survival of 10% is 6.1% conservative as well. The stress scatter index $T_{S,1e7}$ of the model results was evaluated by Equation (2). At this stage, only six specimens from the HIP P2 testing series were available for validation; however, the model lead to sound fatigue results so far for both, Cast-M specimens, as well as Cast-S specimen, where no neural network was involved for prior fatigue strength assessment. The experimental and model-based fatigue strength results of the validation series HIP P2 are compared in Table 7.

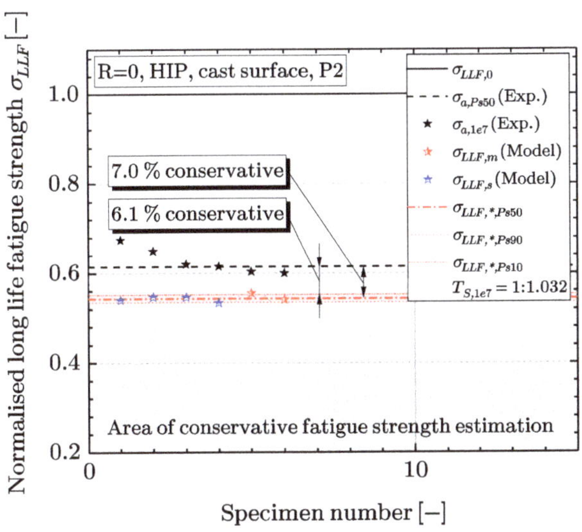

Figure 10. Long life fatigue strength assessment result of the validation series HIP P2.

To prove the fatigue assessment model's applicability to HIP-treated cast surfaces even further, the specimen series HIP P1 was used for additional validation. Figure 11 shows the calculated fatigue assessment results. Again, Cast-S specimens with crack initiation at a surface pit due to the surface roughness are marked in blue, and Cast-M specimens representing a combinatorial defect case with surface roughness and surface layer porosity interaction are marked in red colour. The model's long life fatigue strength at 10%, 50% and 90% probability of survival as well as the evaluated stress scatter index $T_{S,1e7}$ and the experimental fatigue strength of the associated S/N-curve $\sigma_{a,Ps50}$ represented by the dashed black line, are again diagrammed. Both Cast-S as well as Cast-M specimens are well assessed, resulting in an overall 9.3% conservative long life fatigue strength design with a probability of survival of 50%. The stress scatter index increased compared to HIP P2, but still shows sound results as it is below the value of the associated S/N-curve, see Figure 5. An overview of the validation results of the HIP P1 series is also given in Table 7.

Table 7 also lists the validation results of the specimens series HIP P1(2) utilising 16 specimens. Further, as the assessment of Cast-S as well as of Cast-M specimens has been adapted, the validation data set of [36] with 14 specimens has been re-evaluated and is also given in Table 7, labelled as T6 P1. Utilising the modified fatigue assessment model, the result of T6 P1 becomes slightly more conservative compared to the results in [36] and the stress scatter index increased. However, the overall applicability in terms of sand cast aluminium surface layers with different heat treatment conditions, possessing cast surface textures and surface layer porosity is confirmed by this comprehensive validations sets. Summing up, the results of the estimated fatigue strength are about 6% to 9% conservative. Therefore, the introduced model supports an engineering feasible local fatigue assessment concept, to assess distinctions in cast surface roughness structures and their effect on cyclic endurance limit.

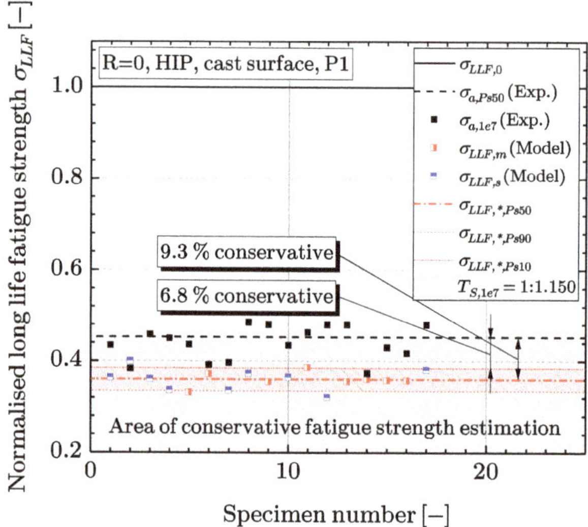

Figure 11. Long life fatigue strength assessment result of the validation series HIP P1.

Table 7. Validation series results and re-evaluation of the T6 P1 validation series from [36].

Validation Series	Experiment $\sigma_{a,Ps50}$ [-]	Model $\sigma_{LLF,*,Ps50}$ [-]	Difference Δ [%]	Model $T_{S,1e7}$ [-]	Sample Size [-]
T6 P1 [36] (re-evaluated)	0.416	0.336	−7.9	1:1.189	14
HIP P1	0.452	0.360	−9.2	1:1.150	17
HIP P1(2)	0.411	0.353	−5.8	1:1.158	16
HIP P2	0.614	0.544	−7.0	1:1.032	6

5. Probabilistic Fatigue Strength Assessment

The proposed surface fatigue assessment model utilises local roughness values evaluated at crack initiation points, identified by means of fracture surface analyses after fatigue testing. However, in engineering design, no a priori knowledge about surface texture is available, instead probabilistic values of the surface layer act as link to the manufacturing process dependent surface layer properties. Moreover, the random variable Sv_{rev}, a parameter based on the distribution of distinctive Sv sub-area values, is an important factor for an appropriate long life fatigue strength calculation. Sv_{rev} should be evaluated with accurateness, necessitating a statistically based recommendation about the sample size of surface measurements.

Thus, this section contributes to the probabilistic assessment of crack initiating surface roughness pits. Previously conducted experiments on Cast-S specimens revealed that for mostly all cases, crack initiation occurred at one of the five deepest surface pits, respectively, one of the five highest Sv values. However, not strictly the maximum value Sv_{max} of all evaluated sub-areas initiates a crack, as the notch root radius interacts in terms of notch stress effect. Concluding, the authors suggest to consider the five highest Sv values of each investigated surface to be representative for statistical surface roughness effect. Subsequently, these surface valley series of five deepest depths is denoted by Sv_i with $1 \leq i \leq 5$. The following sections discuss the effect of sub-area size, statistical distribution, recommendable sample size of surface measurements and demonstrate finally a fatigue strength assessment based on probabilistic surface values.

5.1. Sub-Area Size Effect

First, for statistical characterisation of cast surface textures based on sub-area values, the required sub-area size has to be chosen. The effect of the selectable sub-area size is depicted in Figure 12. It shows the course of the unified mean value of the five deepest surface pits $Sv_{i,mean}$ over the sub-area size. The five deepest surface pits are normalised against the ultimate valley depth of the surface Sv_{max}. Four randomly selected surfaces have been investigated covering both T6 as well as in HIP heat treatment condition, subsequently labelled as specimens 1 to 4. To study the effect of sub-area size, the surface structures are evaluated for the same scope of each specimen. The three pictures (panels (a–c)) within Figure 12 all have the same dimensions and show the same region of the cast surface of specimen 1. In terms of 1 mm × 1 mm sub-area size (Figure 12c), the roughness pit is covered basically by a single patch. As originally published in [36], the Sv value of a sub-area size A_{sub} is evaluated according to Equation (5).

$$Sv = \left| \min_{A_{sub}} z(x,y) \right| \tag{5}$$

Thus, the five extremal Sv_i valley depth values characterise five different surface pits. Comparing the result to the 0.25 mm × 0.25 mm sized sub-area evaluation in Figure 12a, it is recognisable, that at least two or more of the five patches now capture the same surface roughness pit in an adjacent manner. Therefore, no independent statistical description is achieved as the chosen sub-area regions are related. This effect of increasing characterisation of the same surface pit is also indicated within the diagram in Figure 12, as the unified ratio suddenly increases from 0.5 mm sub-area side length to 0.25 mm sub-area side length. Following this trend of continuously decreasing sub-area size, one would end at a ratio of nearly one, when all five sub-areas reflect the deepest spot within the deepest pit of the surface by their value. Of course, this sub-area characterisation also depends on the location of the sub-areas based on the original definition of the surface measurement frame. On the other hand, if the sub-area size is too big, possible crack initiating pits may get neglected, for example, let us assume three critical surface pits are close to each and they may be covered by only one pattern instead. Thus, also not leading to sufficient extreme value characterisation.

Finally, the recommendable sub-area size can also be linked to the sand grain size used in the mould which are typically within the range of 100 µm to 300 µm according to Campbell et al. [40]. As the sand grain sizes, observable in Figures 8 and 9, are up to several hundred µm, a sub-area side length of 0.25 mm would be too small to reliable characterise a single surface pit. Taking these findings into account, the applied sub-area size of 1 mm × 1 mm is an appropriate and recommendable choice for the investigation of the present sand cast surface textures. At this point it should be mentioned that the measured surface should cover a large area of the cast surface to obtain enough sub-area entries reflecting the casting manufacturing process itself. Based on the chosen sub-area size, an amount of at least 100 patches should be evaluated. Within this study, approximately 240 sub-areas are within the investigated cast surface area per cast T6 or HIP-treated specimen.

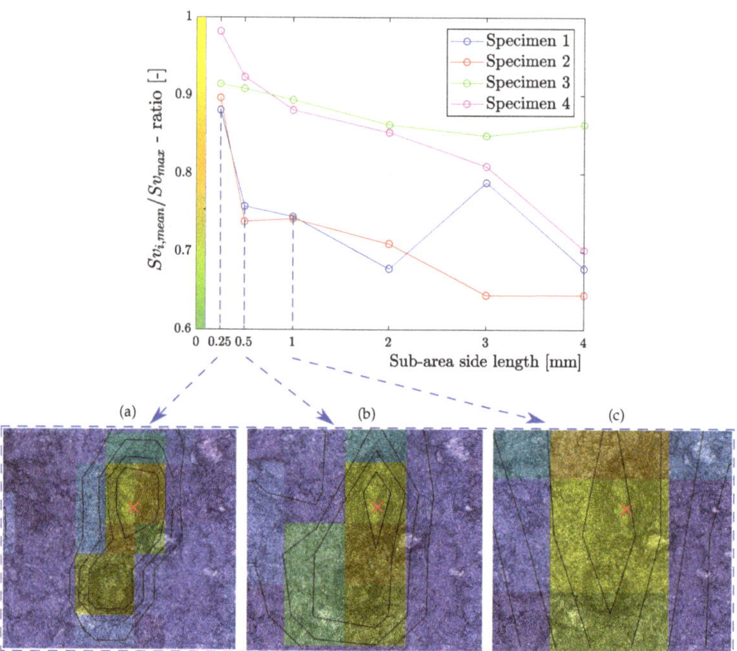

Figure 12. Size effect of (**a**) 0.25 mm, (**b**) 0.5 mm and (**c**) 1.0 mm sub-area side lengths.

5.2. Distribution Parametrisation

For statistical analysis, fatigue-initiating defects can be characterised by means of an extreme value distribution [64]. In terms of limiting extreme value distributions, originally three types have been defined by Gnedenko [65]: the Gumbel distribution (type 1), the Fréchet distribution (type 2) and the Weibull distribution (type 3). The applicable type is thereby determined by the distribution of the basic population from which the extreme value sample has been taken. Jenkinson [66] introduced the General Extreme Value (GEV) distribution, which covers those three types, and is therefore suitable for extreme value statistics. The formulation of the cumulative distribution function of the GEV is given in Equation (6). Therein, δ is the standard deviation (scale parameter), μ is the mean value (location parameter) and ζ is the shape parameter of the distribution. They are often estimated by means of the maximum likelihood method [67,68]. Based on the value of ζ, the type of the distribution is assigned, as the most appropriate GEV type is given by the data itself. Thus, the GEV distribution is frequently used to statistically describe the crack initiating extremal defect size [7,8,14].

$$P(X \leq x) = \int_{-\infty}^{x} \exp\left\{ -\left[1 + \zeta\left(\frac{y - \mu}{\delta} \right) \right]^{-\frac{1}{\zeta}} \right\} dy \tag{6}$$

According, the probability of occurence of an assessment value greater than a chosen threshold value x. is denoted as.

$$P(X \geq x) = 1 - P(X \leq x) \tag{7}$$

As outlined before, the five highest surface pit depth values Sv_i are well suited for parametrisation of a GEV distribution. As both sampling positions as well as the heat treatment of the surface vary, GEV parameters are evaluated to characterise each manufacturing process-based surface texture. Within Figure 13, three examples of GEV distributions are depicted. The green dash-dotted line symbols the GEV distribution of the T6 P1 validation series, whereas the other two HIP series are marked as yellow dashed line for sampling position P1 and as black continuous line for sampling position P2. Additionally, the Sv_i values of each evaluated surface are plotted within the diagram. The HIP P2 series showed the lowest extremal values Sv_i. However, comparing the GEV distributions of the HIP P1 series and the T6 P1 series at the same sampling position, the HIP P1 series exhibited higher values of Sv_i instead. It was observed that the HIP post treatment may affect the extremal Sv_i distribution but keeps the basic population mostly unchanged. It should be noted that the population itself is dependent from the local casting condition and thus no general course of surface valley depth is feasible, but the extremal values can be well parametrised to reflect the local casting process. Table 8 lists the evaluated distribution parameters of the Sv_i GEV distribution as well as the statistical parameter Sv_{rev}. The value Sv_{rev} is based on the associated Sv_i GEV distribution and is calculated as the value with 50% probability of occurrence ($Sv_{rev} = Sv_i(P = 50\%)$). This characteristic value is subsequently used in the derived fatigue strength model and characterises the extremal surface pits in a probabilistic manner.

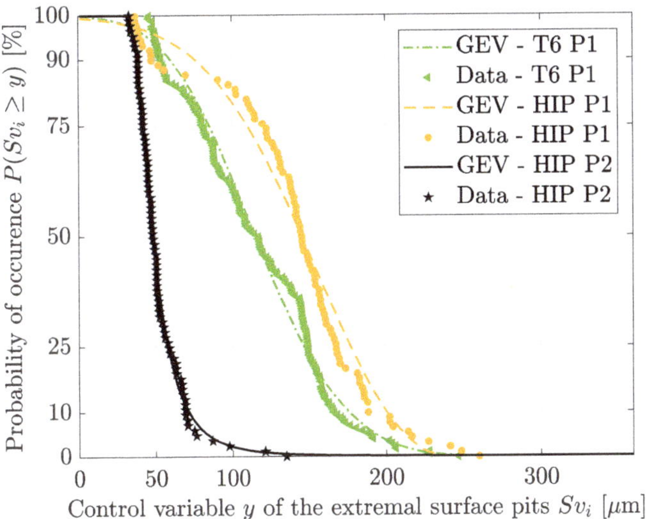

Figure 13. Visualisation of the function $P(Sv_i \geq y)$ of the Sv_i GEV distributions.

Moreover, for a probabilistic fatigue assessment, the distribution of the equivalent notch root radius $\bar{\rho}$ has to be evaluated as well. The $\bar{\rho}$ values are taken from the identical sub-areas as the Sv_i values, thus characterising the notch root radius of the five deepest, most critical surface pits, and thus subsequently denoted by $\bar{\rho}_i$. In order to check for a linear dependency of the population of Sv_i and associated $\bar{\rho}_i$, the coefficient of determination R^2, which is the squared Pearson correlation coefficient (SPCC) [69], was calculated as a measure for the strength of an assumed linear relationship. It is defined as the ratio of the covariance of two random variables A and B to their standard deviations S_A and S_B, see Equation (8).

$$R^2(A,B) = \frac{cov^2(A,B)}{S_A^2 S_B^2} \tag{8}$$

The results in a coefficient of $R^2(Sv_i, \bar{\rho}_i) = 0.01$, which deduces, almost no dependency between Sv_i and $\bar{\rho}_i$. Thus, they are treated as two independent random variables. The evaluated GEV distributions of $\bar{\rho}_i$ are diagrammed in Figure 14 and the fitted distribution parameters are also listed in Table 8. As, in terms of targeted fatigue strength, small notch root radii values are more crucial, the probability of occurrence has to be plotted inversely. Thereby, the T6 P1 series shows the lowest, respectively most critical notch root radii, while the HIP P2 series seems to possess mostly shallow notch curvatures.

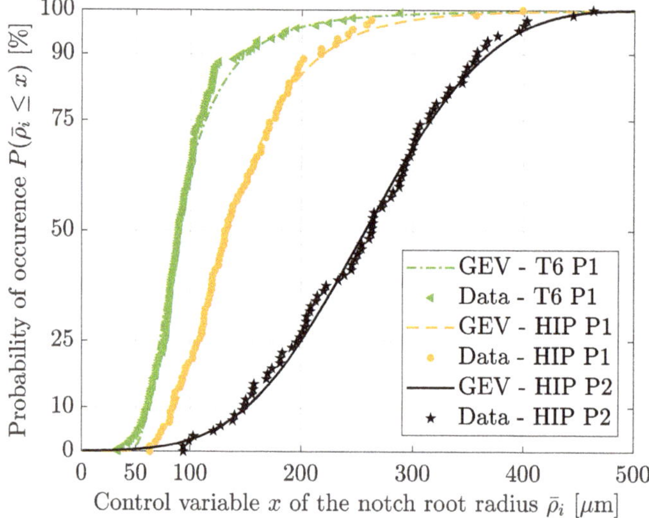

Figure 14. Visualisation of the function $P(\bar{\rho}_i \leq x)$ of the $\bar{\rho}_i$ General Extreme Value (GEV) distributions.

Table 8. Statistical parameter Sv_{rev} and evaluated GEV distribution parameters of Sv_i and $\bar{\rho}_i$.

Specimen Series	Sv_i GEV			Sv_{rev} [µm]	$\bar{\rho}_i$ GEV		
	ζ [-]	μ [µm]	δ [µm]	$Sv_i(P = 50\%)$	ζ [-]	μ [µm]	δ [µm]
T6 P1 [36]	−0.22	99.78	43.04	114	0.27	79.26	23.41
HIP P1	−0.45	128.10	51.11	145	0.07	119.43	40.91
HIP P1(2)	−0.26	126.15	60.95	148	0.03	84.23	29.90
HIP P2	0.24	45.07	8.94	48	−0.28	228.20	82.94

5.3. Impact of Sample Size

For engineering feasibility it is essential for the design engineer to know how many surfaces should be assessed in order to receive statistically reliable information about the cast surface texture. To evaluate an adequate sample size of surface measurements for an assumed basic population, the following methodical procedure is suggested by the authors. The overall workflow is depicted in Figure 15, exemplified for the Sv_i GEV distribution by means of the T6 P1 specimen series. The Sv_i GEV distribution of the T6 P1 specimen series, presented in Figure 13, has been already obtained by means of 34 cast surfaces resulting in 170 Sv_i values in total (five values for each specimen). The evaluated GEV distribution parameters ζ, μ and δ in Table 8 are subsequently treated as main population parameters and support the generation of synthetic, random sample sizes. The dataset S1 acts as reference set as it is based on the original distribution parameters, while the set S2 is randomly derived. Both datasets are parametrised as GEV distributions, implying a stepwise evaluation of probability of 0.5%, leading to 200 equally distanced values. The two datasets $S1$ and $S2$ are assessed by means of the coefficient of determination R^2.

Exemplary, let the synthetic sample size be one, and thus five random Sv_i values will be generated. This synthetically generated random values simulate new samples and thus are applicable for comparison. In the next step, the five random Sv_i values are fitted by a GEV distribution resulting in another ζ_2, μ_2 and δ_2 of the synthetic sample set. Based on that distribution, the synthetic dataset $S2$ is computed by calculation of n = 200 Sv_i values at equally distanced, stepwise (0.5% per step) increased probability of occurrence ($0.5\% \leq P \leq 99.5\%$). Finally, the two datasets $S1$ and $S2$ can be opposed and assessed by the value of $R^2(S1, S2)$. If, in this exemplary case, the GEV distributions of $S1$ (based on the originally evaluated basic population inheriting a quantity of 34 samples) and $S2$ (based on only one sample) would match, the coefficient of determination would be one. To obtain statistically reliable correlation measures, the procedure of randomly calculating and subsequent evaluation of $S2$ is repeated several times. In detail, this $R^2(S1, S2)$ evaluation procedure has been conducted 100 times for sample size one before the sample size is stepwise increased as well.

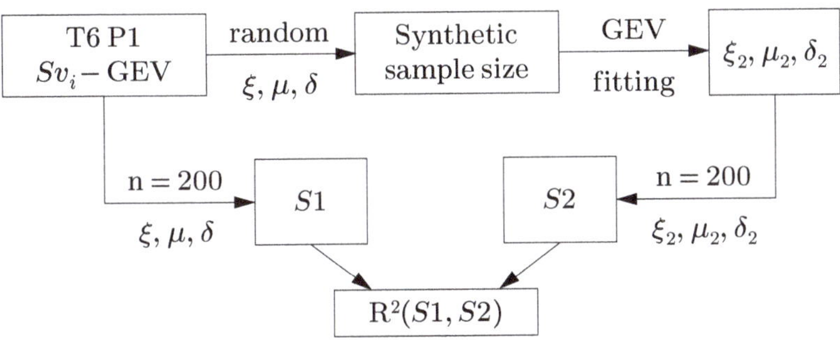

Figure 15. Workflow of the $R^2(S1, S2)$ evaluation exemplified for the T6 P1 Sv_i-GEV distribution.

The result of the sample size effect is depicted in Figure 16a for the distribution of Sv_i of the T6 P1 specimen series. Therein, the mean values of the coefficient of determination $R^2_{mean}(S1, S2)$ are given for each sample size. Furthermore, the area of $R^2(S1, S2)$ values with a probability of occurrence higher than 10% is highlighted by the red area. Once the user-defined criteria ($R^2(S1, S2)$ with $P \geq 0.1) \geq 0.99$ is fulfilled, a satisfying correlation between the sets $S1$ and $S2$, respectively, between the GEV distributions of the original and the synthetic samples, is achieved. This area is marked in grey within Figure 16a and was reached for sample size of nineteen in this case. It is clearly visible that the scatter of the $R^2(S1, S2)$ value decreases as the sample size increases. Thus, the original sample size of 34 investigated specimens has most likely already lead to a basically stable distribution.

The confidence interval of the distribution can be evaluated as well as a measure for change in mean Sv_i values. As for probabilistic fatigue assessment, both Sv_{rev} and Sv_i rely on the distribution, a tight confidence interval has to be aspired. As Sv_{rev} is defined as $Sv_i(P = 0.5)$, the 80% confidence interval at $P = 0.5$ has been studied. For each sample size the evaluation loop of 100 repetitions lead to a certain scattering of the Sv_i values. Figure 16b shows the confidence intervals behaviour of the distributions based on the synthetic sample sets. The mean values of the upper and lower Sv_i confidence bounds are marked as red triangles per sample size, as well as their overall 10–90% $Sv_i(P = 0.5)$ area bordered by the black dotted line. The Sv_{rev} value of 114, as listed in Table 8, is represented by the bold continuous black line. The evaluated sample size threshold from Figure 16a is also drawn, as well as the grey highlighted area whereat the user defined criterion is fulfilled. This indicates that, at the current sample size threshold, the Sv_{rev} value is within $100\ \mu m \leq Sv_{rev} \leq 128\ \mu m$. This refers to a scatter index of $\pm 12\%$ for sample size nineteen, respectively $T_{Sv_i} = 1 : 1.28$. If a tighter confidence interval is aspired, the sample size of surface measurements has to be increased.

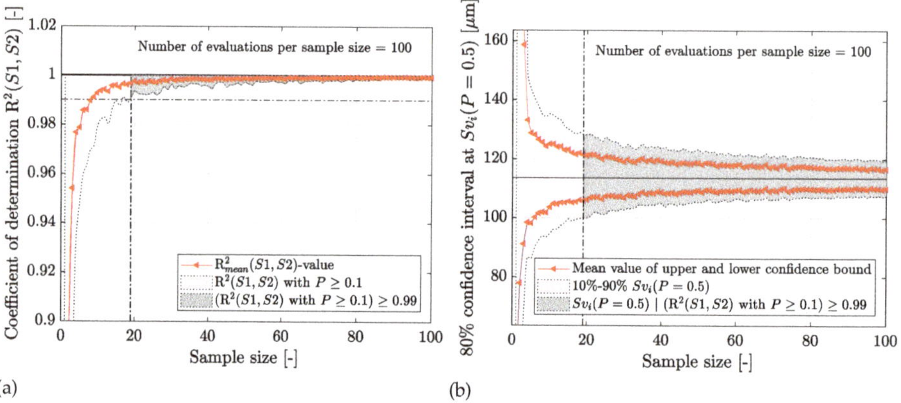

Figure 16. Sample size effect on (**a**) Sv_i-GEV distribution. (**b**) Confidence interval.

5.4. Fatigue Strength Assessment

Utilising both the Sv_i GEV distribution of the roughness parameter Sv and the $\bar{\rho}_i$ GEV distribution of the notch root radius, a bivariate distribution can be evaluated. As it has been proven that Sv_i and $\bar{\rho}_i$ can be handled as independent random variables, the combined probability of occurrence $P(Sv_i \geq y, \bar{\rho}_i \leq x)$ can be calculated by multiplication of the two single probability functions following Equation (9).

$$P(Sv_i \geq y, \bar{\rho}_i \leq x) = (1 - P(Sv_i \leq y)) \cdot P(\bar{\rho}_i \leq x) \tag{9}$$

The bivariate cumulative distribution function is exemplary diagrammed in Figure 17 for the T6 P1 specimen series. Figures A1–A3, depicting the bivariate cumulative distribution function of the other investigated specimens series, are added in the Appendix A. The projected GEV distributions of Sv_i and $\bar{\rho}_i$ are additionally plotted as red lines and the surface mesh color depends on the value of $P(Sv_i \geq y, \bar{\rho}_i \leq x)$. The available Cast-S specimens of this series are marked in blue.

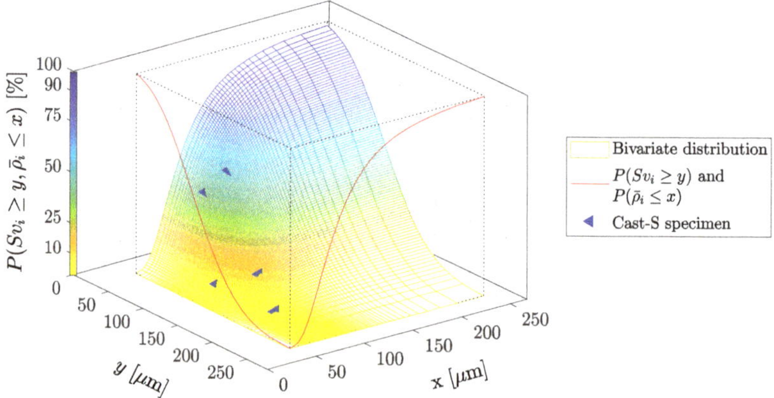

Figure 17. T6 P1 bivariate distribtion of Sv_i GEV distribution and $\bar{\rho}_i$ GEV distribution.

Concluding, the surface fatigue assessment model can be applied utilising this probabilistic surface texture values. The result is the probabilistic cast surface long life fatigue strength $\sigma_{LLF,s}(P(Sv_i \geq$

$y, \bar{\rho}_i \leq x))$, which is normalised to the near defect-free long life fatigue strength $\sigma_{LLF,0}$, as diagrammed in Figure 18. The mesh colour again highlights the combined probability of occurrence $P(Sv_i \geq y, \bar{\rho}_i \leq x)$. The 3D-view is additionally depicted from all three axis projections (Figure 18a–c) to show the model's fatigue life dependency. By comparison of the $\bar{\rho}_i$-plot (Figure 18a) with the Sv_i-plot (Figure 18b) it is recognisable that, although both parameters do have an effect on the fatigue strength result, the notch depth, respectively, surface pit depth Sv, is more pronounced.

This probabilistic fatigue assessment procedure facilitates an engineering feasible fatigue design by providing reliable calculation of the long life fatigue strength, utilising the probability of occurrence of the two statistical model parameters Sv_i and $\bar{\rho}_i$. Based on the fatigue design safety requirements, the designer can obtain the long life fatigue strength by comparably low effort in surface measurements and surface texture evaluation.

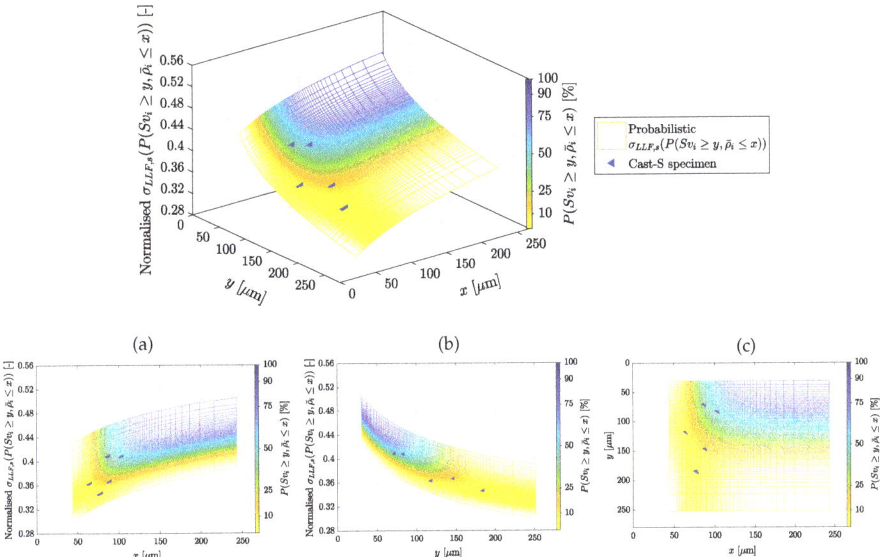

Figure 18. T6 P1 probabilistic cast surface long life fatigue strength $\sigma_{LLF,s}(P(Sv_i \geq y, \bar{\rho}_i \leq x))$ with views: (a) $\bar{\rho}_i$ - $\sigma_{LLF,s}$ (b) Sv_i - $\sigma_{LLF,s}$ (c) $\bar{\rho}_i$ - Sv_i.

6. Discussion

The presented fatigue strength assessment approach, reparametrising the model introduced by Pomberger et al. [36] for sand cast aluminium surface layers with T6 heat treatment, has been applied on two HIP-treated testing series at similar sampling positions (HIP P1 and HIP P1(2)). Furthermore, another cast surface with significantly reduced surface roughness has been investigated (HIP P2). By means of the reparametrised exponents a_1 and a_2, the HIPped validation specimen series, refer to Table 7, as well as the validation series from [36] (T6 P1) showed sound results in terms of long life fatigue strength estimation. Due to the diversification of measured surface roughness data, this concept is now valid to a wider range of cast surfaces. Moreover, current research investigations apply the introduced method also on additively manufactures surface textures.

The HIP-treated cast surfaces revealed that the surface layer pores have partly not being closed. This is caused by the high amount of broached cavities, as depicted in Figures 8 and 9. Metallographic analyses revealed shrinkage pores as cavities reaching depths of up to one millimetre. Therefore, when machining the cast surface, it should be considered that surface layer pores may be broached by the machining process and may result in similar long life fatigue strength reduction as observed for the cast surface texture.

To assess not only surface initiating cracks as in Cast-S specimens, but also surface layer porosity, the neural network has been retrained by means of four input variables on four neurons. The input variables are now the pore size $\sqrt{area}$, the local surface pit depth Sv_{local}, the equivalent notch root radius $\bar{\rho}$ and the statistical surface roughness parameter Sv_{rev}. As this methodology may not be available to designers, and therefore is not easily engineering feasible, the authors study also on the simplified deduction of the interaction coefficient. The first results indicate that the application of the mean values presented in Table 6 for the whole associated sampling series leads to sufficient approximations.

Regarding a probabilistic Cast-S specimen fatigue assessment, the five biggest values Sv_i of the surface pit depth Sv are valid for extreme value statistics. The evaluated distributions of the probabilistic fatigue model parameters Sv_i and $\bar{\rho}_i$ depend on the selected sub-area size. As illustrated, a too small sub area sized may not contain sufficient information about the amount and depth of critical surface pits, but is increasingly characterising only the most critical one and would lead to more conservative assessment. Within this study, it was found that for the investigated sand cast surface textures, a sub-area size of 1 mm × 1 mm is valid. It should be noted that this recommended value is a multiple of the sand grain size. Manufacturing processes resulting in finer surface structures, such as additively manufacturing, may use 0.5 mm × 0.5 mm sub-areas instead. As the presented surface long life fatigue assessment model uses the statistical surface roughness parameter Sv_{rev}, the population of the data for distribution fitting should be substantial enough in order to facilitate sound distribution conformity. Within this study, it was found that evaluating about twenty specimens, with five Sv_i values each, leads to stable distribution parameters of the extremal values. As both fatigue model parameters extremal notch valley depth and averaged notch root radius can be handled as independent random variables, their probabilities of both distributions can be multiplicatively combined, leading to a combined probability of occurrence $P(Sv_i \geq y, \bar{\rho}_i \leq x)$ and subsequently to the probabilistic cast surface long life fatigue strength $\sigma_{LLF,s}(P(Sv_i \geq y, \bar{\rho}_i \leq x))$. Thus, the presented methodology provides a statistically applicable design tool to assess the cast surface effect on the local fatigue strength.

7. Conclusions

Based on the results presented in this paper, the following conclusions can be drawn.

- The presented surface layer fatigue assessment model is valid for aluminium sand cast surfaces in T6 and HIP treatment condition within the investigated range of Sv = 30 µm to 260 µm. Long life fatigue strength estimation results are approximately 6% to 9% conservative.
- The HIP process does not reliably close surface layer pores within the first millimetre of surface layer depth. Therefore, in the investigated manufacturing showcase, the machining process has to remove at least one millimetre of the surface layer to increase the endurable long life fatigue strength by remove surface layer porosity.
- Extremal surface roughness pits may become deeper, respectively, more critical, due to the HIP process. However, this is not compulsory for all investigated surfaces.
- For probabilistic fatigue strength assessment, the sub-area size is meaningful. Sub-area side lengths have to be chosen properly according to the present cast surface texture. For the investigated sand cast aluminium surfaces, a sub-area size of 1 mm × 1 mm is valid. For statistical characterisation, the measured cast surface texture should cover about one-hundred sub-areas at least.
- The statistically assessed surface texture parameters, used in the cast surface fatigue strength assessment model, Sv_i and $\bar{\rho}_i$ are independent variables and can both be statistically described by a GEV distribution. To reliably fit the distribution, at least 20 specimens should be measured, resulting in 100 Sv_i and $\bar{\rho}_i$ values. By means of a bivariate distribution, a probabilistic cast surface long life fatigue strength $\sigma_{LLF,s}(P(Sv_i \geq y, \bar{\rho}_i \leq x))$ can be subsequently calculated and used in fatigue design applications.

Author Contributions: Conceptualisation, S.P., M.O. and M.L.; methodology, S.P., M.O. and M.S.; software, S.P. and M.S.; validation, S.P. and M.L.; formal analysis, S.P. and J.T.; investigation, S.P.; resources, M.S.; data curation, S.P.; writing—original draft preparation, S.P.; writing—review and editing, M.O., M.L., J.T. and M.S.; visualisation, S.P.; supervision, M.L. and M.S.; project administration, M.S.; funding acquisition, M.S.; All authors have read and agreed to the published version of the manuscript.

Funding: This research was funded by the Austrian Federal Ministry for Digital and Economic Affairs and the National Foundation for Research, Technology and Development.

Acknowledgments: The financial support by the Austrian Federal Ministry for Digital and Economic Affairs and the National Foundation for Research, Technology and Development is gratefully acknowledged. Furthermore, the authors would like to thank the industrial partners BMW AG and Nemak Dillingen GmbH for the excellent mutual scientific cooperation within the CD-laboratory framework. Special thanks go to Michael Simon Pegritz, coworker within the Christian Doppler Laboratory for Manufacturing Process based Component Design, for supporting the developement of the microstructural characterisation.

Conflicts of Interest: The authors declare no conflict of interest. The funders had no role in the design of the study; in the collection, analyses, or interpretation of data; in the writing of the manuscript; or in the decision to publish the results.

Abbreviations

The following abbreviations are used in this manuscript:

$\bar{\rho}$	Equivalent notch root radius of a sub-area A_{sub}
$\bar{\rho}_i$	Notch root radii at extremal surface pit depths Sv_i with $1 \leq i \leq 5$
$\sqrt{area}$	Defect size of Murakami's approach
σ_a	Stress amplitude
$\sigma_{a,1e7}$	Experimental SN-point extrapolated to 1e7 load cycles
$\sigma_{LLF,0}$	Near defect-free long life fatigue strength
$\sigma_{a,Ps*}$	Experimental long life fatigue strength with *% probability of survival
$\sigma_{LLF,*,Ps*}$	Estimated long life fatigue strength with *% probability of survival
$\sigma_{LLF,s}(P(Sv_i \geq y, \bar{\rho}_i \leq x))$	Probabilistic cast surface long life fatigue strength
ψ	Interaction coefficient
ψ_{mean}	Mean value of the interaction coefficient
ψ_{std}	Standard deviation of the interaction coefficient
ψ_{min}	Minimum value of the interaction coefficient
ψ_{max}	Maximum value of the interaction coefficient
ζ	Shape parameter of the GEV distribution
δ	Scale parameter of the GEV distribution
μ	Location parameter of the GEV distribution
A	Elongation at rupture
A_{sub}	Sub-area size
a_1, a_2	Exponents in modified stress concentration factor
E	Young's modulus
e_{min}, e_{max}, α	Pore elongation and location parameters
$HV10$	Vickers hardness
k_1, k_2	Inverse slopes of the S/N-curve
$K_{t,mod}$	Modified stress concentration factor
$K_{f,s}$	Surface fatigue notch factor
$K_{f,p}$	Fatigue strength reduction factor
$K_{f,m}$	Mixed fatigue strength reduction factor
N_T	Transition knee point of the S/N-curve
$N_{rupture}$	Load cycles at rupture
$P(Sv_i \geq y); P(\bar{\rho}_i \leq x)$	Probability of occurrence
$P(Sv_i \geq y, \bar{\rho}_i \leq x)$	Combined probability of occurrence
R_m	Ultimate tensile strength
$R_{P0.2}$	0.2% offset yield strength
Sa_{mean}	Mean value of the global arithmetical mean height Sa

Sv	Maximum pit height of the scale limited surface
Sv_{local}	Local maximum pit height of the sub-area A_{sub}
Sv_{max}	Maximum surface pit depth value of the investigated surface
Sv_i	Extremal surface pit depth values of the investigated surface with $1 \leq i \leq 5$
$Sv_{i,mean}$	Mean value of Sv_i
$Sv_{rev}(Sv_i(P = 50\%))$	Sv_i GEV distribution based value with 50% probability of occurrence
$T_{S,1e7}$	Stress scatter index
CPS	Core Package System
DAS	Secondary dendrite arm spacing
GEV	Generalised Extreme Value
HIP	Hot isostatic pressing
NN	Neural network
SPCC	Squared Pearson correlation coefficient
R	Load stress ratio
$R^2(A, B)$	Coefficient of determination of two random variables A and B
$R^2_{mean}(A, B)$	Mean value of $R^2(A, B)$
SGS	Surface geometrical structure
SEM	Scanning electron microscopy

Appendix A

The following figures depict the bivariate distributions of the specimen series HIP P1 (Figure A1), HIP P1(2) (Figure A2) and HIP P2 (Figure A3).

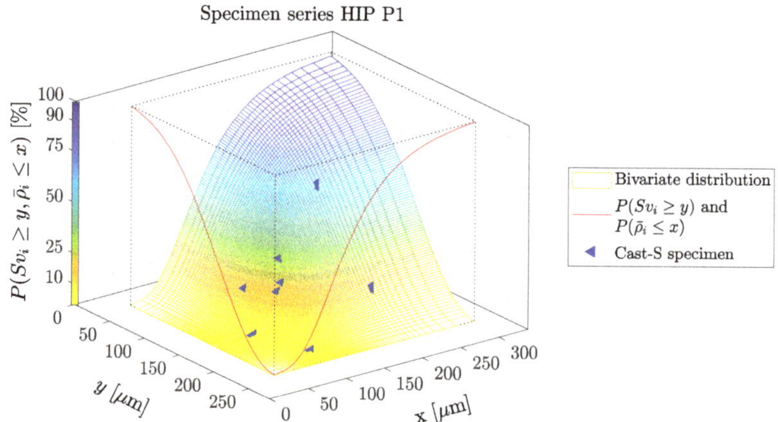

Figure A1. HIP P1 bivariate distribution of Sv_i GEV distribution and $\bar{\rho}_i$ GEV distribution.

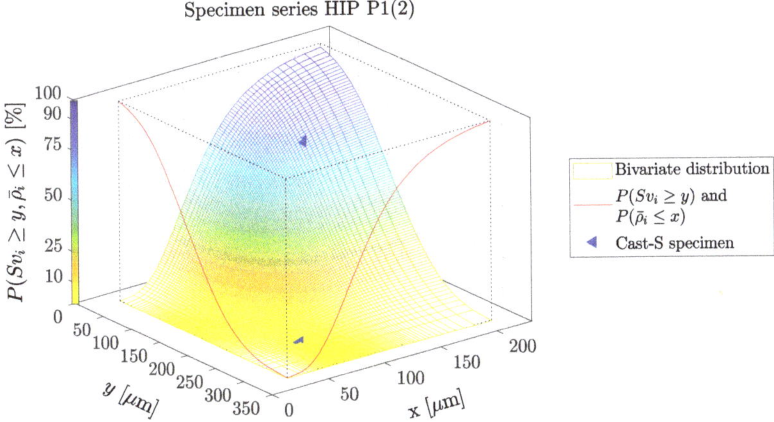

Figure A2. HIP P1(2) bivariate distribution of Sv_i GEV distribution and $\bar{\rho}_i$ GEV distribution.

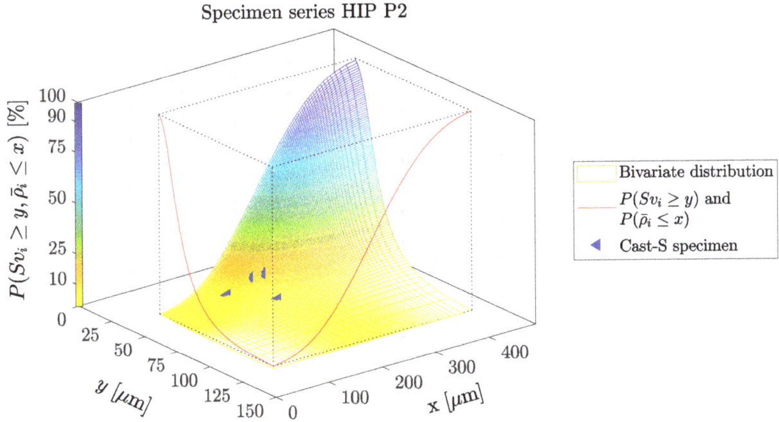

Figure A3. HIP P2 bivariate distribution of Sv_i GEV distribution and $\bar{\rho}_i$ GEV distribution.

References

1. Hirsch, J. Aluminium in Innovative Light-Weight Car Design. *Mater. Trans.* **2011**, *52*, 818–824. [CrossRef]
2. Ueno, A.; Miyakawa, S.; Yamada, K.; Sugiyama, T. Fatigue behavior of die casting aluminum alloys in air and vacuum. *Procedia Eng.* **2010**, *2*, 1937–1943. [CrossRef]
3. Aigner, R.; Garb, C.; Leitner, M.; Stoschka, M.; Grün, F. Application of a $\sqrt{}$ area -Approach for Fatigue Assessment of Cast Aluminum Alloys at Elevated Temperature. *Metals* **2018**, *8*, 1033. [CrossRef]
4. Li, P.; Lee, P.D.; Maijer, D.M.; Lindley, T.C. Quantification of the interaction within defect populations on fatigue behavior in an aluminum alloy. *Acta Mater.* **2009**, *57*, 3539–3548. [CrossRef]
5. Hidalgo, R.; Esnaola, J.A.; Llavori, I.; Larrañaga, M.; Hurtado, I.; Herrero-Dorca, N. Fatigue life estimation of cast aluminium alloys considering the effect of porosity on initiation and propagation phases. *Int. J. Fatigue* **2019**, *125*, 468–478. [CrossRef]
6. Sonsino, C.M.; Ziese, J. Fatigue strength and applications of cast aluminium alloys with different degrees of porosity. *Int. J. Fatigue* **1993**, *15*, 75–84. [CrossRef]
7. Tiryakioğlu, M. Statistical distributions for the size of fatigue-initiating defects in Al–7%Si–0.3%Mg alloy castings: A comparative study. *Mater. Sci. Eng. A* **2008**, *497*, 119–125. [CrossRef]

8. Tiryakioğlu, M. On the size distribution of fracture-initiating defects in Al- and Mg-alloy castings. *Mater. Sci. Eng. A* **2008**, *476*, 174–177. [CrossRef]
9. Tiryakioğlu, M. Relationship between Defect Size and Fatigue Life Distributions in Al-7 Pct Si-Mg Alloy Castings. *Metall. Mater. Trans. A* **2009**, *40*, 1623–1630. [CrossRef]
10. Tijani, Y.; Heinrietz, A.; Stets, W.; Voigt, P. Detection and Influence of Shrinkage Pores and Nonmetallic Inclusions on Fatigue Life of Cast Aluminum Alloys. *Metall. Mater. Trans. A* **2013**, *44*, 5408–5415. [CrossRef]
11. Le, V.D.; Saintier, N.; Morel, F.; Bellett, D.; Osmond, P. Investigation of the effect of porosity on the high cycle fatigue behaviour of cast Al-Si alloy by X-ray micro-tomography. *Int. J. Fatigue* **2018**, *106*, 24–37. [CrossRef]
12. Leitner, M.; Garb, C.; Remes, H.; Stoschka, M. Microporosity and statistical size effect on the fatigue strength of cast aluminium alloys EN AC-45500 and 46200. *Mater. Sci. Eng. A* **2017**, *707*, 567–575. [CrossRef]
13. Aigner, R.; Leitner, M.; Stoschka, M. Fatigue strength characterization of Al-Si cast material incorporating statistical size effect. *MATEC Web Conf.* **2018**, *165*, 14002. [CrossRef]
14. Aigner, R.; Pusterhofer, S.; Pomberger, S.; Leitner, M.; Stoschka, M. A probabilistic Kitagawa-Takahashi diagram for fatigue strength assessment of cast aluminium alloys. *Mater. Sci. Eng. A* **2019**, *745*, 326–334. [CrossRef]
15. Buffiere, J.Y. Fatigue Crack Initiation And Propagation From Defects In Metals: Is 3D Characterization Important? *Procedia Struct. Integr.* **2017**, *7*, 27–32. [CrossRef]
16. Serrano-Munoz, I.; Buffiere, J.Y.; Mokso, R.; Verdu, C.; Nadot, Y. Location, location &size: Defects close to surfaces dominate fatigue crack initiation. *Sci. Rep.* **2017**, *7*, 45239. [CrossRef] [PubMed]
17. Tajiri, A.; Nozaki, T.; Uematsu, Y.; Kakiuchi, T.; Nakajima, M.; Nakamura, Y.; Tanaka, H. Fatigue Limit Prediction of Large Scale Cast Aluminum Alloy A356. *Procedia Mater. Sci.* **2014**, *3*, 924–929. [CrossRef]
18. Aigner, R.; Leitner, M.; Stoschka, M.; Hannesschläger, C.; Wabro, T.; Ehart, R. Modification of a Defect-Based Fatigue Assessment Model for Al-Si-Cu Cast Alloys. *Materials* **2018**, *11*, 2546. [CrossRef]
19. Murakami, Y. Material defects as the basis of fatigue design. *Int. J. Fatigue* **2012**, *41*, 2–10. [CrossRef]
20. Aigner, R.; Leitner, M.; Stoschka, M. On the mean stress sensitivity of cast aluminium considering imperfections. *Mater. Sci. Eng. A* **2019**, *758*, 172–184. [CrossRef]
21. Garb, C.; Leitner, M.; Tauscher, M.; Weidt, M.; Brunner, R. Statistical analysis of micropore size distributions in Al–Si castings evaluated by X-ray computed tomography. *Int. J. Mater. Res.* **2018**, *109*, 889–899. [CrossRef]
22. Murakami, Y. *Metal Fatigue: Effects of Small Defects and Nonmetallic Inclusions*, 1st ed.; Elsevier: Amsterdam, The Netherlands, 2002.
23. Kitagawa, H.; Takahashi, S. Applicability of fracture mechanics to very small cracks or the cracks in the early stage. In Proceedings of the Second International Conference on Mechanical Behavior of Materials, Boston, MA, USA, 16–20 August 1976; pp. 627–631.
24. McKelvey, S.A.; Fatemi, A. Surface finish effect on fatigue behavior of forged steel. *Int. J. Fatigue* **2012**, *36*, 130–145. [CrossRef]
25. McKelvey, S.A.; Lee, Y.L.; Barkey, M.E. Stress-Based Uniaxial Fatigue Analysis Using Methods Described in FKM-Guideline. *J. Failure Anal. Prev.* **2012**, *12*, 445–484. [CrossRef]
26. Rennert, R.; Kullig, E.; Vormwald, M.; Esderts, A.; Siegele, D. *Analytical Strength Assessment of Components Made of Steel, Cast Iron and Aluminium Materials in Mechanical Engineering*, 6th ed.; FKM-Guideline; VDMA-Verl: Frankfurt am Main, Germany, 2012.
27. Cheng, Z.; Liao, R.; Lu, W.; Wang, D. Fatigue notch factors prediction of rough specimen by the theory of critical distance. *Int. J. Fatigue* **2017**, *104*, 195–205. [CrossRef]
28. Cheng, Z.; Liao, R.; Lu, W. Surface stress concentration factor via Fourier representation and its application for machined surfaces. *Int. J. Solids Struct.* **2017**, *113–114*, 108–117. [CrossRef]
29. Abroug, F.; Pessard, E.; Germain, G.; Morel, F. A probabilistic approach to study the effect of machined surface states on HCF behavior of a AA7050 alloy. *Int. J. Fatigue* **2018**, *116*, 473–489. [CrossRef]
30. Neuber, H. *Kerbspannungslehre*; Springer: Berlin/Heidelberg, Germany, 1958. [CrossRef]
31. Pilkey, W.D.; Pilkey, D.F. *Peterson's Stress Concentration Factors*, 3rd ed.; John Wiley & Sons Incorporated: Chichester, UK, 2008.
32. Suraratchai, M.; Limido, J.; Mabru, C.; Chieragatti, R. Modelling the influence of machined surface roughness on the fatigue life of aluminium alloy. *Int. J. Fatigue* **2008**, *30*, 2119–2126. [CrossRef]
33. Arola, D.; Ramulu, M. An Examination of the Effects from Surface Texture on the Strength of Fiber Reinforced Plastics. *J. Compos. Mater.* **1999**, *33*, 102–123. [CrossRef]

34. Arola, D.; Williams, C. Estimating the fatigue stress concentration factor of machined surfaces. *Int. J. Fatigue* **2002**, *24*, 923–930. [CrossRef]

35. Cheng, Z.; Liao, R. Effect of surface topography on stress concentration factor. *Chinese J. Mech. Eng.* **2015**, *28*, 1141–1148. [CrossRef]

36. Pomberger, S.; Stoschka, M.; Aigner, R.; Leitner, M.; Ehart, R. Areal fatigue strength assessment of cast aluminium surface layers. *Int. J. Fatigue* **2020**, *133*, 105423. [CrossRef]

37. Pomberger, S.; Stoschka, M.; Leitner, M. Cast surface texture characterisation via areal roughness. *Precis. Eng.* **2019**, *60*, 465–481. [CrossRef]

38. Pomberger, S.; Leitner, M.; Stoschka, M. Evaluation of surface roughness parameters and their impact on fatigue strength of Al-Si cast material. *Mater. Today Proc.* **2019**, *12*, 225–234. [CrossRef]

39. Schneller, W.; Leitner, M.; Springer, S.; Grün, F.; Taschauer, M. Effect of HIP Treatment on Microstructure and Fatigue Strength of Selectively Laser Melted AlSi10Mg. *J. Manuf. Mate. Proc.* **2019**, *3*, 16. [CrossRef]

40. Campbell, J. *Complete Casting Handbook: Metal Casting Processes, Techniques and Design*; Elsevier Butterworth-Heinemann: Oxford, UK, 2011.

41. Ostermann, F. *Anwendungstechnologie Aluminium*, 2nd ed.; Springer: Berlin/Heidelberg, Germany, 2007.

42. Ernst, F.; Kube, D.; Klaus, G.; Nemak Dillingen GmbH. Al-Kurbelgehäuse mit thermisch gespritzter Eisenbasisbeschichtung: Gießtechnische Anforderungen. *Giesserei* **2013**, *100*, 44–51.

43. Krause, G. Weiterentwicklungen bei Gehäusen von Elektromotoren: Weiterentwicklungen bei Gehäusen von Elektromotoren. *Giesserei* **2017**, *104*, 108–115.

44. Lei, C.S.C.; Frazier, W.E.; Lee, E.W. The effect of hot isostatic pressing on cast aluminum. *JOM* **1997**, *49*, 38–39. [CrossRef]

45. Lee, M.H.; Kim, J.J.; Kim, K.H.; Kim, N.J.; Lee, S.; Lee, E.W. Effects of HIPping on high-cycle fatigue properties of investment cast A356 aluminum alloys. *Mater. Sci. Eng. A* **2003**, *340*, 123–129. [CrossRef]

46. Nayhumwa, C.; Green, N.R.; Campbell, J. Influence of casting technique and hot isostatic pressing on the fatigue of an Al-7Si-Mg alloy. *Metall. Mater. Trans. A* **2001**, *32*, 349–358. [CrossRef]

47. Staley, J.T.; Tiryakioğlu, M.; Campbell, J. The effect of hot isostatic pressing (HIP) on the fatigue life of A206-T71 aluminum castings. *Mater. Sci. Eng. A* **2007**, *465*, 136–145. [CrossRef]

48. Staley, J.T.; Tiryakioglu, M.; Campbell, J. Applicability of fracture mechanics to very small cracks or the cracks in the early stage. In Proceedings of the Effect of Various hip Conditions on Bifilms and Mechanical Properties in Aluminum Castings, Orlando, FL, USA, 25 February–1 March 2007; Crepeau, P.N., Tiryakioglu, M., Campbell, J., Eds.; TMS (The Minerals, Metals and Materials Society): Warrendale, PA, USA, 2007; pp. 159–166

49. Aigner, R. Characterising the Fatigue Strength of Alminium Castings by Applied Statistical Evaluation of Imperfections. Ph.D. Thesis, Montanuniversität Leoben, Leoben, Austria, 2019.

50. Ceschini, L.; Morri, A.; Sambogna, G. The effect of hot isostatic pressing on the fatigue behaviour of sand cast A356-T6 and A204-T6 aluminum alloys. *J. Mater. Proc. Technol.* **2008**, *204*, 231–238. [CrossRef]

51. Vandersluis, E.; Ravindran, C. Comparison of Measurement Methods for Secondary Dendrite Arm Spacing. *Metall. Microst. Anal.* **2017**, *6*, 89–94. [CrossRef]

52. Haibach, E. *Betriebsfestigkeit: Verfahren und Daten zur Bauteilberechnung*; VDI-Buch; Springer: Berlin, Germany, 2006. [CrossRef]

53. Leitner, H. Simulation of the Fatigue Behaviour of Aluminium Cast Alloys. Ph.D. Thesis, Montanuniversität Leoben, Leoben, Austria, 2001.

54. Dengel, D.; Harig, H. Estimation of the fatigue limit by progressively-increasing load tests. *Fatigue Fract. Eng. Mater. Struct.* **1980**, *3*, 113–128. [CrossRef]

55. Sonsino, C.M. Course of SN-curves especially in the high-cycle fatigue regime with regard to component design and safety. *Int. J. Fatigue* **2007**, *29*, 2246–2258. [CrossRef]

56. Atkinson, H.V.; Davies, S. Fundamental aspects of hot isostatic pressing: An overview. *Metall. Mater. Trans. A* **2000**, *31*, 2981–3000. [CrossRef]

57. Masuo, H.; Tanaka, Y.; Morokoshi, S.; Yagura, H.; Uchida, T.; Yamamoto, Y.; Murakami, Y. Effects of Defects, Surface Roughness and HIP on Fatigue Strength of Ti-6Al-4V manufactured by Additive Manufacturing. *Procedia Struct. Integr.* **2017**, *7*, 19–26. [CrossRef]

58. Schneller, W.; Leitner, M.; Springer, S.; Beter, F.; Grün, F. Influencing factors on the fatigue strength of selectively laser melted structures. *Procedia Struct. Integr.* **2019**, *19*, 556–565. [CrossRef]

59. Schneller, W.; Leitner, M.; Pomberger, S.; Springer, S.; Beter, F.; Grün, F. Effect of Post Treatment on the Microstructure, Surface Roughness and Residual Stress Regarding the Fatigue Strength of Selectively Laser Melted AlSi10Mg Structures. *J. Manuf. Mater. Process.* **2019**, *3*, 89. [CrossRef]

60. Leitner, M.; Stoschka, M.; Fröschl, J.; Wiebesiek, J. Surface Topography Effects on the Fatigue Strength of Cast Aluminum Alloy AlSi8Cu3. *Mater. Perform. Charact.* **2018**, *7*, 20170127. [CrossRef]

61. Campbell, J. Entrainment defects. *Mater. Sci. Technol.* **2013**, *22*, 127–145. [CrossRef]

62. Wang, Q.G.; Crepeau, P.N.; Davidson, C.J.; Griffiths, J.R. Oxide films, pores and the fatigue lives of cast aluminum alloys. *Metall. Mater. Trans. B* **2006**, *37*, 887–895. [CrossRef]

63. Yousefian, P.; Tiryakioğlu, M. Pore Formation During Solidification of Aluminum: Reconciliation of Experimental Observations, Modeling Assumptions, and Classical Nucleation Theory. *Metall. Mater. Trans. A* **2018**, *49*, 563–575. [CrossRef]

64. Gumbel, E.J. *Statistics of Extremes*; Columbia University Press: New York, NY, USA, 1958.

65. Gnedenko, B. Sur la distribution limite du terme maximum d'une serie aleatoire. *Annals Math.* **1943**, *44*, 423–453. [CrossRef]

66. Jenkinson, A.F. The frequency distribution of the annual maximum (or minimum) values of meteorological elements. *Q. J. R. Meteorol. Soc.* **1955**, *87*, 145–158. [CrossRef]

67. Beretta, S.; Murakami, Y. Statistical analysis of defects for fatigue strength prediction and quality control of materials. *Fatigue Fract. Eng. Mater. Struct.* **1998**, *21*, 1049–1065. [CrossRef]

68. Raynal-Villasenor, J.A.; Raynal-Gutierrez, M.E. Estimation procedures for the GEV distribution for the minima. *J. Hydrol.* **2014**, *519*, 512–522. [CrossRef]

69. Benesty, J.; Chen, J.; Huang, Y.; Cohen, I. Pearson Correlation Coefficient. In *Noise Reduction in Speech Processing*; Cohen, I., Huang, Y., Chen, J., Benesty, J., Eds.; Springer: New York, NY, USA, 2009; Volume 2; pp. 1–4. [CrossRef]

Article

Validation Study on the Statistical Size Effect in Cast Aluminium

Matthias Oberreiter *, Sebastian Pomberger, Martin Leitner and Michael Stoschka

Christian Doppler Laboratory for Manufacturing Process Based Component Design, Chair of Mechanical Engineering, Montanuniversität Leoben, Franz-Josef-Straße 18, 8700 Leoben, Austria; sebastian.pomberger@unileoben.ac.at (S.P.); martin.leitner@unileoben.ac.at (M.L.); michael.stoschka@unileoben.ac.at (M.S.)
* Correspondence: matthias.oberreiter@unileoben.ac.at; Tel.: +43-3842-402-1474

Received: 29 April 2020; Accepted: 22 May 2020; Published: 27 May 2020

Abstract: Imperfections due to the manufacturing process can significantly affect the local fatigue strength of the bulk material in cast aluminium alloys. Most components possess several sections of varying microstructure, whereat each of them may inherit a different highly-stressed volume (HSV). Even in cases of homogeneous local casting conditions, the statistical distribution parameters of failure causing defect sizes change significantly, since for a larger highly-stressed volume the probability for enlarged critical defects gets elevated. This impact of differing highly-stressed volume is commonly referred as statistical size effect. In this paper, the study of the statistical size effect on cast material considering partial highly-stressed volumes is based on the comparison of a reference volume V_0 and an arbitrary enlarged, but disconnected volume V_α utilizing another specimen geometry. Thus, the behaviour of disconnected highly-stressed volumes within one component in terms of fatigue strength and resulting defect distributions can be assessed. The experimental results show that doubling of the highly-stressed volume leads to a decrease in fatigue strength of 5% and shifts the defect distribution towards larger defect sizes. The highly-stressed volume is numerically determined whereat the applicable element size is gained by a parametric study. Finally, the validation with a prior developed fatigue strength assessment model by R. Aigner et al. leads to a conservative fatigue design with a deviation of only about 0.3% for cast aluminium alloy.

Keywords: aluminium casting; fatigue assessment; shrinkage porosity; statistical size effect; extreme value statistics; highly-stressed volume

---

1. Introduction

Complex cast aluminium parts possess a severely heterogeneous microstructure and therefore it is essential to consider its interaction with the highly stressed volume (HSV). The result of elevated highly stressed volumes in terms of cyclic loading is generally a reduced the fatigue strength. According to References [1,2], size effects can be classified into technological, geometrical, statistical and surface technology size effects. Larger components, respectively larger HSV, increase the probability of critical defect sizes, thus lessening the endurable fatigue strength. The aim of this work is the validation of the statistical size effect with consideration of the microstructural properties, as introduced as probabilistic design method for aluminium castings in References [3–5]. In general, the local fatigue strength correlates well with the dedicated microstructure because of the statistical distribution of the defects, apparent in preliminary studies [6–11]. Therefore, it is essential to consider the local pore size distribution in the fatigue design process. Fatigue initiating defects in cast parts can be described well with extreme value statistics, like the generalized extreme value distribution (GEV) or the Gumbel distribution [12–14]. Further methodologies to assess the statistical size effect with regard to volumetric dependencies and highly stressed surface models are given in References [15–22]. In these latter cases, a highly stressed volume, which is defined as the volume with a particular percentage of the maximum stress node, is taken into account. One of these approaches is the volumetric model of Sonsino [16], who invokes the 90% highly-stressed volume V_{90} and the Weibull exponent κ to assess the size effect related fatigue strength, represented in Equation (1).

$$\frac{\sigma_{LLF,0}}{\sigma_{LLF,1}} = \left(\frac{V_{90,1}}{V_{90,0}}\right)^{\frac{1}{\kappa}} \tag{1}$$

In this equation, $\sigma_{LLF,0}$ and $\sigma_{LLF,1}$ represent the long-life fatigue strength of the highly-stressed volumes $V_{90,0}$ and $V_{90,1}$. The material dependent Weibull exponent κ specifies the slope in the double logarithmic σ_{LLF}-V_α-plot and therefore the reduction of the fatigue strength against the highly-stressed volume. Its value and can be taken either by a common guideline [23], which defines the parameter as $\kappa = 10$ for aluminium castings, or be calculated dependent on the probability distribution of the fatigue data, represented by T_S [17], see Equation (2). In this equation, T_S is the scatter index of the high cycle fatigue region at ten million load cycles, defined as the stress ratio between a 10% and 90% probability of survival. In Reference [24], Sonsino proposed a threshold volume $V_\infty = 8000$ mm^3 for cast aluminium material, implying that no further noticeable decrease in fatigue strength may be observed.

$$\kappa = \frac{1.3151}{log(T_s)}. \tag{2}$$

The weakest link model of Weibull [25] as well as the discussed volumetric model are in good accordance to the experimental fatigue data [26]. Studies on artificial defects in References [27,28] exhibit that the highly-stressed volume approach is more suitable to investigate the statistical size effect. Both, the common engineering guideline [23] and short-crack growth findings in Reference [1] recommend a highly stressed surface model but refer additionally to highly-stressed volume models. Hence, the model of Sonsino [16] is used in this study for the validation of the statistical size effect. Kitagawa and Takahashi recommended in Reference [29] that the long life fatigue strength σ_{LLF} can be related to a dedicated crack length a, respectively to equivalent defect size, which can be defined as equivalent circle diameter (ECD) or by the equivalent edge length of a square ($\sqrt{area}$), see Equation (3). The sound applicability of the model from Kitagawa and Takahashi has been proven in several studies, see References [3,4,7,30–36].

$$\Delta\sigma_{LLF} = \frac{\Delta K_{th,lc}}{Y\sqrt{\pi a}}. \tag{3}$$

In this equation, $\Delta K_{th,lc}$ is the long crack threshold and Y a geometry factor depending on the geometrical shape and location of the defects, as discussed in preliminary studies such as that in Reference [37]. The design strength is limited on the one hand by the long life fatigue strength of the near defect free material $\Delta\sigma_0$, evaluated at specimens with hot isostatic pressed (HIP) condition with T6 heat treatment (HIP + T6). Otherwise, the fracture mechanical approach takes into account the long crack threshold value $\Delta K_{th,lc}$ and the effective crack threshold value $\Delta K_{th,eff}$. These crack threshold values come into effect for flaw sizes becoming larger than a intrinsic crack length $a_{0,eff}$, respectively $a_{0,lc}$. Further improvements of the Kitagawa Takahashi diagram by El Haddad [38,39] and Chapetti [40] are considering the crack resistance curve. A schematic representation of the Kitagawa Takahaschi diagram (KTD) and its modifications are given in Figure 1.

The crack extension from the intrinsic threshold $\Delta K_{th,eff}$ to the long crack threshold $\Delta K_{th,lc}$ can be represented by applying the cyclic crack resistance curve (R-cureve), as introduced by Reference [41]. The build-up of the crack resistance from the intrinsic $\Delta K_{th,eff}$ to the long crack threshold $\Delta K_{th,lc}$ with elevating crack length is caused by crack closure effects [42,43], whereat a premature contact of the crack faces generally leads to a minor real effective load ΔK_{eff} for further crack propagation, see Equation (4).

$$\Delta K_{eff} = K_{max} - K_{op}. \tag{4}$$

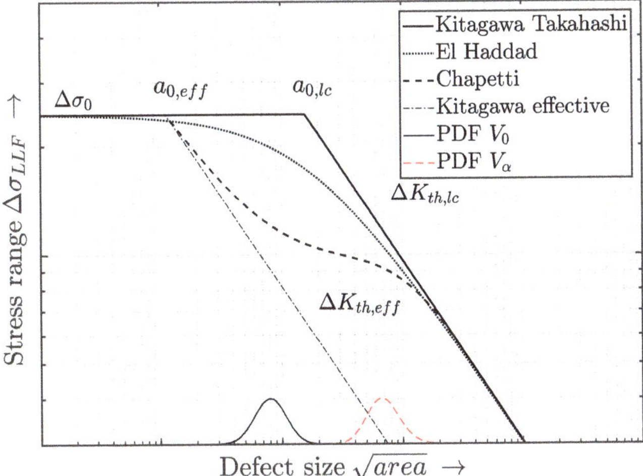

Figure 1. Schematic set up of the Kitagawa Takahashi diagram with its modifications and exemplary defect distributions of a volume V_0 and an enhanced volume V_α.

Crack closure effects can be classified into plasticity-, roughness-, and oxide-induced crack closure fractions as the most pronounced ones, whereat an explicit separation of these effects is not possible [44–49]. Maierhofer recommended in Reference [41] a procedure to describe the R-curve in a unified manner as given in Equation (5).

$$\Delta K_{th,\Delta a} = \Delta K_{th,eff} + \left(\Delta K_{th,lc} - \Delta K_{th,eff}\right)\left[1 - \sum_{i=0}^{n} v_i \cdot \exp\left(-\frac{\Delta a}{l_i}\right)\right], \tag{5}$$

with

$$\sum_{i=0}^{n} v_i \equiv 1.$$

In this equation, the crack closure effects are considered using the parameters v_i and l_i, implying that if the crack reaches length l_i the corresponding closure effect v_i is completely developed. By inserting Equation (5) in Equation (3), the cyclic R-curve can be implemented in the KTD whereby this extension of the KTD is useful to assess both, physically short and long, cracks. Therein, the crack length a is substituted by the equivalent defect size $\sqrt{area}$. Murakami introduced in Reference [50] the $\sqrt{area}$-parameter, which is the cross section of a defect in respect to the load direction. According to a study in Reference [51], the stress field surrounding the defect correlates well with the $\sqrt{area}$-parameter. Hence, this parameter is used to assess the crack initiating defects. Preliminary studies [5,52,53] contributed to the measurement methods of defects in cast aluminium alloys, as also applied within this study.

In References [12–14,54] it was shown that the statistical distribution of defect sizes follow an extreme value distribution. The Generalized Extreme Value (GEV) distribution includes the Frechet, Gumbel and Weibull distribution [55] and is applicable for characterizing crack initiating defect sizes [12]. Its cumulative distribution function (CDF), see Equation (6), is defined by three parameters, named as location μ, scale δ and shape ξ parameter which can be estimated by using the maximum likelihood method, applied in the studies [12,53,56,57]. The shape parameter ξ determines the type of extreme value distribution, differentiating between three cases: $\xi \rightarrow 0$ indicates a Gumbel, $\xi < 0$ a Weibull and $\xi > 0$ a Fréchet distribution, see Reference [13].

As published in a previous study [3], the CDF of an α-times enlarged volume V_α of the defect distribution P^α can be derived based on the distribution of the reference volume V_0 according to Reference [58], expressed in Equations (6)–(12).

$$V_0 \sim P(\sqrt{area}; \mu, \delta, \xi) = \exp\left\{-\left[1 + \xi\left(\frac{\sqrt{area} - \mu}{\delta}\right)\right]^{-\frac{1}{\xi}}\right\} \tag{6}$$

$$V_\alpha \sim P^\alpha, \tag{7}$$

with

$$\xi_\alpha = \xi \tag{8}$$

$$\delta_\alpha = \delta \cdot \alpha^\xi \tag{9}$$

$$\mu_\alpha = \mu + \frac{\delta}{\xi} \cdot \left(\alpha^\xi - 1\right), \tag{10}$$

which leads to

$$P^\alpha = \exp\left\{-\left[1 + \xi\left(\frac{\sqrt{area} - (\mu + \frac{\delta}{\xi}(\alpha^\xi - 1))}{\delta\alpha^\xi}\right)\right]^{-\frac{1}{\xi}}\right\} \tag{11}$$

$$V_\alpha \sim P\left(\sqrt{area}; \mu + \frac{\delta}{\xi}\left(\alpha^\xi - 1\right), \delta\alpha^\xi, \xi\right). \tag{12}$$

Detailed methodologies to calculate the maximum defect in geometries with enlarged HSV are given in References [59,60], whereat it is shown that the most extremal defects are commonly Gumbel distributed, applied in Equation (13), using the location parameter μ and scale parameter δ.

$$P(\sqrt{area}) = exp\left\{-exp\left[-\frac{\sqrt{area} - \mu}{\delta}\right]\right\}. \tag{13}$$

Now the size of a critical defect in an enlarged control volume V_α, which is considered by the ratio of the enlarged volume V_α divided by the reference volume V_0, can be calculated by Equation (14).

$$\sqrt{area}(\alpha) = \mu - \delta \cdot ln\left[-ln\left(1 - \frac{1}{\alpha}\right)\right] \tag{14}$$

with the return perid α denoted as:

$$\alpha = \frac{V_\alpha}{V_0}. \tag{15}$$

Complex components exhibit various HSVs whereas, mostly, each of them features differences in microstructure due to dependency on local casting process conditions. Thus, a local fatigue assessment considering the microstructural characteristics is beneficial. Even in case of the same microstructure, respectively basic defect distribution, the HSV depends on the component geometry and load condition, which lead to the question if single HSVs may be added together, resulting in a HSV of the whole part and fatigue strength design according to Equation (1), or if each HSV has to be considered individually for all unconnected ones. This paper clarifies this task regarding size effect based fatigue strength design in cast aluminium. Summing up, this paper scientifically contributes to the following points:

- The influence of disconnected highly-stressed volumes as statistical size effect based on accumulated highly-stressed volumes.

- The impact of the highly-stressed volume on the defect distribution and its associated parameters is verified for samples with not-yet investigated casting process conditions. This enhances the existing database and strengthens the a priori established model framework of probabilistic fatigue strength design.
- The effect of the element size during numerical evaluation of the highly-stressed volume is studied and supports recommendations for engineering applicability.
- The work validates the prior developed statistical size effect approach which depends not only on the return period of the highly-stressed volume but takes also the defect distribution of the fractograpic analysis and the material resistance as probabilistic values into account.

2. Investigated Alloy

The material is taken out of a gravity cast automotive part, manufactured using the core package system casting process [61,62]. The components are made of EN AC-46200 with T6 heat treatment [63], whose nominal chemical composition is given in Table 1. In general, applied steps for T6 heat treatment at aluminium alloys are solution treatment, quenching and age hardening, following defined temperature and time conditions [61,64,65]. First, solution treatment is conducted at high temperatures of approximately 490 °C to 510 °C for about 0.5 h to 8 h to dissolve Cu-rich particles [64–68]. The following quenching in water at ambient temperature, or at 60 °C, leads to a over-saturated solid solution [64,69]. In the third step, the age hardening process is conducted at temperatures from 160 °C to 210 °C for about 4 h to 18 h, whereat in case of higher temperatures a reduced time span is needed to reach the peak hardness, which is the overall aim of T6 treatment [64,67,70–74]. Furthermore, the peak hardness decreases with increasing age hardening temperature [64]. The specimens are manufactured from two different sampling positions, denoted as A and B, where A possesses a highly-stressed volume V_0 and B an increased highly-stressed volume V_1. Further information about these positions and its local microstructural and mechanical properties are given in detail in References [3,53,75,76]. Within these preliminary studies, the fundamental KTD was built up.

Table 1. Nominal chemical composition of the investigated cast alloy in weight percent [63].

Alloy	Si [%]	Cu [%]	Fe [%]	Mn [%]	Mg [%]	Ti [%]	Al [-]
EN AC-46200	7.5–8.5	2.0–3.5	0.8	0.15–0.65	0.05–0.55	0.25	balance

The secondary dendrite arm spacing (SDAS) in position A and B is almost identical and differ by only five percent, resulting in a negligible technological size effect between these two positions. The SDAS was evaluated through an automated procedure described in Reference [77] for linking the microstructural properties to quasi-static [78–80] and fatigue properties [7,73,81,82]. Thus, the chosen positions A and B feature specimens of varying geometric sizes but with almost identical microstructural and mechanical properties. The investigated samples possess the same basic circular cross section, but their total length differ. For clarification, specimen A is taken from position A and specimen B is manufactured out of position B. Subsequently, only the specimens are denoted as A and B. To reduce the stress concentration factor within the cross section transition region, the specimens have been numerically shape optimized resulting in a stress concentration factor of only 1.04.

The difference between specimen A and specimen B is, that in case of specimen B, the basic geometry of sample A has been invoked two times in a row. Thus, it is the same as two specimens of type A. Figures 2 and 3 depict the two specimen geometries for high cycle fatigue testing under uniaxial tension load.

The highly-stressed volume of specimen geometry B is roughly doubled in comparison to geometry A. Thus, considering the sum of both sections, a noticeable statistical size effect is expected. To determine the return period of the highly-stressed volumes more accurately, a numerical study regarding the applicable element seed is conducted. A linear elastic finite element analysis has been set-up featuring an uni-axial tension load with couplings to match the experimental clamping conditions.

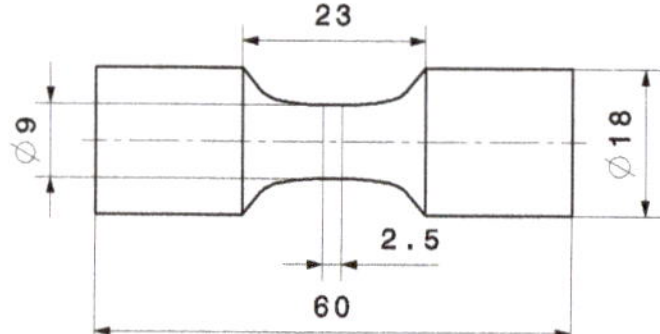

Figure 2. High cycle fatigue (HCF) specimen A with dimensions in [mm].

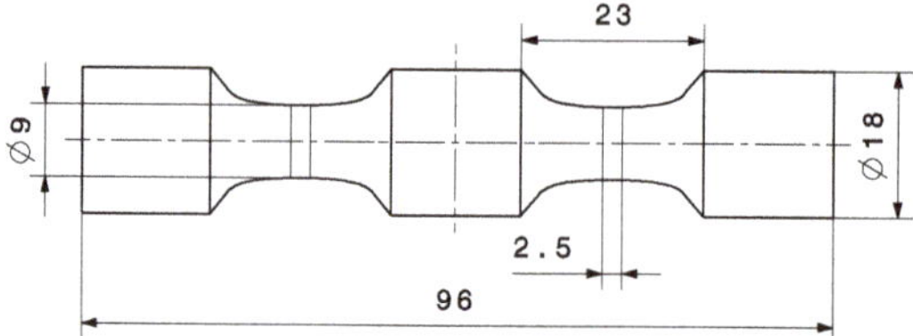

Figure 3. HCF specimen B with dimensions in [mm].

The element types employed are 20-node quadratic brick C3D20R and 10-node quadratic tetrahedron C3D10 elements with 8 up to 116 elements on each circumference. Additionally, axisymmetric CAX8R elements are used with the same element dimensions to significantly reduce the simulation time. This results in an average element size of approximately 0.24 mm to 3.5 mm in the HSV-region, see Figure 4. Another possibility to define the element seed, respectively number of elements per unit length, is the deviation factor, which is defined as the ratio between height h of the segment and the chord length L with n as element number on the circumference, see Equation (16).

$$\frac{h}{L} = \frac{1}{2} \cdot tan\left(\frac{\pi}{2 \cdot n}\right).$$

(16)

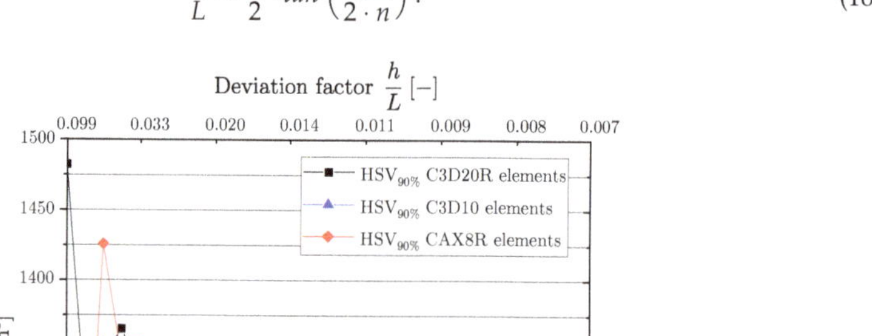

Figure 4. Effect of element seed on numerically determined highly-stressed volume (HSV).

Thus, a number of about 32 elements on circumference, or a deviation factor of 0.03, leads to a sound compromise between simulation time and accuracy. The numerically evaluated volume results in a value of $V_{0,90\%} = 647$ mm^3 for specimen A and $V_{1,90\%} = 1284$ mm^3 for specimen B, see Figure 5. Concluding, a 90 % highly-stressed volume ratio of $\alpha = 1.98$ is obtained for specimen A and B.

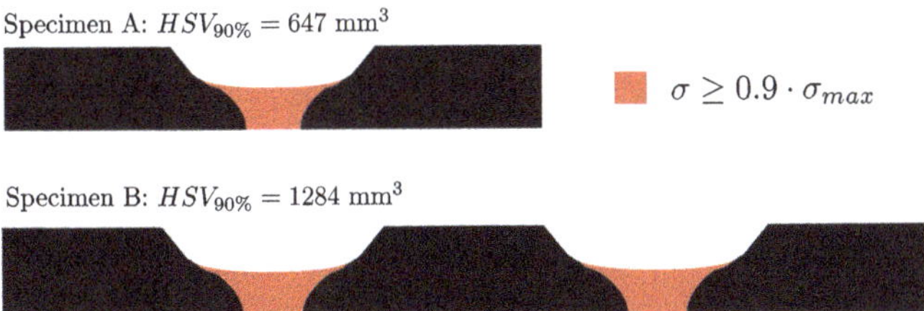

Figure 5. Finite Element (FE) analysis of specimens A and B with 90% HSV determined with C3D20R elements.

In the first phase of the testing procedure the experiment is carried out by clamping part 1 and 3 of the entire specimen with subsequent high cycle fatigue testing until rupture, either at the upper (section 2–3) or the lower (section 1–2) specimen fraction, as depicted in Figure 6. Next, the fractured part is removed (shorter specimen part of section 1 or section 3). Subsequently, the specimen is clamped at the middle part (section 2), see secondary clamping in Figure 6, and the test is continued at the same load level until rupture of the remaining short specimen. It should be highlighted that this shortened specimen possess a HSV which is equivalent to specimen geometry A. The result of this testing procedure are two points in the S/N diagram, which will be discussed in more detail in Section 3.

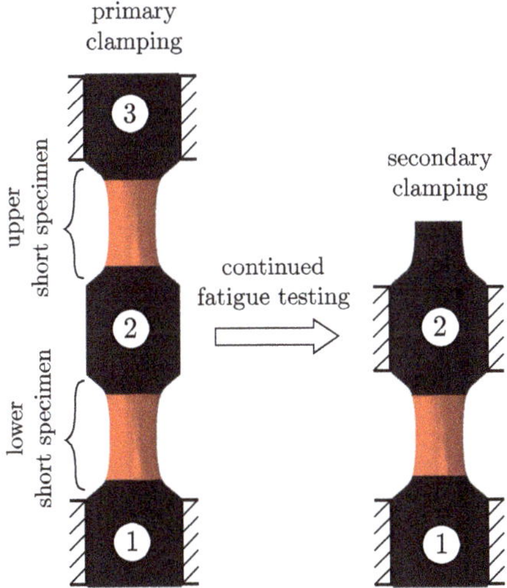

Figure 6. First and second testing of specimen type B.

In order to ensure a homogeneously distributed surface quality with prevention of human influence by polishing, the specimens are polished by a vibratory finishing process. After the CNC machining process, the components are placed in an oscillating bowl containing polishing media. Thereby, the specimens are precision grinded and polished with different abrasive media for several hours until the required surface quality is obtained.

3. Experimental Results

3.1. Fatigue Strength

The fatigue strength of the material is determined at a resonant testing machine with a testing frequency of about 108 Hz with compression/tension loading at a stress ratio of R = −1. In order to focus on the long life fatigue region, the run-out number was set to ten million load cycles. Previous investigations [3] indicated that the transition knee point is close to about two million load cycles for such unnotched samples made of aluminium alloy. As proposed in Reference [83] and applied in preliminary studies [3,76,83,84], the slope of the S/N-curve in the long life region k_2 scales with the slope in the finite life region k_1 and therefore it is assigned with $k_2 = 5 \cdot k_1$. The S/N curve in the finite life region is evaluated by the statistical procedure given in the standard [85]. The long life region is assessed by the $arcsin\sqrt{P}$ methodology, as proposed in Reference [86]. In the following, the long life fatigue strength of specimen A, taken out of position A, at ten million load cycles and at a probability of survival $P_S = 50\,\%$ is used as unifying reference value. Figure 7 presents the statistically evaluated S/N curve of specimen A series including the 90% and 10% scatter band. Be aware that specimen A inherit the highly-stressed volume V_0.

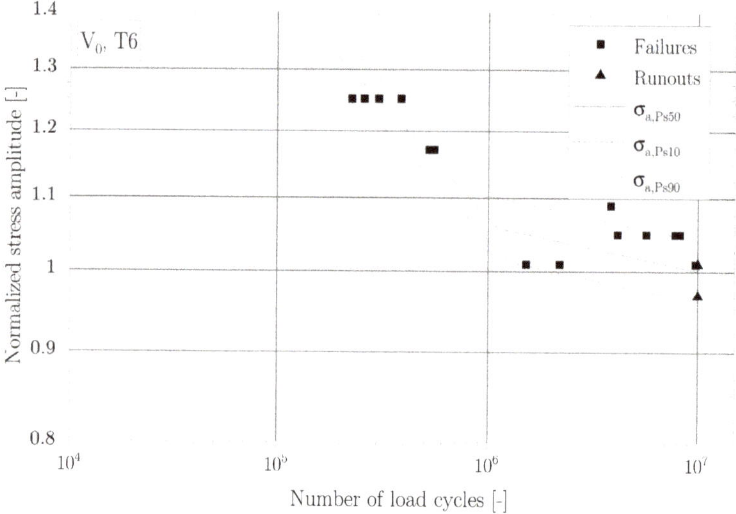

Figure 7. S/N curve of specimen A.

Next, the evaluated fatigue data of specimen B at first failure is depicted in Figure 8, again with the 90% and 10% scatter band of the high cycle fatigue region. Thus, the mean long-life fatigue strength σ_{LLF} of position B decreased by approximately five percent compared to specimen A. The doubling of the highly-stressed volume in position B reveals an evaluable decrease in fatigue strength contributed as statistical size effect.

The evaluated slope k_1 at position B in the finite life region is somewhat higher with respect to position A. Additionally, the number of load cycles N_T of the transition knee-point is slightly enhanced. Comparing the scatter indices T_S of the positions A and B, an increase at disconnected highly-stressed volumes is observed. The evaluated long life fatigue strength σ_{LLF} is listed in Table 2, where all fatigue strengths are normalized by position A with a probability of survival $Ps = 50\,\%$. Furthermore, the slope k_1 of the finite life region, the number of load cycles for the transition knee-point N_T and the statistically evaluated fatigue scatter index T_S are given in Table 2.

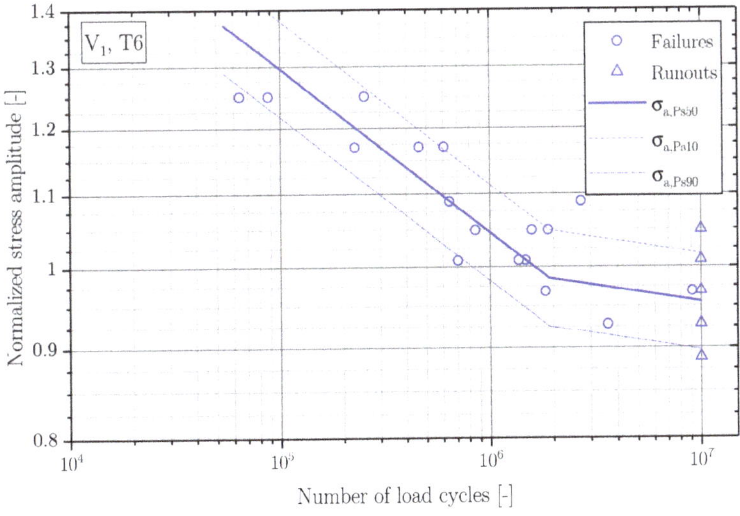

Figure 8. S/N curve of specimen B (evaluation only first failures).

Table 2. Results of the fatigue tests of specimen A and B.

Specimen	HT	Volume	k_1 [-]	$\sigma_{LLF,50\%}$ [-]	N_T [-]	T_S [-]
A	T6	V_0	7.84	1.00	1,100,000	1:1.08
B	T6	V_1	10.73	0.96	1,900,000	1:1.23

Thus, the statistical size effect may be clearly identified for such samples possessing an increased highly-stressed volume, even though this volume is not coherent as shown in Figure 5. On the other hand, if the highly-stressed volume would be considered separately, which means that no statistical size effect occurs in case of non-coherent highly-stressed volume, both S/N curves in Figures 7 and 8 must coincide. Therefore, the experimental point $\sigma_{B,Ps50}$ should be congruent with the point $\sigma_{A,Ps50}$ for the same connected highly-stressed volume. But the experimental point of specimen B (V_1) with two separated highly-stressed volumes V_0 is below the fatigue strength of specimen A (V_0).

Therefore, as main finding based on the presented experiments, the entire highly-stressed volume has to be considered for the statistical size effect. Thereby, the entire highly-stressed volume V_1 is calculated by the sum of the separated, non-coherent highly-stressed volumes whereat the failure of one single highly-stressed volume leads to a collapse of the specimen. The working hypothesis for first, and second, fatigue failure of specimen B and a theoretical discussion is given in detail in Appendix A.

3.2. Fractography

The crack initiating defect sizes of the HCF specimens are evaluated subsequently to the fatigue testing utilizing a digital optical microscopy for macroscopic inspection and scanning electron microscopy respectively for magnification enhanced, local analysis. According to previous investigations [5,53], defect sizes are evaluated by their precise contour in contrast to the coarser method proposed by Murakami in Reference [59], where a smooth hull contour, which envelopes the original shape, is utilized. This measurement methodology leads to smaller, but more precisely evaluated defect sizes, and it minimizes the distortive effects of projected pore shape onto the statistical evaluation of defect sizes. Therefore, a spline is drawn manually at the contour of the defect using the software Fiji, allowing to calculate the enclosed area. The analysis of the initiating cracks of the specimen A and B revealed that in most cases the technical crack initiates at surface near defects, as depicted in Figure 9. Thus, the increased stress intensity of surface-intersecting defects and surface near defects lead to a lowered crack initiation phase compared to closed defects within the bulk volume [53]. With the existence of superior internal defects in a few samples, in the majority of them specimen B cases, origin of fracture is shifted into the centre of the specimen, exemplary see Figure 10.

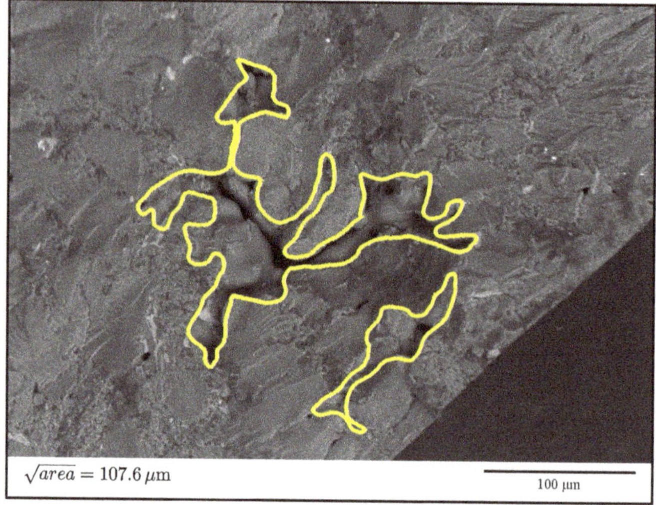

Figure 9. Fracture initiating defect at specimen A.

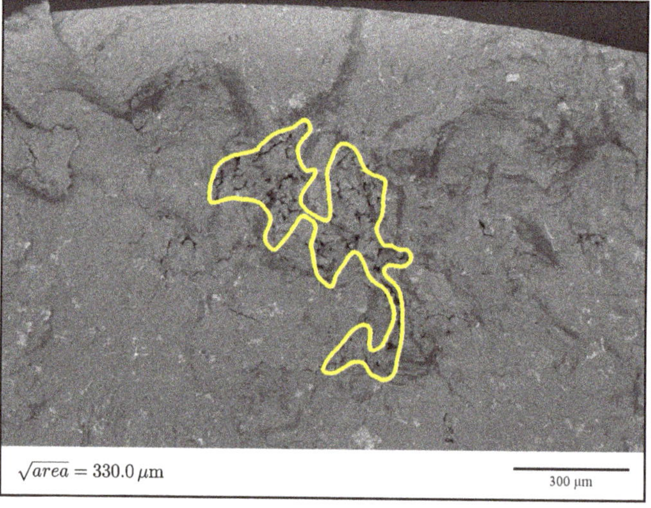

Figure 10. Fracture initiating defect at specimen B.

As depicted in Figure 11, in a few cases another failure mechanism is recognizable. According to previous studies [5,87] large slip plane areas can operate as failure reason for load amplitudes within the finite life region. This is more likely to happen for increasing loads. It is stated in References [88,89], that in fine microstructures with a small SDAS, the dislocations are able to move across the cell boundaries of the dendrites, since there are no particles to block them. This is in contrast to larger SDAS values by means of coarse microstructures where the dendrite cell is isolated by a thick eutectic wall blocking the dislocations. Due to that, a critical defect size exists, below that the crack initiates at slip bands instead of interdendritic shrinkage pores. From preliminary studies [5,53] it can be assumed that this failure mechanism only occurs at specimens with lower SDAS and quite high load levels of the S/N-curve.

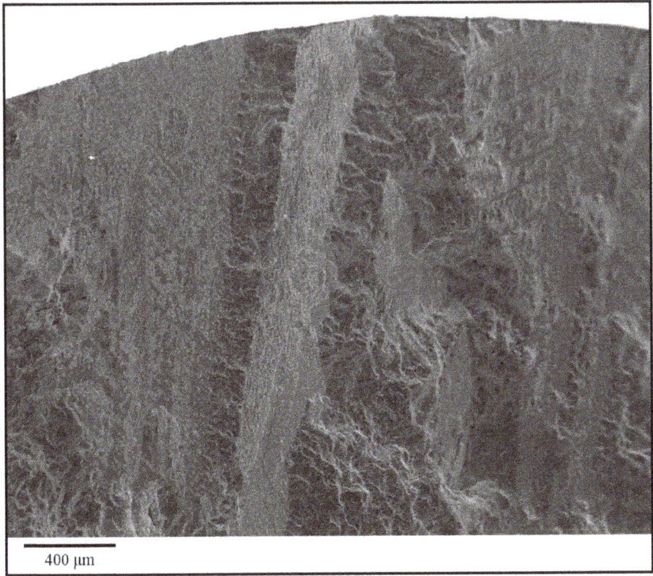

Figure 11. Slip plane area that occurs in both positions.

Summing up the experimental work, the fractographic analysis revealed that in most cases the crack initiation starts at interacting shrinkage porosity near the surface, see Figure 9. Therefore, defects are regarded as interacting if the distance between two defects is less than the size of the smaller defect, as proposed in Reference [90]. For the subsequent statistical evaluation of the critical defect sizes the generalized extreme value distribution is applied, following the proposal of Reference [13]. The associated cumulative distribution function (CDF) is given in Equation (6). Following Reference [91], a Kolmogorov-Smirnov (KS) test is conducted to evaluate the goodness of fit for the statistical assessment of the distribution. A perfect compliance for the fit is given with a value of $p_{KS} = 1.00$ in the KS-test.

The evaluated probability of occurrence P_{Occ} of casting defects for the reference volume V_0 as well as the two times enlarged volume V_1, reflecting specimen A and B, is drawn in Figure 12. Assuming that the failure of one section causes the failure of the whole component, only the first fracture and its associated flaw size are utilized for the evaluation of the distribution parameters. In addition, the parameters for the distributions in Figure 12 are statistically evaluated using the maximum likelihood estimation, as proposed in Reference [56]. The evaluated parameters of the distributions from specimen A and B, the result of the Kolmogorov-Smirnov test and the evaluated defect size with a probability of occurrence of 50% are listed for comparison in Table 3.

Table 3. Statistically evaluated distribution parameters of the generalized extreme value distribution (GEV).

Position	Volume	μ [µm]	δ [µm]	ζ [-]	$\sqrt{area}(P_{Occ=0.5})$ [µm]	p_{ks} [-]
A	V_0	95.1	20.1	0.43	103	0.93
B	V_1	118.4	28.1	0.36	129	0.69
B (model)	V_1	111.3	27.1	0.43	122	0.58

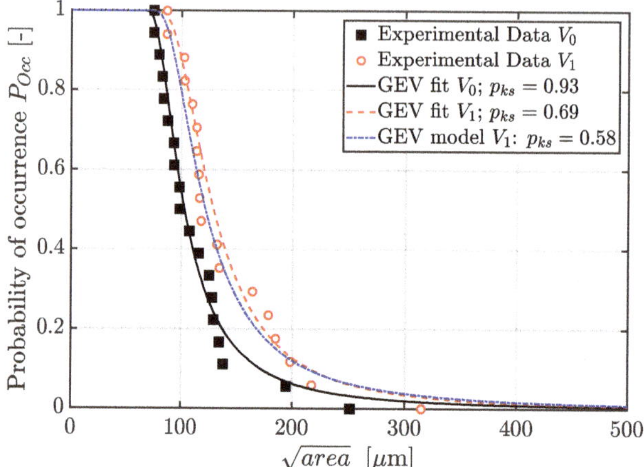

Figure 12. Probability of occurrence of critical defect sizes in specimen A and B.

As mentioned before, in this study only the first failures per specimen are considered for the validation of the statistical size effect. The probability of occurrence P_{Occ} of a critical defect for an α-times enlarged volume V_α can be estimated based on the statistical distribution of the reference volume as reasoned in Reference [3] and shown in Equation (17). The evaluation of the location μ_α, shape ζ_α and scale δ_α parameter for the distribution P^α is given in Equations (8)–(10). The parameters of the distribution P^α are listed in Table 3.

$$P^\alpha = \exp \left\{ - \left[1 + \zeta_\alpha \left(\frac{\sqrt{area} - \mu_\alpha}{\delta_\alpha} \right) \right]^{-\frac{1}{\zeta_\alpha}} \right\}. \tag{17}$$

4. Verification of Size-Effect Related Fatigue Strength

In order to study the size effect as influence of the highly stressed volume, the probabilistic model of the preliminary work [3] has to be applied to evaluate the local Weibull factor κ. It depends on the return period α of the highly-stressed volume and the local defect population μ_0. As the same aluminium alloy with T6 heat treatment was used also in the previous model development regarding fatigue strengths, the diagram can be easily rebuilt for the varying return period, respective defect population within the highly-stressed volume. The local Weibull factor $\kappa(\mu_0, \alpha)$ can be obtained by transforming Equation (1). This results in a value of $\kappa_{\mu_0,\alpha} = 15.27$ based on the experimental results for the α-times enlarged volume in case of specimen B, see Equation (18).

$$\kappa(\mu_0, \alpha) = \frac{\log(\alpha)}{\log(\Delta\sigma_{\mathrm{LLF},V_0}) - \log(\Delta\sigma_{\mathrm{LLF},V_\alpha})}. \tag{18}$$

In the fundamental work of Reference [3], the Kitagawa-Takahashi diagram (KTD) was used to assess the fatigue strength $\Delta\sigma_{\mathrm{LLF},V_0}$ and $\Delta\sigma_{\mathrm{LLF},V_\alpha}$ depending on defects, respective microcracks. Therein, crack propagation tests have been conducted with specimens manufactured from the identical positions as used in this study to minimize microstructural deviations. To extend the KTD for physically short and long cracks, the crack-resistance curve was implemented [4]. A summary of the fracture mechanical variables, determined by crack propagation tests from previous studies [4], is given in Table 4 for the investigated alloy. No statistically feasible difference in fracture mechanical material properties of position A and B has emerged.

Table 4. Parameters resulting from crack propagation tests in position A and B for a probability of occurrence of $P_{Occ} = 50\%$.

$\Delta K_{th,lc}$ [MPa$\sqrt{\text{m}}$]	$\Delta K_{th,eff}$ [MPa$\sqrt{\text{m}}$]	v_1 [-]	v_2 [-]	l_1 [mm]	l_2 [mm]
3.95	1.06	0.4	0.6	0.03	0.75

The long life fatigue strength of the near defect free material $\Delta\sigma_0$, which defines the upper limit of the left side of the KTD, was evaluated with specimens in HIP treatment condition at the same position. In this model the fatigue strength $\Delta\sigma_{LLF,V_0}$ is determined using the R-curve extension [40] for a defect size represented by the size of an defect a_m of the reference volume V_0 implying a probability of occurrence of $P_{Occ} = 50\%$. The critical defect size for an enhanced volume can be estimated by application of Equations (6)–(12). This results in a fatigue strength $\Delta\sigma_{LLF,V_\alpha}$ for an enhanced volume V_α using the given defect distribution $(\mu_\alpha, \delta_\alpha)$ with a specific defect size $a_{m,\alpha}$.

$$\Delta\sigma_{LLF,V_0} = \frac{\Delta K_{th,\Delta a}}{Y \cdot \sqrt{\pi \cdot a_m}}, \tag{19}$$

with

$$\Delta K_{th,\Delta a} = \Delta K_{th,eff} + \left(\Delta K_{th,lc} - \Delta K_{th,eff}\right)\left[1 - \sum_{i=0}^{n} v_i \cdot \exp\left(-\frac{\Delta a}{l_i}\right)\right], \tag{20}$$

$$a_m = \mu_0 + \delta_0\left(-log(-log(P))\right), \tag{21}$$

$$\Delta a = a_m - a_{0,eff}, \tag{22}$$

$$a_{0,eff} = \frac{\Delta K_{th,eff}}{(Y \cdot \Delta\sigma_0)^2} \cdot \frac{1}{\pi}. \tag{23}$$

Thus, the long life fatigue strength of the reference volume V_0 with a certain defect distribution can be calculated using Equations (19)–(23). Moreover, the fatigue strength of an enlarged volume V_α can be determined by means of Equation (24) to (27), (20) and (23), as exemplified in Reference [3].

$$\Delta\sigma_{LLF,V_\alpha} = \frac{\Delta K_{th,\Delta a}}{Y \cdot \sqrt{\pi \cdot a_{m,\alpha}}}, \tag{24}$$

with

$$a_{m,\alpha} = \mu_\alpha + \delta_\alpha\left(-log(-log(P))\right), \tag{25}$$

$$\Delta a = a_{m,\alpha} - a_{0,eff}, \tag{26}$$

$$\mu_\alpha = \mu_0 + log(\alpha) \cdot \delta_0, \tag{27}$$

$$\delta_\alpha = \delta_0. \tag{28}$$

Now, the local Weibull factor $\kappa(\mu_0, \alpha)$ can be derived as a function of inhomogeneity population represented by its location parameter μ_0 in a control volume V_α. The course of the local Weibull factor κ is plotted in dependence of α and μ_0 in Figure 13. It is evident that κ increases with rising return period α and defect population μ_0. This relationship is a significant improvement compared to the common guideline [23], where the Weibull factor is determined with a constant value of ten.

Hence, this generalized model of Aigner et al. [3] can be used to check on the size effect of the return period α, thereby validating the influence of disconnected highly-stressed volumes, as discussed in Section 3.2. Therefore the evaluated defect distribution for the reference volume V_0 in Section 3.2 and the return period of $\alpha = 1.98$ are utilized and leading to a model-based local Weibull factor of $\kappa(\mu, \alpha) = 13.8$, as depicted in Figure 13 as red marked triangle.

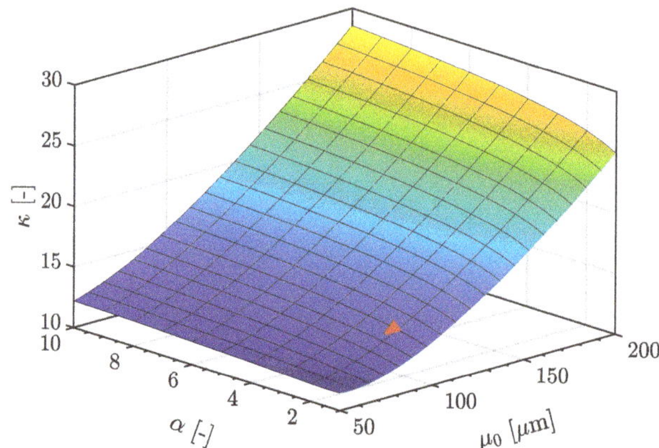

Figure 13. Weibull factor κ depending on the return period α and the defect population μ_0 and evaluated point of the current test series of specimen B.

By applying the common guideline [23], respectively, the volumetric model of Sonsino [16], (Equation (1)), the fatigue strength of an elevated HSV with return period α, defined as V_α, can be calculated as a function of a reference volume V_0 and its associated parameters. The calculation is done for the HSV of specimen B.

Aside from the discussed volumetric approaches, the estimation of the Weibull factor can be related to the scatter index of the experimental fatigue strength distribution only [17], see Equation (2). Substantiated by the high manufacturing quality of the samples and quite homogeneous manufacturing process conditions within the HSV, a comparably small fatigue scatter index T_S in the long-life fatigue region is obtained. This approach leads to a value of $\kappa = 39.3$, resulting in non-conservative fatigue data. This is depicted as dash-dotted line in Figure 14.

In Figure 14, all three different approaches [3,17,23] are compared, where each of them leads to different Weibull factors κ resulting in differing fatigue strength values. Table 5 lists the normalized fatigue strength results from the three different κ-values. The fatigue assessment model proposed in Reference [3] fits the experimental data with a value of $\kappa = 13.8$ best, plotted as continuous line in Figure 14. The common guideline (dotted line in Figure 14) leads to a more conservative fatigue design compared to the experimental results, because a constant weibull factor κ is defined for groups of materials. The model published by Reference [17] leads to an improper, non-conservative fatigue design due to the small scatter band of the fatigue data.

Table 5. Comparison of the normalized fatigue strength resulting from different Weibull parameters κ using a return period of $\alpha = 1.98$ (specimen B in this study).

$\sigma_{LLF,50\%}$ [-]	Δ	κ [-]	Model
0.96	Reference	-	Experiment
0.93	-2.28%	10.0	[23]
0.98	$+2.84\%$	39.3	[17]
0.95	-0.42%	13.8	[3]

To summarize, the fatigue assessment model of Reference [3] is validated for EN AC-46200 in sand cast condition for volume ratios up to a value of α about two, leading to an enhanced fatigue assessment avoiding an over-conservative design. The updated size effect model, utilizing a highly stressed volume of 90%, now covers a return period of about two and up to six [3]. Nevertheless, other return periods shall be investigated to approve the statistical method even further.

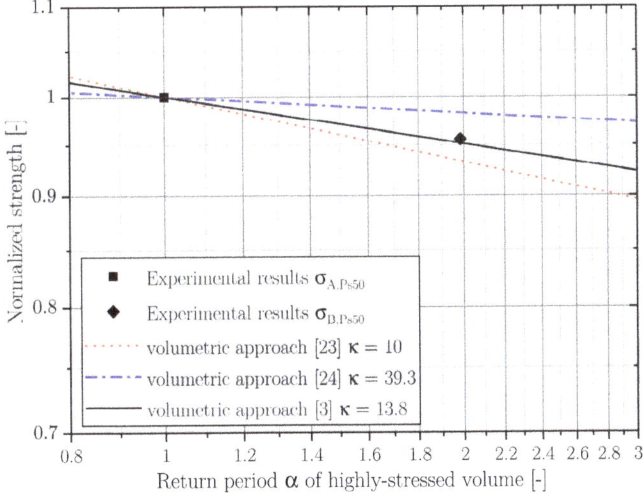

Figure 14. Comparison of HSV-based fatigue assessment models for S/N-results of specimen A and B.

5. Conclusions

This paper evaluates the size-effect based fatigue strength design of EN AC-46200 in T6 heat treatment condition. Therefore, a special specimen geometry was designed, which possess a non-coherent highly-stressed volume. Volumetric approaches are reviewed and their applicability for conservative fatigue designs is discussed. Overall, the following conclusions can be drawn:

- Based on a numerical parameter study, a deviation factor of about 0.03 is recommendable for numerical evaluation of the highly-stressed volume (HSV) in engineering applications.
- If several independent HSVs with the same microstructural properties are attached as one component and loaded simultaneously, the failure of each HSV leads to failure of the whole component. Hence, the aggregated sum of disconnected HSVs has to be considered as size effect in fatigue strength design. But in the case of varying microstructures between the individual highly-stressed volumes, the local microstructure has to be considered as well.
- The conducted validation of the aforesaid defect based probabilistic fatigue assessment model, originally published in Reference [3], is based on samples with a return period of about two. The results confirm that the model assesses the fatigue strength in terms of statistical size effect best by applying the local Weibull factor κ depending on the return period α and defect population μ_0. Thus, the verified probabilistic approach is recommendable for engineering design of complex parts, whereat the HSV has to be linked to the local microstructural properties for proper fatigue strength design.

Current work focuses on the design strength related interaction between HSV and associated microstructure in cast aluminium alloys, especially in case of service load cases which enforces locally varying HSV and subsequent feasible damage sum calculations. Moreover, the applicability of the design concept for notched components considering different load cases and local stress gradients will be investigated.

Author Contributions: Conceptualization, M.O., S.P., M.L. and M.S.; methodology, M.O., S.P. and M.S.; software, M.O., M.S. and S.P.; validation, M.O. and M.L.; formal analysis, M.O. and S.P.; investigation, M.O.; resources, M.S.; data curation, M.O.; writing–original draft preparation, M.O.; writing–review and editing, M.O., M.L., S.P. and M.S.; visualization, M.O.; supervision, M.L. and M.S.; project administration, M.S.; funding acquisition, M.S.; All authors have read and agreed to the published version of the manuscript.

Funding: This research was funded by the Austrian Federal Ministry for Digital and Economic Affairs and the National Foundation for Research, Technology and Development.

Acknowledgments: The financial support by the Austrian Federal Ministry for Digital and Economic Affairs and the National Foundation for Research, Technology and Development is gratefully acknowledged. Furthermore, the authors would like to thank the industrial partners BMW AG and Nemak Dillingen GmbH for the excellent mutual scientific cooperation within the CD-laboratory framework.

Conflicts of Interest: The authors declare no conflict of interest.

Abbreviations

The following abbreviations are used in this manuscript:

$\sqrt{area}$	Defect size of Murakami's approach
α	Return period of the highly-stressed volume
κ	Weibull factor
σ_{LLF}	Long life fatigue strength
σ_{LLF,V_0}	Long life fatigue strength of the reference volume V_0
σ_{LLF,V_α}	Long life fatigue strength of the α-times enlarged volume V_α
$\sigma_{LLF,50}$	Estimated long life fatigue strength with 50% probability of survival
$\sigma_{*,Ps50}$	Experimental long life fatigue strength at position * with 50% probability of survival
Δ	Deviation of model to experiment
$\Delta\sigma_0$	Fatigue range of near defect free material
δ	Scale parameter of the GEV distribution
δ_0	Scale parameter of the GEV distribution for the reference volume V_0
δ_α	Scale parameter of the GEV distribution for the α-times enlarged volume V_α
μ	Location parameter of the GEV distribution
μ_0	Location parameter of the GEV distribution for the reference volume V_0
μ_α	Location parameter of the GEV distribution for the α-times enlarged volume V_α
ξ	Shape parameter of the GEV distribution
ξ_α	Shape parameter of the GEV distribution for the α-times enlarged volume V_α
ν_i	Weighting factor for crack closure effect i
l_i	Crack elongation, where the crack closure effect ν_i is completely build-up
$\Delta K_{th,lc}$	Long crack threshold range
$\Delta K_{th,\Delta a}$	Crack threshold range in respect to the crack extension
$\Delta K_{th,eff}$	Effective crack threshold range
ΔK_{eff}	Effective stress intensity factor range
K_{max}	Maximum stress intensity factor
K_{op}	Opening stress intensity factor
Δa	Crack extension
a	Crack length
$a_{0,eff}$	Intrinsic crack length
$a_{0,lc}$	Crack length at the transition to long crack behaviour
a_m	Crack length of the reference volume V_0 for a probability of occurrence of 50%
$a_{m,\alpha}$	Crack length of the reference volume V_α for a probability of occurrence of 50%
h	Segment height of a circle
L	Chord length of the segment
n	Number of elements on circumference
P	Probability
P_{Occ}	Probability of occurrence
P_S	Probability of survival
P^α	Defect distribution of α-times enlarged volume V_α
V_0, V_1	Highly stressed volume of specimen A and B
$V_{90,0}, V_{90,1}$	90% highly stressed volume of specimen A (V_0) and B (V_1)
V_∞	Threshold volume

V_α	α-times enlarged highly stressed volume
p_{ks}	p-value of the Kolmogorov-Smirnov test
Y	Geometry factor
k_1	Inverse slope of the S/N-curve in finite life region
k_2	Inverse slope of the S/N-curve in long life region
T_S	Fatigue scatter band of the S/N-curve
N_T	Transition knee point of the S/N-curve
R	Load ratio
R-curve	Cyclic crack resistance curve
HSV	Highly stressed volume
SDAS	Secondary dendrite arm spacing
GEV	Generalized extreme value distribution
CDF	Cumulative distribution function
KTD	Kitagawa Takahashi diagram
ECD	Equivalent circle diameter
FE	Finite element
HCF	High cycle fatigue
HIP	Hot isostatic pressing

Appendix A. Fatigue Failure Hypothesis

Lets assume that there is a cube containing a homogeneous defect distribution. Therefore, specimens manufactured from this cube, containing a certain highly stressed volume V_0, named specimen geometry A in this hypothesis, see Figure A1. This homogeneous distribution of defects results in a fatigue strength σ_{LLF,V_0}, inheriting a defect distribution GEV_{V_0}, evaluated by means of a fractographic analysis. Next, specimens possessing a connected doubled highly-stressed volume V_1 are manufactured from the same cube, which results in a fatigue strength σ_{LLF,V_1} with associated defect distribution GEV_{V_1}. According to [3], this defect distribution GEV_{V_1} is shifted to larger defect sizes compared to the GEV_{V_0} caused by the increased probability for larger, extremal defects in an increased highly-stressed volume. In the third step, two cubes containing a highly stressed volume V_0 are linked together as specimen B, to get a disconnected highly-stressed volume, which is two times V_0, see Figure 3. This results also in a lowered fatigue strength σ_{LLF,V_1} considering only the first failure of each specimen.

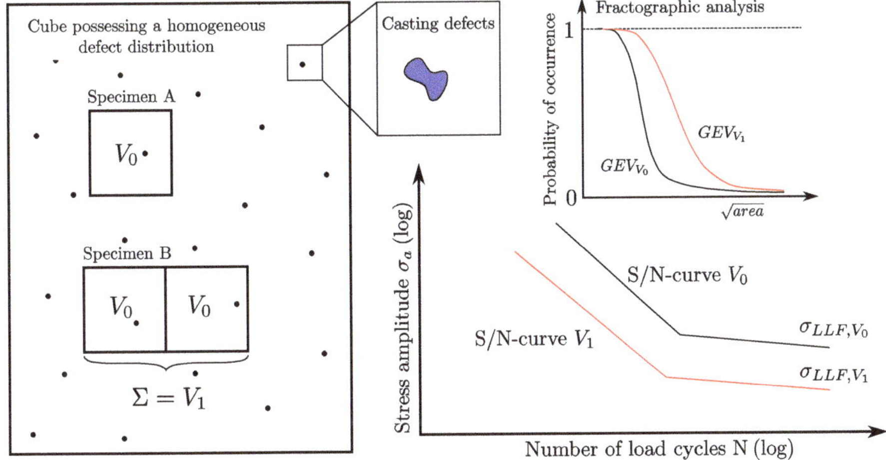

Figure A1. Schematic representation of the specimens manufactured from a cube possessing a homogeneous defect distribution and sketch of expected fatigue strength results.

Additionally, the fatigue strength of the second failures should result towards the higher value σ_{LLF,V_0}. Considering the defect distributions in the third case, the fractographical evaluated defect distribution of the first failures is supposed to coincide with GEV_{V_1} and the defect distribution of the second failures should coincide with GEV_{V_0}.

References

1. Hertel, O.; Vormwald, M. Statistical and geometrical size effects in notched members based on weakest-link and short-crack modelling. *Eng. Fract. Mech.* **2012**, *95*, 72–83, doi:10.1016/j.engfracmech.2011.10.017. [CrossRef]
2. Tomaszewski, T.; Sempruch, J. Size effect in high-cycle fatigue. *J. Mach. Constr. Maint. Probl. Eksploat.* **2017**, *1*, 29–35.
3. Aigner, R.; Pomberger, S.; Leitner, M.; Stoschka, M. On the Statistical Size Effect of Cast Aluminium. *Materials* **2019**, *12*, 1578, doi:10.3390/ma12101578. [CrossRef]
4. Aigner, R.; Pusterhofer, S.; Pomberger, S.; Leitner, M.; Stoschka, M. A probabilistic Kitagawa-Takahashi diagram for fatigue strength assessment of cast aluminium alloys. *Mater. Sci. Eng. A* **2019**, *745*, 326–334, doi:10.1016/j.msea.2018.12.108. [CrossRef]
5. Aigner, R.; Leitner, M.; Stoschka, M. Fatigue strength characterization of Al-Si cast material incorporating statistical size effect. In Proceedings of the MATEC Web Conference, Poitiers, France, 27 May—1 June 2018; Volume 165, p. 14002. [CrossRef]
6. Yi, J.Z.; Gao, Y.X.; Lee, P.D.; Lindley, T.C. Microstructure-based fatigue life prediction for cast A356-T6 aluminum-silicon alloys. *Metall. Mater. Trans. B* **2006**, *37*, 301–311. [CrossRef]
7. Iben Houria, M.; Nadot, Y.; Fathallah, R.; Roy, M.; Maijer, D.M. Influence of casting defect and SDAS on the multiaxial fatigue behaviour of A356-T6 alloy including mean stress effect. *Int. J. Fatigue* **2015**, *80*, 90–102, doi:10.1016/j.ijfatigue.2015.05.012. [CrossRef]
8. Stroppe, H. Calculation of the S-N curve for cast aluminium alloys based on static tensile test and dendrite arm spacing: Berechnung der Wöhler–Linie für Aluminium–Gusslegierungen aus dem statischen Zugversuch und dem Dendritenarmabstand. *Mater. Werkst.* **2009**, *40*, 738–742. [CrossRef]
9. Brueckner-Foit, A.; Luetje, M.; Wicke, M.; Bacaicoa, I.; Geisert, A.; Fehlbier, M. On the role of internal defects in the fatigue damage process of a cast Al-Si-Cu alloy. *Int. J. Fatigue* **2018**, *116*, 562–571, doi:10.1016/j.ijfatigue.2018.07.012. [CrossRef]
10. Brueckner-Foit, A.; Luetje, M.; Bacaicoa, I.; Geisert, A.; Fehlbier, M. On the role of internal defects in the fatigue damage process of a cast Al-Si-Cu alloy. *Procedia Struct. Integr.* **2017**, *7*, 36–43, doi:10.1016/j.prostr.2017.11.058. [CrossRef]
11. Atxaga, G.; Pelayo, A.; Irisarri, A.M. Effect of microstructure on fatigue behaviour of cast Al–7Si–Mg alloy. *Mater. Sci. Technol.* **2013**, *17*, 446–450, doi:10.1179/026708301101510023. [CrossRef]
12. Tiryakioğlu, M. On the size distribution of fracture-initiating defects in Al- and Mg-alloy castings. *Mater. Sci. Eng. A* **2008**, *476*, 174–177, doi:10.1016/j.msea.2007.04.088. [CrossRef]
13. Tiryakioğlu, M. Statistical distributions for the size of fatigue-initiating defects in Al–7%Si–0.3%Mg alloy castings: A comparative study. *Mater. Sci. Eng. A* **2008**, *497*, 119–125, doi:10.1016/j.msea.2008.06.023. [CrossRef]
14. Anderson, K.V.; Daniewicz, S.R. Statistical analysis of the influence of defects on fatigue life using a Gumbel distribution. *Int. J. Fatigue* **2018**, *112*, 78–83, doi:10.1016/j.ijfatigue.2018.03.008. [CrossRef]
15. Kuguel, R. A relation between theoretical stress concentration factor and fatigue notch factor deduced from the concept of highly stressed volume. *Proc. ASTM* **1961**, *61*, 732–748.
16. Sonsino, C.M. Zur Bewertung des Schwingfestigkeitsverhaltens von Bauteilen mit Hilfe örtlicher Beanspruchungen. *Konstruktion* **1993**, *45*, 25–33.
17. Diemar, A.; Thumser, R.; Bergmann, J.W. Statistischer Größeneinfluss und Bauteilfestigkeit. *Mater. Test.* **2004**, *46*, 16–21.
18. Böhm, J.; Heckel, K. Die Vorhersage der Dauerschwingfestigkeit unter Berücksichtigung des statistischen Größeneinflusses. *Mater. Werkst.* **1982**, *13*, 120–128, doi:10.1002/mawe.19820130408. [CrossRef]

19. Kuwazuru, O.; Murata, Y.; Hangai, Y.; Utsunomiya, T.; Kithara, S.; Yoshikawa, N. X-Ray CT Inspection for Porosities and Its Effect on Fatigue of Die Cast Aluminium Alloy. *J. Solid Mech. Mater. Eng.* **2008**, *2*, 1220–1231, doi:10.1299/jmmp.2.1220. [CrossRef]

20. Rotella, A.; Nadot, Y.; Piellard, M.; Augustin, R.; Fleuriot, M. Fatigue limit of a cast Al-Si-Mg alloy (A357-T6) with natural casting shrinkages using ASTM standard X-ray inspection. *Int. J. Fatigue* **2018**, *114*, 177–188, doi:10.1016/j.ijfatigue.2018.05.026. [CrossRef]

21. González, R.; González, A.; Talamantes-Silva, J.; Valtierra, S.; Mercado-Solís, R.D.; Garza-Montes-de Oca, N.F.; Colás, R. Fatigue of an aluminium cast alloy used in the manufacture of automotive engine blocks. *Int. J. Fatigue* **2013**, *54*, 118–126, doi:10.1016/j.ijfatigue.2013.03.018. [CrossRef]

22. González, R.; Martínez, D.I.; González, J.A.; Talamantes, J.; Valtierra, S.; Colás, R. Experimental investigation for fatigue strength of a cast aluminium alloy. *Int. J. Fatigue* **2011**, *33*, 273–278, doi:10.1016/j.ijfatigue.2010.09.002. [CrossRef]

23. *Rechnerischer Festigkeitsnachweis für Maschinenbauteile aus Stahl, Eisenguss- und Aluminiumwerkstoffen*, 6th ed.; FKM-Richtlinie, VDMA-Verl.: Frankfurt am Main, Germany, 2012; Volume 6.

24. Sonsino, C.M.; Ziese, J. Fatigue strength and applications of cast aluminium alloys with different degrees of porosity. *Int. J. Fatigue* **1993**, *15*, 75–84, doi:10.1016/0142-1123(93)90001-7. [CrossRef]

25. Weibull, W. Zur Abhängigkeit der Festigkeit von der Probengröße. *Ingenieur-Archiv* **1959**, *28*, 360–362. [CrossRef]

26. Gänser, H.P. Some notes on gradient, volumetric and weakest link concepts in fatigue. *Comput. Mater. Sci.* **2008**, *44*, 230–239, doi:10.1016/j.commatsci.2008.03.021. [CrossRef]

27. Abroug, F.; Pessard, E.; Germain, G.; Morel, F.; Hénaff, G. Fatigue size effect due to defects in an AA7050 alloy. In Proceedings of the MATEC Web Conference, Poitiers, France, 27 May—1 June 2018; Volume 165, p. 14015. [CrossRef]

28. Abroug, F.; Pessard, E.; Germain, G.; Morel, F. A probabilistic approach to study the effect of machined surface states on HCF behavior of a AA7050 alloy. *Int. J. Fatigue* **2018**, *116*, 473–489, doi:10.1016/j.ijfatigue.2018.06.048. [CrossRef]

29. Kitagawa, H.; Takahashi, S. Applicability of fracture mechanics to very small cracks or the cracks in the early stage. In Proceedings of the Second International Conference on Mechanical Behavior of Materials, Boston, MA, USA, 16–20 August 1976; pp. 627–631.

30. Garb, C.; Leitner, M.; Stauder, B.; Schnubel, D.; Grün, F. Application of modified Kitagawa-Takahashi diagram for fatigue strength assessment of cast Al-Si-Cu alloys. *Int. J. Fatigue* **2018**, doi:10.1016/j.ijfatigue.2018.01.030. [CrossRef]

31. Tenkamp, J.; Koch, A.; Knorre, S.; Krupp, U.; Michels, W.; Walther, F. Defect-correlated fatigue assessment of A356-T6 aluminum cast alloy using computed tomography based Kitagawa-Takahashi diagrams. *Int. J. Fatigue* **2018**, *108*, 25–34, doi:10.1016/j.ijfatigue.2017.11.003. [CrossRef]

32. Roy, M.J.; Nadot, Y.; Nadot-Martin, C.; Bardin, P.G.; Maijer, D.M. Multiaxial Kitagawa analysis of A356-T6. *Int. J. Fatigue* **2011**, *33*, 823–832, doi:10.1016/j.ijfatigue.2010.12.011. [CrossRef]

33. Beretta, S.; Romano, S. A comparison of fatigue strength sensitivity to defects for materials manufactured by AM or traditional processes. *Int. J. Fatigue* **2017**, *94*, 178–191, doi:10.1016/j.ijfatigue.2016.06.020. [CrossRef]

34. Benedetti, M.; Santus, C. Building the Kitagawa-Takahashi diagram of flawed materials and components using an optimized V-notched cylindrical specimen. *Eng. Fract. Mech.* **2020**, *224*, 106810, doi:10.1016/j.engfracmech.2019.106810. [CrossRef]

35. Patriarca, L.; Beretta, S.; Foletti, S.; Riva, A.; Parodi, S. A probabilistic framework to define the design stress and acceptable defects under combined-cycle fatigue conditions. *Eng. Fract. Mech.* **2020**, *224*, 106784, doi:10.1016/j.engfracmech.2019.106784. [CrossRef]

36. Poulin, J.R.; Kreitcberg, A.; Terriault, P.; Brailovski, V. Fatigue strength prediction of laser powder bed fusion processed Inconel 625 specimens with intentionally-seeded porosity: Feasibility study. *Int. J. Fatigue* **2020**, *132*, 105394, doi:10.1016/j.ijfatigue.2019.105394. [CrossRef]

37. Radaj, D. Geometry correction for stress intesity at elliptical cracks. *Weld. Cut.* **1977**, *29*, 198–402.

38. El Haddad, M.H.; Smith, K.N.; Topper, T.H. Fatigue Crack Propagation of Short Cracks. *J. Eng. Mater. Technol.* **1979**, *101*, 42, doi:10.1115/1.3443647. [CrossRef]

39. El Haddad, M.H.; Topper, T.H.; Smith, K.N. Prediction of non propagating cracks. *Eng. Fract. Mech.* **1979**, *11*, 573–584, doi:10.1016/0013-7944(79)90081-X. [CrossRef]

40. Chapetti, M.D. Fatigue propagation threshold of short cracks under constant amplitude loading. *Int. J. Fatigue* **2003**, *25*, 1319–1326, doi:10.1016/S0142-1123(03)00065-3. [CrossRef]

41. Maierhofer, J.; Pippan, R.; Gänser, H.P. Modified NASGRO equation for physically short cracks. *Int. J. Fatigue* **2014**, *59*, 200–207, doi:10.1016/j.ijfatigue.2013.08.019. [CrossRef]

42. Zerbst, U.; Vormwald, M.; Pippan, R.; Gänser, H.P.; Sarrazin-Baudoux, C.; Madia, M. About the fatigue crack propagation threshold of metals as a design criterion—A review. *Eng. Fract. Mech.* **2016**, *153*, 190–243, doi:10.1016/j.engfracmech.2015.12.002. [CrossRef]

43. Pippan, R.; Hohenwarter, A. Fatigue crack closure: A review of the physical phenomena. *Fatigue Fract. Eng. Mater. Struct.* **2017**, *40*, 471–495, doi:10.1111/ffe.12578. [CrossRef]

44. Suresh, S.; Ritchie, R.O. A geometric model for fatigue crack closure induced by fracture surface roughness. *Metall. Trans. A* **1982**, *13*, 1627–1631, doi:10.1007/BF02644803. [CrossRef]

45. Wasén, J.; Heier, E. Fatigue crack growth thresholds—The influence of Young's modulus and fracture surface roughness. *Int. J. Fatigue* **1998**, *20*, 737–742, doi:10.1016/S0142-1123(98)00034-6. [CrossRef]

46. Kim, J.H.; Lee, S.B. Behavior of plasticity-induced crack closure and roughness-induced crack closure in aluminum alloy. *Int. J. Fatigue* **2001**, *23*, 247–251, doi:10.1016/S0142-1123(01)00155-4. [CrossRef]

47. Suresh, S.; Zamiski, G.F.; Ritchie, D.R.O. Oxide-Induced Crack Closure: An Explanation for Near-Threshold Corrosion Fatigue Crack Growth Behavior. *Metall. Mater. Trans. A* **1981**, *12*, 1435–1443, doi:10.1007/BF02643688. [CrossRef]

48. Newman, J.A.; Piascik, R.S. Interactions of plasticity and oxide crack closure mechanisms near the fatigue crack growth threshold. *Int. J. Fatigue* **2004**, *26*, 923–927, doi:10.1016/j.ijfatigue.2004.02.001. [CrossRef]

49. Lados, D.; Apelian, D.; Paris, P.; Donald, J. Closure mechanisms in Al–Si–Mg cast alloys and long-crack to small-crack corrections. *Int. J. Fatigue* **2005**, *27*, 1463–1472, doi:10.1016/j.ijfatigue.2005.06.013. [CrossRef]

50. Murakami, Y.; Endo, M. Effects of defects, inclusions and inhomogeneities on fatigue strength. *Int. J. Fatigue* **1994**, *16*, 163–182, doi:10.1016/0142-1123(94)90001-9. [CrossRef]

51. Li, P.; Lee, P.D.; Maijer, D.M.; Lindley, T.C. Quantification of the interaction within defect populations on fatigue behavior in an aluminum alloy. *Acta Mater.* **2009**, *57*, 3539–3548, doi:10.1016/j.actamat.2009.04.008. [CrossRef]

52. Aigner, R.; Garb, C.; Leitner, M.; Stoschka, M.; Grün, F. Application of a $\sqrt{area}$ -Approach for Fatigue Assessment of Cast Aluminum Alloys at Elevated Temperature. *Metals* **2018**, *8*, 1033, doi:10.3390/met8121033. [CrossRef]

53. Aigner, R.; Leitner, M.; Stoschka, M.; Hannesschläger, C.; Wabro, T.; Ehart, R. Modification of a Defect-Based Fatigue Assessment Model for Al-Si-Cu Cast Alloys. *Materials* **2018**, *11*, 2546, doi:10.3390/ma11122546. [CrossRef]

54. Gnedenko, B. Sur la distribution limite du terme maximum d'une serie aleatoire. *Ann. Math.* **1943**, *44*, 423–453.

55. Jenkinson, A.F. The frequency distribution of the annual maximum (or minimum) values of meteorological elements. *Q. J. R. Meteorol. Soc.* **1955**, *87*, 145–158, doi:10.1002/qj.49708134804. [CrossRef]

56. Beretta, S.; Murakami, Y. Statistical analysis of defects for fatigue strength prediction and quality control of materials. *Fatigue Fract. Eng. Mater. Struct.* **1998**, *21*, 1049–1065, doi:10.1046/j.1460-2695.1998.00104.x. [CrossRef]

57. Mahdi.; Cenac, S.; Myrtene. Estimating Parameters of Gumbel Distribution using the Methods of Moments, probability weighted moments and maximum likelihood. *Rev. Mat. Teor. Apl.* **2005**, *12*, 151–156.

58. Gumbel, E.J. *Statistics of Extremes*; Columbia University Press: New York, NY, USA, 1958.

59. Murakami, Y. *Metal Fatigue: Effects of Small Defects and Nonmetallic Inclusions*, 1th ed.; Elsevier: Amsterdam, The Netherlands, 2002.

60. Murakami, Y. Material defects as the basis of fatigue design. *Int. J. Fatigue* **2012**, *41*, 2–10, doi:10.1016/j.ijfatigue.2011.12.001. [CrossRef]

61. Feikus, F.J.; Bernsteiner, P.; Gutiérrez, R.F.; Łuszczak, M. Weiterentwicklungen bei Gehäusen von Elektromotoren. *MTZ-Motortechnische Zeitschrift* **2020**, *81*, 42–47, doi:10.1007/s35146-019-0180-5. [CrossRef]

62. Campbell, J. *Complete Casting Handbook//Complete Casting Handbook: Metal Casting Processes, Metallurgy, Techniques and Design*, 2nd ed.; Elsevier: Amsterdam, The Netherlands, 2015; doi:10.1016/C2014-0-01548-1. [CrossRef]

63. DIN EN 1706. *Aluminium and Aluminium Alloys–Castings—Chemical Composition and Mechanical Properties*; German version EN 1706:2010; Beuth: Berlin, Germany

64. Sjölander, E.; Seifeddine, S. The heat treatment of Al–Si–Cu–Mg casting alloys. *J. Mater. Process. Technol.* **2010**, *210*, 1249–1259, doi:10.1016/j.jmatprotec.2010.03.020. [CrossRef]

65. Yang, H.; Ji, S.; Fan, Z. Effect of heat treatment and Fe content on the microstructure and mechanical properties of die-cast Al–Si–Cu alloys. *Mater. Des.* **2015**, *85*, 823–832, doi:10.1016/j.matdes.2015.07.074. [CrossRef]

66. Costa, A.T.; Dias, M.; Gomes, G.L.; Rocha, O.L.; Garcia, A. Effect of solution time in T6 heat treatment on microstructure and hardness of a directionally solidified Al–Si–Cu alloy. *J. Alloys Compd.* **2016**, *683*, 485–494, doi:10.1016/j.jallcom.2016.05.099. [CrossRef]

67. Toschi, S. Optimization of A354 Al-Si-Cu-Mg Alloy Heat Treatment: Effect on Microstructure, Hardness, and Tensile Properties of Peak Aged and Overaged Alloy. *Metals* **2018**, *8*, 961, doi:10.3390/met8110961. [CrossRef]

68. Han, Y.; Samuel, A.M.; Doty, H.W.; Valtierra, S.; Samuel, F.H. Optimizing the tensile properties of Al–Si–Cu–Mg 319-type alloys: Role of solution heat treatment. *Mater. Des.* **2014**, *58*, 426–438, doi:10.1016/j.matdes.2014.01.060. [CrossRef]

69. Ceschini, L.; Morri, A.; Toschi, S.; Seifeddine, S. Room and high temperature fatigue behaviour of the A354 and C355 (Al–Si–Cu–Mg) alloys: Role of microstructure and heat treatment. *Mater. Sci. Eng. A* **2016**, *653*, 129–138, doi:10.1016/j.msea.2015.12.015. [CrossRef]

70. Samuel, A.M.; Doty, H.W.; Valtierra, S.; Samuel, F.H. Relationship between tensile and impact properties in Al–Si–Cu–Mg cast alloys and their fracture mechanisms. *Mater. Des.* **2014**, *53*, 938–946, doi:10.1016/j.matdes.2013.07.021. [CrossRef]

71. do Lee, C. Effect of T6 heat treatment on the defect susceptibility of fatigue properties to microporosity variations in a low-pressure die-cast A356 alloy. *Mater. Sci. Eng. A* **2013**, *559*, 496–505, doi:10.1016/j.msea.2012.08.131. [CrossRef]

72. Zhu, M.; Jian, Z.; Yang, G.; Zhou, Y. Effects of T6 heat treatment on the microstructure, tensile properties, and fracture behavior of the modified A356 alloys. *Mater. Des.* **2012**, *36*, 243–249, doi:10.1016/j.matdes.2011.11.018. [CrossRef]

73. Boileau, J.M.; Allison, J.E. The effect of solidification time and heat treatment on the fatigue properties of a cast 319 aluminum alloy. *Metall. Trans.* **2003**, *34*, 1807–1820, doi:10.1007/s11661-003-0147-4. [CrossRef]

74. Fabrizi, A.; Capuzzi, S.; de Mori, A.; Timelli, G. Effect of T6 Heat Treatment on the Microstructure and Hardness of Secondary AlSi9Cu3(Fe) Alloys Produced by Semi-Solid SEED Process. *Metals* **2018**, *8*, 750, doi:10.3390/met8100750. [CrossRef]

75. Aigner, R.; Leitner, M.; Stoschka, M. On the mean stress sensitivity of cast aluminium considering imperfections. *Mater. Sci. Eng. A* **2019**, *758*, 172–184, doi:10.1016/j.msea.2019.04.119. [CrossRef]

76. Leitner, M.; Garb, C.; Remes, H.; Stoschka, M. Microporosity and statistical size effect on the fatigue strength of cast aluminium alloys EN AC-45500 and 46200. *Mater. Sci. Eng. A* **2017**, *707*, 567–575, doi:10.1016/j.msea.2017.09.023. [CrossRef]

77. Vandersluis, E.; Ravindran, C. Comparison of Measurement Methods for Secondary Dendrite Arm Spacing. *Metallogr. Microstruct. Anal.* **2017**, *6*, 89–94, doi:10.1007/s13632-016-0331-8. [CrossRef]

78. Boileau, J.M.; Zindel, J.W.; Allison, J.E. The Effect of Solidification Time on the Mechanical Properties in a Cast A356-T6 Aluminum Alloy. *SAE Trans.* **1997**, *106*, 63–74.

79. Zhang, L.Y.; Jiang, Y.H.; Ma, Z.; Shan, S.F.; Jia, Y.Z.; Fan, C.Z.; Wang, W.K. Effect of cooling rate on solidified microstructure and mechanical properties of aluminium-A356 alloy. *J. Mater. Process. Technol.* **2008**, *207*, 107–111, doi:10.1016/j.jmatprotec.2007.12.059. [CrossRef]

80. Ceschini, L.; Boromei, I.; Morri, A.; Seifeddine, S.; Svensson, I.L. Microstructure, tensile and fatigue properties of the Al–10%Si–2%Cu alloy with different Fe and Mn content cast under controlled conditions. *J. Mater. Process. Technol.* **2009**, *209*, 5669–5679, doi:10.1016/j.jmatprotec.2009.05.030. [CrossRef]

81. Zhang, B.; Chen, W.; Poirier, D.R. Effect of solidification cooling rate on the fatigue life of A356.2-T6 cast aluminium alloy. *Fatigue Fract. Eng. Mater. Struct.* **2000**, *23*, 417–423, doi:10.1046/j.1460-2695.2000.00299.x. [CrossRef]

82. Gerbe, S.; Krupp, U.; Michels, W. Influence of secondary dendrite arm spacing (SDAS) on the fatigue properties of different conventional automotive aluminum cast alloys. *Frat. Integrità Strutt.* **2019**, *13*, 105–115, doi:10.3221/IGF-ESIS.48.13. [CrossRef]

83. Leitner, H. Simulation des Ermüdungsverhaltens von Aluminiumgusslegierungen. Ph.D. Thesis, Montanuniversität Leoben, Leoben, Austria, 2001.

84. Garb, C.; Leitner, M.; Grün, F. Application of $\sqrt{area}$-concept to assess fatigue strength of AlSi7Cu0.5Mg casted components. *Eng. Fract. Mech.* **2017**, *185*, 61–71. doi:10.1016/j.engfracmech.2017.03.018. [CrossRef]

85. ASTM International E 739. Standard Practice for Statistical Analysis of Linear or Linearized Stress-Life (S-N) and Strain Life (E-N) Fatigue Data, ASTM International: West Conshohocken, PA, 1998.

86. Dengel, D.; Harig, H. Estimation of the fatigue limit by progressively-increasing load tests. *Fatigue Fract. Eng. Mater. Struct.* **1980**, *3*, 113–128, doi:10.1111/j.1460-2695.1980.tb01108.x. [CrossRef]

87. Garb, C.; Leitner, M.; Grün, F. Effect of elevated temperature on the fatigue strength of casted AlSi8Cu3 aluminium alloys. *Procedia Struct. Integr.* **2017**, *7*, 497–504, doi:10.1016/j.prostr.2017.11.118. [CrossRef]

88. Wang, Q.; Apelian, D.; Lados, D. Fatigue behavior of A356/357 aluminum cast alloys. Part II—Effect of microstructural constituents. *J. Light Met.* **2001**, *1*, 85–97, doi:10.1016/S1471-5317(00)00009-2. [CrossRef]

89. Wang, Q.; Apelian, D.; Lados, D. Fatigue behavior of A356-T6 aluminum cast alloys. Part I—Effect of casting defects. *J. Light Met.* **2001**, *1*, 73–84, doi:10.1016/S1471-5317(00)00008-0. [CrossRef]

90. Åman, M.; Okazaki, S.; Matsunaga, H.; Marquis, G.B.; Remes, H. The effect of interacting small defects on the fatigue limit of a medium carbon steel. *Procedia Struct. Integr.* **2016**, *2*, 3322–3329, doi:10.1016/j.prostr.2016.06.414. [CrossRef]

91. Massey, F.J., Jr. The Kolmogorov-Smirnov test for goodness of fit. *J. Am. Stat. Assoc.* **1951**, *46*, 68–78.

Article

Fatigue Assessment of Wire and Arc Additively Manufactured Ti-6Al-4V

Sebastian Springer [1,*], Martin Leitner [2], Thomas Gruber [3], Bernd Oberwinkler [3], Michael Lasnik [3] and Florian Grün [1]

[1] Chair of Mechanical Engineering, Montanuniversität Leoben, 8700 Leoben, Austria; florian.gruen@unileoben.ac.at
[2] Institute of Structural Durability and Railway Technology, Graz University of Technology, 8010 Graz, Austria; martin.leitner@tugraz.at
[3] voestalpine BÖHLER Aerospace GmbH & Co KG, 8605 Kapfenberg, Austria; thomas.gruber@voestalpine.com (T.G.); bernd.oberwinkler@voestalpine.com (B.O.); michael.lasnik@voestalpine.com (M.L.)
* Correspondence: sebastian.springer@unileoben.ac.at; Tel.: +43-3842-402-1430

Abstract: Wire and arc additively manufactured (WAAM) parts and structures often present internal defects, such as gas pores, and cause irregularities in the manufacturing process. In order to describe and assess the effect of internal defects in fatigue design, this research study investigates the fatigue strength of wire arc additive manufactured structures covering the influence of imperfections, particularly gas pores. Single pass WAAM structures are manufactured using titanium alloy Ti-6Al-4V and round fatigue, tensile specimen are extracted. Tensile tests and uniaxial fatigue tests with a load stress ratio of $R = 0.1$ were carried out, whereby fatigue test results are used for further assessments. An extensive fractographic and metallographic fracture surface analysis is utilized to characterize and measure crack-initiating defects. As surface pores as well as bulk pores are detected, a stress intensity equivalent ΔK_{eqv} transformation approach is presented in this study. Thereby, the defect size of the surface pore is transformed to an increased defect size, which is equivalent to a bulk pore. Subsequently, the fatigue strength assessment method by Tiryakioğlu, commonly used for casting processes, is applied. For this method, a cumulative Gumbel extreme value distribution is utilized to statistically describe the defect size. The fitted distribution with modified data reveals a better agreement with the experimental data than unmodified. Additionally, the validation of the model shows that the usage of the ΔK modified data demonstrates better results, with a slight underestimation of up to about -7%, compared to unmodified data, with an overestimation of up to about 14%, comparing the number of load cycles until failure. Hence, the presented approach applying a stress intensity equivalent transformation of surface to bulk pores facilitates a sound fatigue strength assessment of WAAM Ti-6Al-4V structures.

Keywords: wire arc additive manufacturing; fatigue assessment; Ti-6Al-4V; defects; statistical distribution

Citation: Springer, S.; Leitner, M.; Gruber, T.; Oberwinkler, B.; Lasnik, M.; Grün, F. Fatigue Assessment of Wire and Arc Additively Manufactured Ti-6Al-4V. *Metals* **2022**, *12*, 795. https://doi.org/10.3390/met12050795

Academic Editor: Yongho Sohn

Received: 28 March 2022
Accepted: 28 April 2022
Published: 4 May 2022

1. Introduction

Novel, innovative manufacturing technologies, such as additive manufacturing, have the potential to become a time- and cost-efficient method of producing more or less complex high-tech parts and structures using expensive and hard-to-manufacture materials [1–4]. In contrast to commonly used powder-bed processes, such as selective laser melting, wire-based technologies offer high deposition rates and high material utilization. Wire arc additive manufacturing is one possible wire-based AM technique particularly used for large components, where the deposition material is fed as a wire, melted by means of an electrical arc and added layer-by-layer on a substrate.

Due to its benefits regarding a buy-to-fly optimization and lightweight design potential, the titanium alloy Ti-6Al-4V is of interest in the aerospace industry [5–8]. Despite the main advantages and the potentials of WAAM, technological challenges are currently under investigation. For example, some challenges are process stability, component design, formation and simulative prediction of residual stresses and distortion and the mechanical material behavior itself [8–10]. Reproducible quasi-static mechanical and fatigue properties are an essential requirement for the use of additively manufactured components in aerospace industry and, therefore, need to be investigated in detail. Studies on the fatigue performance of additively manufactured parts pointed out a comparably large scatter in fatigue life and a reduced fatigue strength compared to conventional manufactured, e.g., forged parts, due to process induced defects, such as pores, as-built surfaces and a different microstructure [8,11–13]. Failure critical aerospace components are always machined to avoid a critical, rough surface. Due to the machining of the rough surface, crack initiating defects are found to be on inner areas or surfaces near pores [14].

Preliminary studies showed that the fatigue strength of additively manufactured structures is reduced with the presence of defects to the same extent as traditionally manufactured structures. Basically, the origin of crack initiation does not differ to conventional processes and fatigue failure should occur at the largest defect in the tested volume. Fatigue assessment methods developed for traditional processes, such as casting and using the statistics of extremes, seem to be applicable for AM parts. Therefore, concepts for conventional manufacturing processes using extreme value statistics can be adopted and used for determination of the fatigue strength of AM parts [15,16]. In [12,17–19], different statistical extreme value distributions are utilized for the occurring defects in additively manufactured parts and structures.

The objective of this study is to investigate the finite fatigue strength of additively manufactured Ti-6Al-4V structures with the presence of process induced defects. Three structures are manufactured and fatigue tests are carried out of this material followed by a holistic fracture surface analysis. The Gumbel extreme value distribution is fitted to the size of the observed failure critical pores/defects. In order to compare surface defects to inner defects, a concept for the transformation of surface pores to inner pores is utilized using a stress intensity equivalent approach. Finally, the proposed assessment method of Tiryakioğlu is applied and validated with experimental data from fatigue tests out of wire arc additive manufacturing structures.

The scientific contribution of this paper is to extensively investigate the impact of porosity on the finite fatigue strength of additively manufactured Ti-6Al-4V structures, the transformation of surface pores to inner pores by an stress intensity equivalent approach and the application and validation of the fatigue assessment methodology by Tiryakioğlu for material out of a Ti-6Al-4V WAAM structure.

2. Materials and Methods

2.1. Material and Manufacturing

The investigated material within this study is the titanium alloy Ti-6Al-4V, which is commonly used for parts and structures in aerospace industry. Ti-6Al-4V is used for the wire as well as for the substrate. The substrates material condition was conventionally processed by hot-rolling. The nominal chemical composition out of the suppliers data sheets for the wire and the substrate is given in Table 1.

Table 1. Nominal chemical composition of the used material in weight %, comparable to [20,21].

	Al	V	Fe	O	N	C	H	Ti
Substrate	5.50–6.75	3.50–4.50	<0.30	<0.20	<0.05	<0.08	<0.015	Balance
Wire	6.00	4.00	<0.15	0.18	<0.03	<0.05	<0.01	Balance

In structure manufacturing, a diameter 1.2 mm wire was added layer-by-layer on a bolt clamped substrate with the dimensions of $250 \times 150 \times 12.4$ mm^3. The process is carried out using a Fronius Cold Metal Transfer welding machine combined with a

Yaskawa Motoman welding robot, see Figure 1. In order to protect the weld pool and the heat affected zone from oxidation and other undesirable gaseous elements, a local trailing shield purged with 99.9999% pure Argon was used. In Table 2, the process parameters used for the manufacture of the pieces are shown. Dwell times between layers are defined in order to ensure a constant interpass temperature of about 150 °C.

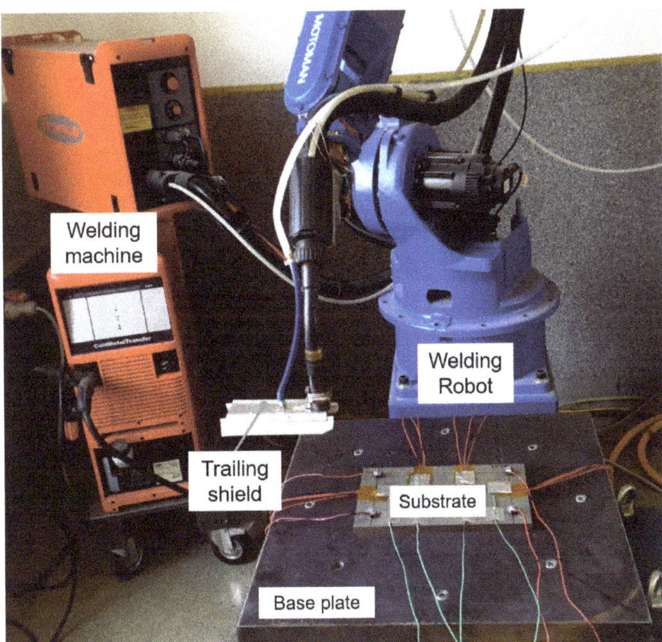

Figure 1. Experimental setup for additive manufacturing.

Table 2. Used WAAM building parameters.

Process Parameter	Unit	Value
Current	A	92.0
Voltage	V	18.1
Travel speed	mm/s	2.5
Wire feed speed	m/min	5.0
Wire diameter	mm	1.2

In total, three structures are built, with each structure consisting of twenty layers with a total length of about 130 mm in deposition direction. These walls are manufactured utilizing a single layer strategy without oscillation and with the same deposition direction in each layer. The overall measured height of the WAAM structure is about 110 mm, and the effective wall thickness is measured to be about 8.5 mm. An exemplary twenty layer WAAM structure is shown in Figure 2.

Quasi-static tensile and fatigue test specimens are extracted horizontally, and it is parallel to the deposition direction (see Figure 2). The positions of specimens in building direction are defined in order to neglect any starting effects related to, e.g., heating up the base plate and, hence, obtaining a comparable microstructure within all specimens, starting with extraction above the sixth layer. To statistically evaluate the mechanical material properties and the fatigue strength of this AM material, in total 19 fatigue and six tensile specimen are cut out of the three manufactured WAAM structures. By polishing the surface of the extracted and machined test specimen before testing, any influence of a surface roughness on the experimental investigations can be avoided. Tensile and fatigue tests were conducted at room temperature. The geometry of the tensile test specimens (see Figure 3a), is defined according to the requirements of standard EN ISO 6821 [22] with an initial test length of 25 mm. An uniaxial, servo-hydraulic cylinder from Instron-Schenk with a maximum load capacity of 25 kN is used for the tensile tests. All tests are strain-rate controlled until failure of the specimen and carried out with an touching extensometer and a strain rate of $2.5 \times 10^{-3}\,\mathrm{s}^{-1}$.

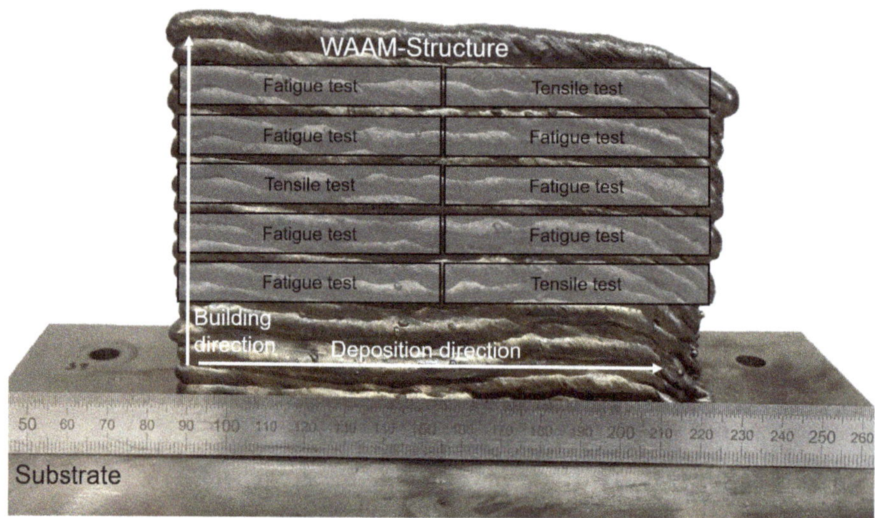

Figure 2. Additively manufactured structure for experiments and schematic specimen positions.

The fatigue tests are utilized on a resonant test machine RUMUL Microtron with a resonance testing frequency of about 130 Hz. The dimensions of the investigated round fatigue specimen geometry are illustrated in Figure 3b, with a diameter of 4 mm in the test area. Specimens are cyclically tested in a pulsating tension load range with a load stress ratio of $R = 0.1$. The abort criterion was set to total fracture of the specimen or a defined number of ten million load-cycles without failure. This study focuses on the finite life region; thus, the applied load for the fatigue tests is relatively high. After fatigue tests, an extensive fracture surface analysis was carried out to evaluate the failure origin using a digital optical microscope.

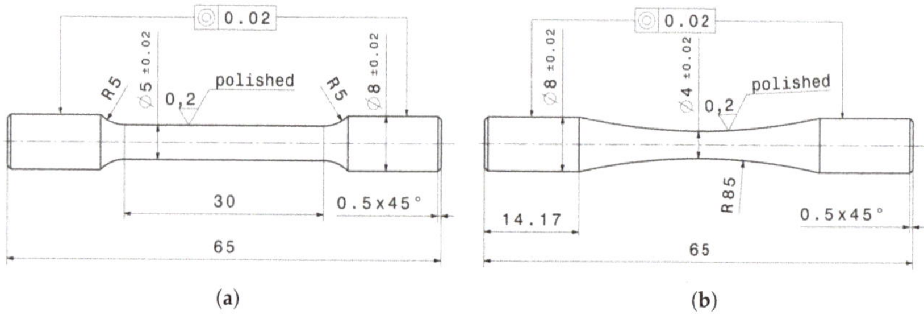

Figure 3. Geometry of test specimen in [mm]. (**a**) Quasi-static tensile test. (**b**) Fatigue test.

2.2. Assessment Methodology

In order to determine the survival probability of components, it is of utmost importance to consider local pore size distributions in the fatigue assessment process. Different statistical distributions were investigated in [23,24] for crack initiating defects in cast materials. These studies revealed that the defect size can be well described by extreme value statistics using the extreme value distribution of type one, which is also called Gumbel distribution and enables the assessment of the largest values of the distribution [25]. Better fit results are found using a Gumbel extreme value distribution compared to commonly used lognormal or Weibull distributions [26,27]. Further studies [28,29] for cast materials revealed a good agreement of the assessed fatigue strength to the experimental data with the proposed methodology according to Tiryakioğlu. To quantify the size of irregularities or pores, the projected area is measured. A relationship between the projected defect area A_{proj} and the equivalent diameter d_{eqv} can be determined using Equation (1), whereby d_{eqv} equals the diameter of a circle, which covers the same area as the defect itself. The size of the critical defects is commonly measured subsequently after fatigue testing by investigations of the fractured surfaces or non-destructive by using X-ray tomography before fatigue testing. At the latter, the minimum detectable defect size has to be considered.

$$d_{eqv} = \sqrt{\frac{4}{\pi} \cdot A_{proj}} \tag{1}$$

As mentioned before, the Gumbel distribution can be used to describe the distribution of defects, with the cumulative Gumbel probability P for a defined equivalent defect diameter is given in Equation (2) [25]. There, λ is a pore size dependent location and δ a scale parameter.

$$P(d_{eqv}) = \exp\left[-\exp\left(\frac{d_{eqv} - \lambda}{\delta}\right)\right] \tag{2}$$

In order to link the cumulative defect distribution of the failure initiating defect size to the finite fatigue life of parts and structures containing defects, Tiryakioğlu suggests a new methodology. In this fatigue assessment methodology, the crack propagation law of Paris–Erodgan [30] for stable crack growth is combined to the cumulative Gumbel defect distribution (see Equation (3)) [23]. Based on this methodology, failure probability P_f can be determined dependent on a specified number of load cycles until failure N_f. Because cracks from structural defects start to grow immediately after the first cycle, the number of cycles to initiate a crack is set to 0 in this study [23,31].

$$P_f(N_f) = 1 - \exp\left\{-\exp\left[\frac{\lambda}{\delta} - \frac{2}{\delta \cdot \sqrt{\pi}}\left(\frac{N_f}{B \cdot \sigma_a^{-m}}\right)^{\frac{2}{2-m}}\right]\right\} \tag{3}$$

Within Equation (3), λ and δ are the previously introduced location and scale parameters of the Gumbel extreme value distribution, σ_a is the nominal stress amplitude and B and m are material dependent constants. Crack propagation slope is represented by m in the stable crack growth region and B is a offset parameter. The survival probability P_s can be calculated by the following.

$$P_s(N_f) = 1 - P_f(N_f) \tag{4}$$

3. Results

3.1. Quasi-Static and Fatigue Test Results

Quasi-static tests are carried out to investigate the mechanical properties of the WAAM material at room temperature. As mentioned in the previous section, all tensile tests are strain controlled. In Figure 4, a representative stress–strain curve of a horizontally extracted tensile test is presented. The exemplary stress–strain curve shows a distinctive plastic deformation with a small amount of work hardening.

All important and necessary mechanical properties are subsequently determined out of the experimental data using a common guideline out of the standard EN ISO 6892-1 for tensile tests [22]. The experimental data of the six static tests are afterward statistically analyzed using simple descriptive statistics. The results of the strain controlled tensile tests are summarized in Table 3. At room temperature, mean YS is found to be about 867 MPa, the mean UTS equals 957 MPa and A equals 6%. A comparison of data from the literature [32–35] with the material tested in this paper shows a good agreement of the mechanical properties and lays in the same range (see Table 3).

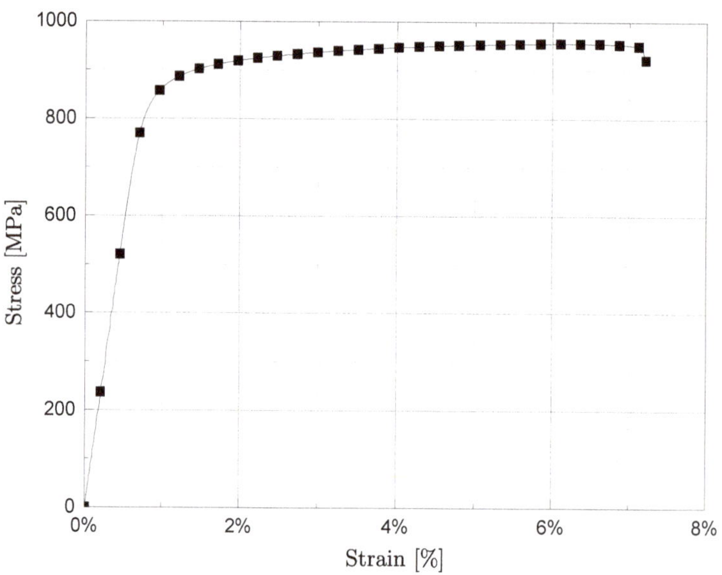

Figure 4. Representative tensile test with specimen extracted in horizontal direction.

Table 3. Summarized mechanical properties of tensile tests extracted in horizontal directions.

Condition	Orientation	UTS [MPa]	YS [MPa]	A (%)	References
As-built	Horizontal	957 ± 4	867 ± 16	6 ± 1	This study
As-built	Horizontal	923–995	840–909	5–11	[32–35]

As mentioned in the previous section, uniaxial fatigue tests are utilized on a resonant testing rig. The stress amplitude σ_a for the fatigue tests is defined in accordance with the stress–strain curves. Additionally, all specimen are tested in the finite life region in order to assess the fatigue strength with the method of Tiryakioğlu. The experimental fatigue test points in the finite life region are statistically evaluated by applying the standard by ASTM E 739 [36] to determine the S-N curves for the survival probabilities of $P_s = 10\%$, $P_s = 50\%$ and $P_s = 90\%$. In Figure 5, the fatigue test results and the statistically evaluated S-N curves for different survival probabilities are presented.

The statistically estimated S-N parameters are given in Table 4, whereby the evaluated S-N curve reveals a stress amplitude of $\sigma_a = 261.6$ MPa for $P_s = 50\%$ at $N_f = 1 \times 10^5$ load cycles. Additionally, the comparably high slope of the S-N curve in the finite life region is found to be about $k = 4.4$, which is in a good agreement with data from the literature [37] and proves the presence of process-induced, crack-initiating defects. In accordance to the common definition in [38], the scatter bands for stress $1:T_S$ and load cycles $1:T_N$ of the S-N curve are evaluated, whereby the ratio of $P_s = 10\%$ to $P_s = 90\%$ is used. Moreover, the evaluated scatter index reveals the presence of defects, comparable to [37].

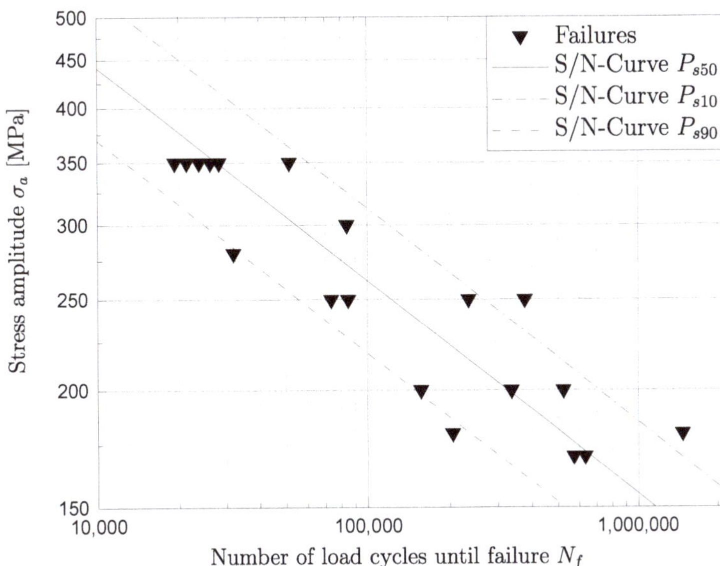

Figure 5. Experimental fatigue test results.

Table 4. Evaluated fatigue test data.

Condition	Orientation	$\sigma_{a,P_{s50},N_1 \times 10^5}$ [MPa]	Slope k	1:T_S	1:T_N
As-built	Horizontal	261.6	4.40	1.43	4.78

3.2. Fractography

After fatigue testing, all specimens were analyzed by means of an extensive fracture surface analysis using a digital light microscope. Within the analysis, the crack initiating defect of each tested specimen is evaluated and additionally relevant geometric defect parameters, such as area, diameter and position are measured and subsequently determined. In cases of all specimens, gas pores were found as failure critical defects. On the one hand, pores were located inside the specimen; on the other hand, pores at the surface were found. In Figure 6, a characteristic gas pore inside of a specimen is presented, and in Figure 7, a representative gas pore at the surface is shown. A summary of all pores is given in Table 5, whereby the position of the pore is presented as well as the projected area A_{proj}, the estimated equivalent diameter d_{eqv}, calculated according to Equation (1), and the stress intensity equivalent defect diameter $d_{\Delta K,eqv}$ are shown.

Crack initiating pores are found in the bulk material as well as at the surface of specimens. In fact, surface near pores are more critical than bulk pores in terms of a higher stress intensity factor [15,39]. Experimental data are often divided into two sample batches dependent on the position of the failure. Furthermore, fatigue assessment methods are applied separately for each sample batch dependent on position of the defect. To assess the fatigue strength of the WAAM material by using one approach for bulk and surface pores, a stress intensity equivalent ΔK_{eqv} transformation is applied. This approach is based on the stress intensity, which can be determined using Equation (5) [40]. In Equation (5), σ_a is the applied stress amplitude, Y is a defect based geometry factor and a is the defect or initial crack size.

$$\Delta K = 2 \cdot \sigma_a \cdot Y \cdot \sqrt{\pi \cdot a} \tag{5}$$

It is assumed that the transformed inner pore has the same stress intensity as the surface pore; therefore, the stress intensity of the surface pore $\Delta K_{Surface,max}$ equals the stress intensity of the inner pore within the bulk material $\Delta K_{Bulk,max}$.

$$\Delta K_{Surface,max} = \Delta K_{Bulk,max} \tag{6}$$

The insertion of Equation (5) into Equation (6) and the reduction in the constant stress amplitude σ_a leads to Equation (7).

$$Y_{Surface,max} \cdot \sqrt{\pi \cdot d_{eqv}} = Y_{Bulk,max} \cdot \sqrt{\pi \cdot d_{\Delta K,eqv}} \tag{7}$$

After rearranging Equation (7), the stress intensity equivalent diameter $d_{\Delta K,eqv}$ can be determined dependent on the square of the ratio between the geometry factor of a surface pore $Y_{Surface,max}$ to the geometry factor of a bulk pore $Y_{Bulk,max}$ as shown in Equation (8).

$$d_{\Delta K,eqv} = \left(\frac{Y_{Surface,max}}{Y_{Bulk,max}} \right)^2 \cdot d_{eqv} \tag{8}$$

The scheme of the stress intensity equivalent ΔK_{eqv} approach for the transformation of a surface pore to an inner pore is illustrated in Figure 8. To sum this methodology up, when a pore at the surface is detected, the projected area A_{proj} is measured and the equivalent diameter d_{eqv} calculated in the same manner as for a bulk pore. Equivalent diameter d_{eqv} is then transformed to $d_{\Delta K,eqv}$ with Equation (8). The geometry factors for the surface pore $Y_{Surface,max}$ and the bulk pore $Y_{Bulk,max}$ are taken out of the literature and are estimated based on the literature data for surface pore $Y_{Surface,max} = 0.75$ and for bulk pore $Y_{Bulk,max} = \frac{2}{\pi}$ [41]. Overall, the transformation factor is determined to be $\left(\frac{Y_{Surface,max}}{Y_{Bulk,max}} \right)^2 = 1.39$.

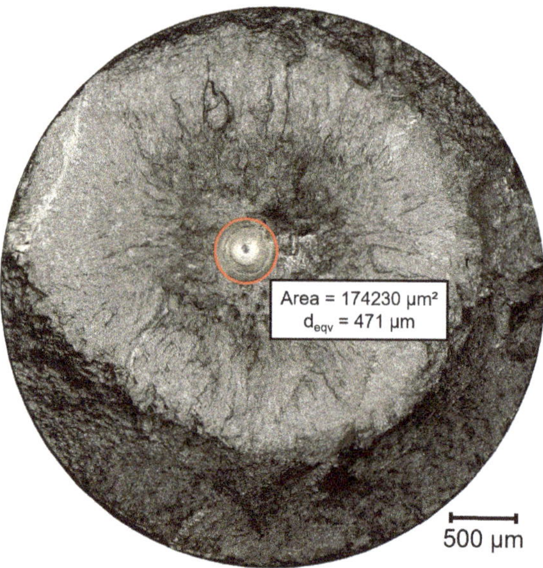

Figure 6. Gas pore within bulk material as representative inner defect.

After extensive characterization of the occurring critical defects, the statistical Gumbel extreme value distribution is fitted to them. To do so, the cumulative extreme value distribution according to Gumbel (see Equation (2)) is fitted to the experimental data and the Gumbel parameters are evaluated by using a maximum likelihood function fit introduced in [42]. In order to evaluate the goodness of the fit and the improvement of the distribution regarding the ΔK_{eqv} modification, two additional fit tests are carried out. On the one hand, Anderson–Darling goodness [43] and Kolmogorov–Smirnov goodness [44] are utilized.

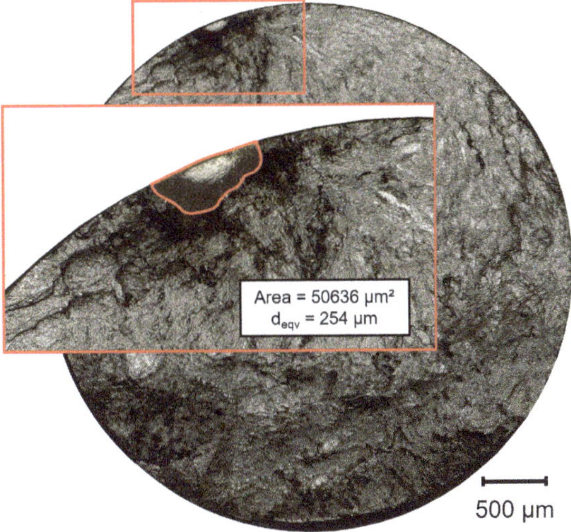

Figure 7. Gas pore at surface as representative surface defect.

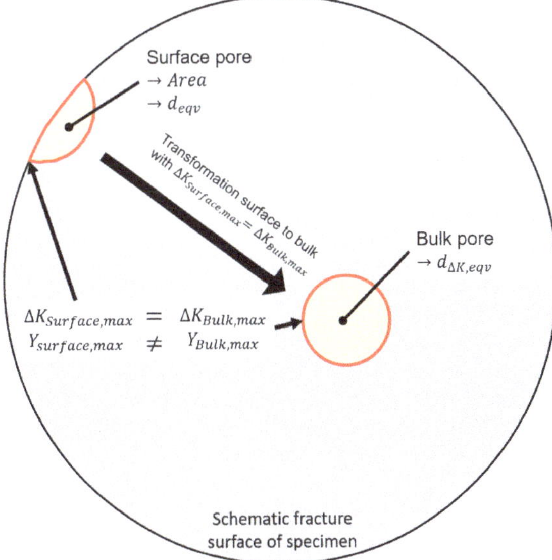

Figure 8. Schematic illustration of the surface to bulk pore transformation.

Additionally, to validate the stress intensity equivalent transformation of surface to bulk pores, the distribution fit was performed for two datasets. The first dataset is the original data without transformation and the second dataset is the ΔK_{eqv} modified set with the transformed surface pores. The scale δ and location parameter λ of the cumulative

Gumbel distribution for the equivalent defect diameter and the Anderson–Darling A^2_{AD} and Kolmogorov–Smirnov p_{kol} goodness-of-fit test results are shown in Table 6. Both goodness of fit tests reveal significantly better results of the Gumbel distribution fit in the case of the ΔK_{eqv} modified dataset. A value of $p_{kol} = 1$ means that the distribution perfectly fits the data. Additionally, if the Anderson–Darling test variable is below $A^2_{AD} < 0.75$, then the distribution fit is valid [45]. Figure 9 outlines the cumulative probability of occurrence for both data sets and the data points for the fits. Thereby, an equivalent defect diameter of about 237 µm will occur with a 50% probability of occurrence in case of the original dataset and an equivalent defect diameter of about 276 µm in case the ΔK_{eqv} modified set. Hence, the equivalent diameter with 50% occurrence probability increases by about 16% with the use of the surface to bulk pore transformation, and additionally better fit results are found.

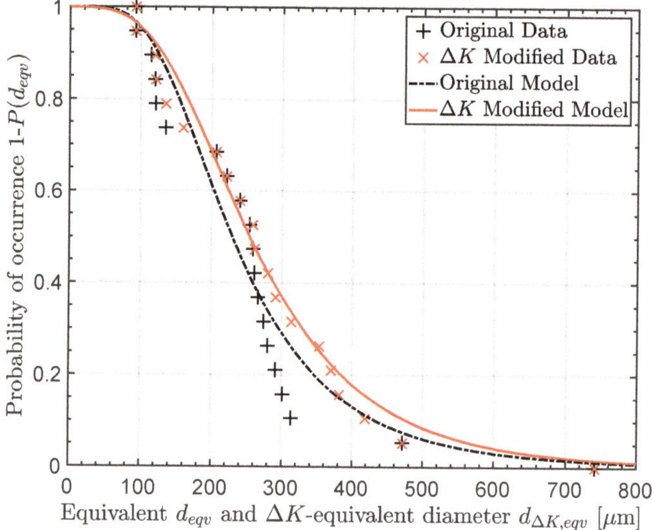

Figure 9. Cumulative probability of occurrence dependent on equivalent and ΔK-equivalent defect size.

Table 5. Position, size and equivalent diameter of the detected pores.

Specimen	Pore Position	Area [µm^2]	d_{eqv} [µm]	$d_{\Delta K,eqv}$ [µm]
1	Bulk	38,360	221	
2	Bulk	14,314	135	
3	Bulk	11,310	120	
4	Bulk	53,502	261	$\Big\}\widehat{=}$
5	Bulk	76,945	313	
6	Bulk	45,239	240	
7	Bulk	52,685	259	
8	Surface	59,155	274	381
9	Surface	50,636	253	353
10	Bulk	61,757	280	$\}\widehat{=}$
11	Bulk	66,508	291	
12	Surface	71,227	301	418
13	Bulk	431,247	741	
14	Bulk	174,234	471	$\}\widehat{=}$
15	Bulk	11,499	121	
16	Surface	55,623	266	370
17	Surface	10,311	114	159
18	Bulk	33,329	206	$\}\widehat{=}$
19	Bulk	6793	93	

Table 6. Estimated Gumbel distribution parameters and goodness of fit test.

Dataset	λ [µm]	δ [µm]	A^2_{AD}	p_{Kol}
Original	202.3	96.3	0.74	0.61
ΔK_{eqv} Modified	232.9	119.8	0.33	0.97

3.3. Fatigue Assessment Methodology

In order to estimate the fatigue life in the finite life region, the assessment model according to Tiryakioğlu was utilized, as introduced in Section 1. The necessary constants for the model in Equation (3) are the fitted Gumbel distribution parameters and the material constants B and m. The constant m is estimated to be 4 based on the literature data [46], and parameter B is evaluated to be about 3×10^{18} by a best-fit approach. The assessment model is used for both datasets to validate the surface to bulk pore transformation, whereby the parameters of the Gumbel distribution are different. To validate assessment methodology, three different load levels and number of load cycles are used and subsequently compared to the experimental S-N data. In cases of a constant stress amplitude, the stress amplitudes are 200 MPa, 300 MPa, and 350 MPa, and in case of a constant number of load cycles, the chosen numbers of load cycles are $1 \times 10^5, 5 \times 10^5$ and 1×10^6.

The results of the assessment for constant amplitudes are presented in Figure 10. Thereby, the probability of survival P_S is calculated in accordance to Equation (3) with a constant amplitude σ_a as a function of the number of load cycles until failure N_f. Additionally, the experimental S-N data are plotted in Figure 10 to validate the assessment model. A comparison of results of the two different datasets with the experimental data is given in Table 7, whereby the number of cycles until failure for a probability of 50% and the scatter band are compared for different stress amplitude levels. The model with the original data overestimates the number of load cycles until failure. By comparison, the ΔK_{eqv}-modified model provides a slight underestimation. When the scatter bands are compared, the model with the modified data delivers a better agreement when compared with the original data. The assumption can be made that the model with the transformed dataset delivers results in good agreement with the experimental data, and it is necessary to use stress intensity-equivalent pore transformation.

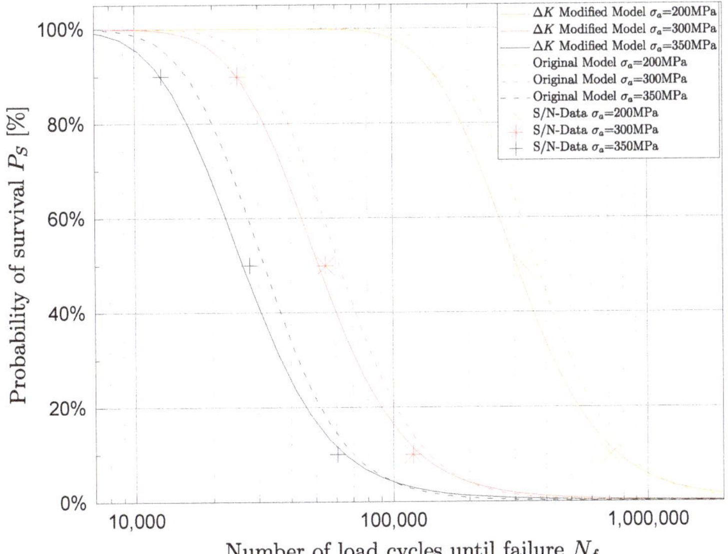

Figure 10. Model validation for constant stress amplitude.

Table 7. Fatigue assessment model validation for constant stress amplitude.

	Dataset	$\sigma_a = 200$ MPa		$\sigma_a = 300$ MPa		$\sigma_a = 350$ MPa	
		$N_{f,P_{s=50\%}}$	$1{:}T_N$	$N_{f,P_{s=50\%}}$	$1{:}T_N$	$N_{f,P_{s=50\%}}$	$1{:}T_N$
Experiment		3.3×10^5	4.78	5.5×10^4	4.78	2.8×10^4	4.78
Model Deviation	Original	3.7×10^5 13%	4.39	6.2×10^4 13%	4.38	3.2×10^4 14%	4.38
Model Deviation	ΔK_{eqv} Modified	3.1×10^5 -6%	5.12	5.1×10^4 -6%	5.12	2.6×10^4 -6%	5.12

The results of the assessment with a constant number of load cycles until failure are shown in Figure 11. For the evaluation of the survival probability in case of a constant number of load cycles, the stress amplitude is used as control variable. A comparison of the results for the two different datasets with the experimental data is given in Table 8, where the stress amplitudes for a survival probability of 50% and the scatter bands are compared for different numbers of load cycles until failure. The comparison of the assessed stress amplitude with the experimental data points out that the modified dataset delivers better results with a slight deviation of about 2%. Similar behavior as in the previous comparison for a constant stress amplitude arises: the model with the original data overestimates the experimental data, and the modified data delivers better results with a slight underestimation.

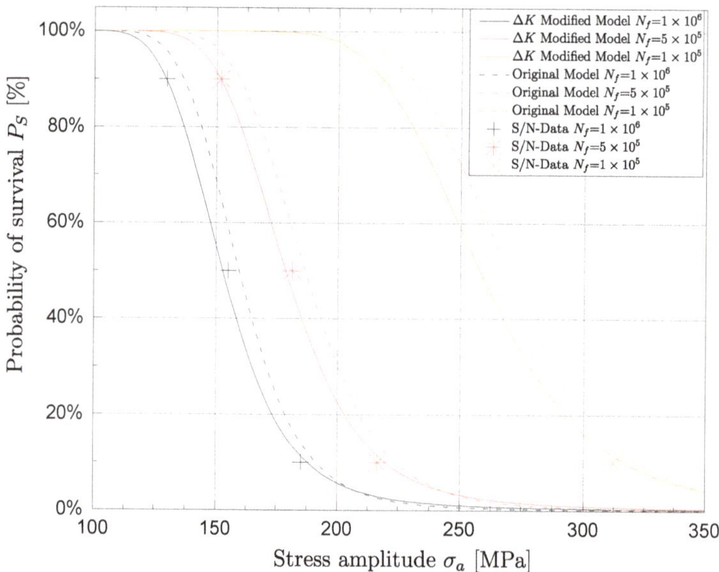

Figure 11. Model validation for constant load cycles.

Table 8. Fatigue assessment model validation defined number of load cycles.

	Dataset	$N_f = 1 \times 10^6$		$N_f = 5 \times 10^5$		$N_f = 1 \times 10^5$	
		$\sigma_{a,P_{s=50\%}}$ [MPa]	$1{:}T_S$	$\sigma_{a,P_{s=50\%}}$ [MPa]	$1{:}T_S$	$\sigma_{a,P_{s=50\%}}$ [MPa]	$1{:}T_S$
Experiment		155	1.43	182	1.43	262	1.43
Model Deviation	Original	159 3%	1.40	187 3%	1.40	269 3%	1.40
Model Deviation	ΔK_{eqv} Modified	153 -2%	1.45	179 -2%	1.45	258 -1%	1.45

4. Discussion

Within fractographic and metallographic investigations, pores are found as failure origins in all specimens. These pores are detected on the surface as well as in the bulk of the specimen. To compare pores independent of their position, a transformation routine was successfully developed. The used Gumbel extreme value distribution is in a good agreement with the experimental measured pore sizes, when the size of surface pores is stress intensity equivalent transformed to a size of an inner pore. For an extended statistical validation of the defect distribution, future work will focus on non-destructive, computer tomographic scans of wire arc additive manufacturing structures. In the presence of defects, the assessment method developed by Tiryakioğlu, validated for cast material, is applicable to additive manufacturing material and delivers well suitable results when the presented surface pore transformation is utilized.

Comparison of assessment results with and without ΔK_{eqv} modification reveals in general an overestimation without modification and an underestimation with modification. In Figure 12, the modeled data are presented dependent on the experimental data. Both evaluations for constant stress amplitude σ_a (Figure 12a) and constant number of load cycles until fracture N_f (Figure 12b) reveal that the original model lies above the ideal coherence and the ΔK_{eqv} modified model lies slightly below or shows almost ideal coherence. The model without modification delivers an overestimation up to 14% and the model with ΔK_{eqv} modification an slight underestimation of up to -7%. This behavior can be described with the lower occurring defect size (50% probability) of about 237 µm without modification compared to 276 µm with ΔK_{eqv} modification.

For the final application of additive manufacturing components in aerospace, the proof reliability is of utmost importance and must be taken into account. The impact of reliability calculations is investigated in [47,48] for additive manufacturing parts in aerospace. There, a new assessment methodology is proposed taking reliability into account in addition to the probabilistic defect size distribution.

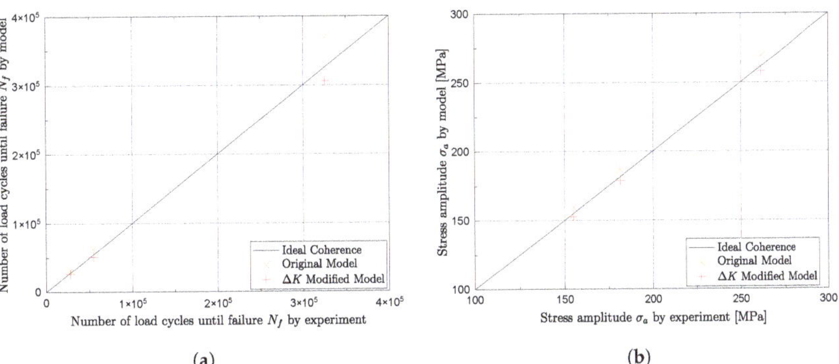

Figure 12. Comparison of results by experimental and model. (**a**) For constant load stress amplitude σ_a. (**b**) For constant number of load cycles N_f.

5. Conclusions

Based on the conducted investigations and presented results, we conclude the following:

- Imperfections, in particular gas pores, significantly influence the fatigue strength of WAAM structures and should be taken into account within fatigue assessment methods.
- A stress intensity equivalent transformation approach of surface pores to inner bulk material pores is presented and successfully validated.
- A cumulative extreme value distribution by Gumbel fit extremal pore sizes. The fitted distribution reveals a sound agreement with the experimental defect size data with a goodness of fit test value of $p_{kol} = 0.97$, when the ΔK_{eqv} transformation of surface to

inner pores is used. In contrast, the distribution fit with the unmodified defect sizes shows a lack of fit to the measured defects sizes with an goodness of fit test value of $p_{kol} = 0.61$. Thus, the ΔK_{eqv} transformation of surface to inner bulk pores leads to a significant improvement in the goodness of the distribution's fit.

- The used fatigue assessment method according to Tiryakioğlu is well applicable for this defect containing, additive manufacturing system structures. By utilizing the ΔK_{eqv} transformation, the fatigue assessment reveals a sound agreement with an underestimation of up to -7% compared to the experiments. Fatigue assessed with the original data reveals an overestimation of up to 14%. Therefore, fatigue estimation can be significantly improved by considering the ΔK_{eqv} transformation of surface pores to inner bulk pores.

Ongoing research focuses on non-destructive investigations of the defect sizes in WAAM structures, whereby computer tomographic scans are planned and will be used for a better statistical generation of the chosen cumulative extreme value distribution. Additional work is focusing on fracture mechanical crack growth tests to evaluate and validate the necessary parameters for the used assessment methodology. Further research is planned on additional effects on the fatigue strength of structures, such as residual stresses and their impact on the effective load stress ratio as well as the effect of different microstructures and the effect of anisotropy due to the directional solidification during additive manufacturing [49,50].

Author Contributions: Conceptualization, S.S. and M.L. (Martin Leitner); methodology, S.S. and M.L. (Martin Leitner); software, S.S.; validation, S.S. and M.L. (Martin Leitner); formal analysis, S.S. and M.L. (Martin Leitner); investigation, S.S.; resources, F.G., T.G., M.L. (Michael Lasnik) and B.O.; data curation, S.S.; writing—original draft preparation, S.S.; writing—review and editing, S.S., M.L. (Martin Leitner), F.G., T.G., M.L. (Michael Lasnik) and B.O.; visualization, S.S.; supervision, M.L. (Martin Leitner), T.G. and F.G.; project administration, S.S. and T.G.; funding acquisition, M.L. (Martin Leitner), M.L. (Michael Lasnik) and T.G. All authors have read and agreed to the published version of the manuscript.

Funding: This research was funded by the Austrian Federal Ministry for Climate Action, Environment, Energy, Mobility, Innovation and Technology (bmk) and the Federal Ministry for Digital and Economic Affairs (bmdw).

Institutional Review Board Statement: Not applicable.

Informed Consent Statement: Not applicable.

Data Availability Statement: Not applicable.

Acknowledgments: Special thanks are given to the Austrian Research Promotion Agency(FFG; project number 32765288), who funded the research project with funds of the Federal Ministry for Climate Action, Environment, Energy, Mobility, Innovation and Technology (bmk) and the Federal Ministry for Digital and Economic Affairs (bmdw).

Conflicts of Interest: The authors declare no conflict of interest. The funders had no role in the design of the study; in the collection, analyses or interpretation of data; in the writing of the manuscript; or in the decision to publish the results.

Abbreviations

The following abbreviations and symbols are used in this manuscript:

A_{proj}	Projected defect area of pore;
A_{AD}^2	Goodness of fit test value of the Anderson-Darling test;
a	Crack length;
B	Offset parameter;
$d_{\Delta K,eqv}$	Stress intensity equivalent diameter of defect;
d_{eqv}	Equivalent diameter of defect;
δ	Scale parameter of the Gumbel distribution;

ΔK	Stress intensity factor;
ΔK_{eqv}	Equivalent stress intensity factor;
$\Delta K_{Surface,max}$	Stress intensity factor of surface pore;
$\Delta K_{Bulk,max}$	Stress intensity factor of bulk pore;
k	Slope of S-N curve in finite life region;
λ	Location parameter of the Gumbel distribution;
m	Slope of crack propagation in stable crack growth region;
N_f	Number of load cycles until failure;
P	Probability;
P_{Occ}	Probability of occurrence;
P_s	Probability of survival;
P_f	Probability of failure;
p_{kol}	Goodness of fit test value of the Kolmogorov–Smirnov test;
R	Load ratio;
σ_a	Stress amplitude;
T_N	Scatter band of load cycles of the S/N-curve;
T_S	Scatter band of stress amplitude of the S/N-curve;
Y	Geometry factor;
$Y_{Surface,max}$	Geometry factor of surface pore;
$Y_{Bulk,max}$	Geometry factor of bulk pore;
A	Elongation at fracture;
UTS	Ultimate tensile strength;
WAAM	Wire arc additive manufacturing;
YS	Yield strength.

References

1. Baufeld, B.; van der Biest, O.; Gault, R. Additive manufacturing of Ti–6Al–4V components by shaped metal deposition: Microstructure and mechanical properties. *Mater. Des.* **2010**, *31*, S106–S111. [CrossRef]
2. Herzog, D.; Seyda, V.; Wycisk, E.; Emmelmann, C. Additive manufacturing of metals. *Acta Mater.* **2016**, *117*, 371–392. [CrossRef]
3. Prakash, K.S.; Nancharaih, T.; Rao, V.S. Additive Manufacturing Techniques in Manufacturing—An Overview. *Mater. Today Proc.* **2018**, *5*, 3873–3882. [CrossRef]
4. Kruth, J.P.; Leu, M.C.; Nakagawa, T. Progress in Additive Manufacturing and Rapid Prototyping. *CIRP Ann.* **1998**, *47*, 525–540. [CrossRef]
5. Singh, S.R.; Khanna, P. Wire arc additive manufacturing (WAAM): A new process to shape engineering materials. *Mater. Today Proc.* **2020**, *67*, 1191. [CrossRef]
6. Rodrigues, T.A.; Duarte, V.; Miranda, R.M.; Santos, T.G.; Oliveira, J.P. Current Status and Perspectives on Wire and Arc Additive Manufacturing (WAAM). *Materials* **2019**, *12*, 1121. [CrossRef] [PubMed]
7. Frazier, W.E. Metal Additive Manufacturing: A Review. *J. Mater. Eng. Perform.* **2014**, *23*, 1917–1928. [CrossRef]
8. Liu, S.; Shin, Y.C. Additive manufacturing of Ti6Al4V alloy: A review. *Mater. Des.* **2019**, *164*, 107552. [CrossRef]
9. Martina, F.; Mehnen, J.; Williams, S.W.; Colegrove, P.; Wang, F. Investigation of the benefits of plasma deposition for the additive layer manufacture of Ti–6Al–4V. *J. Mater. Process. Technol.* **2012**, *212*, 1377–1386. [CrossRef]
10. Springer, S.; Röcklinger, A.; Leitner, M.; Florian, G.; Gruber, T.; Lasnik, M.; Oberwinkler, B. Implementation of a viscoplastic creep model in the thermomechanical simulation of the WAAM process. *Weld. World* **2021**, *66*, 441–453. [CrossRef]
11. de Jesus, J.; Martins Ferreira, J.A.; Borrego, L.; Costa, J.D.; Capela, C. Fatigue Failure from Inner Surfaces of Additive Manufactured Ti-6Al-4V Components. *Materials* **2021**, *14*, 737. [CrossRef] [PubMed]
12. Sandell, V.; Hansson, T.; Roychowdhury, S.; Månsson, T.; Delin, M.; Åkerfeldt, P.; Antti, M.L. Defects in Electron Beam Melted Ti-6Al-4V: Fatigue Life Prediction Using Experimental Data and Extreme Value Statistics. *Materials* **2021**, *14*, 640. [CrossRef] [PubMed]
13. Greitemeier, D.; Palm, F.; Syassen, F.; Melz, T. Fatigue performance of additive manufactured TiAl6V4 using electron and laser beam melting. *Int. J. Fatigue* **2017**, *94*, 211–217. [CrossRef]
14. Akgun, E.; Zhang, X.; Biswal, R.; Zhang, Y.; Doré, M. Fatigue of wire+arc additive manufactured Ti-6Al-4V in presence of process-induced porosity defects. *Int. J. Fatigue* **2021**, *150*, 106315. [CrossRef]
15. Beretta, S.; Romano, S. A comparison of fatigue strength sensitivity to defects for materials manufactured by AM or traditional processes. *Int. J. Fatigue* **2017**, *94*, 178–191. [CrossRef]
16. Beretta, S.; Gargourimotlagh, M.; Foletti, S.; Du Plessis, A.; Riccio, M. Fatigue strength assessment of "as built" AlSi10Mg manufactured by SLM with different build orientations. *Int. J. Fatigue* **2020**, *139*, 105737. [CrossRef]
17. Romano, S.; Brandão, A.; Gumpinger, J.; Gschweitl, M.; Beretta, S. Qualification of AM parts: Extreme value statistics applied to tomographic measurements. *Mater. Des.* **2017**, *131*, 32–48. [CrossRef]

18. Romano, S.; Brückner-Foit, A.; Brandão, A.; Gumpinger, J.; Ghidini, T.; Beretta, S. Fatigue properties of AlSi10Mg obtained by additive manufacturing: Defect-based modelling and prediction of fatigue strength. *Eng. Fract. Mech.* **2018**, *187*, 165–189. [CrossRef]

19. Romano, S.; Nezhadfar, P.D.; Shamsaei, N.; Seifi, M.; Beretta, S. High cycle fatigue behavior and life prediction for additively manufactured 17-4 PH stainless steel: Effect of sub-surface porosity and surface roughness. *Theor. Appl. Fract. Mech.* **2020**, *106*, 102477. [CrossRef]

20. Valbruna Edel Inox GmbH. *Material Datasheet Ti-Grade5/Ti-6Al-4V*; Valbruna Edel Inox GmbH: Dormagen, Germany, 2021.

21. Voestalpine Böhler Welding GmbH. *Material Datasheet 3Dprint AM Ti Grade 5/Ti-6Al-4V*; Voestalpine Böhler Welding GmbH: Düsseldorf, Germany, 2021.

22. *EN ISO 6892-1*; Metallic Materials-Tensile Testing—Part 1: Method of Test at Room Temperature. Committee for Standardization: Brussels, Belgium, 2016.

23. Tiryakioğlu, M. Statistical distributions for the size of fatigue-initiating defects in Al–7%Si–0.3%Mg alloy castings: A comparative study. *Mater. Sci. Eng. A* **2008**, *497*, 119–125. [CrossRef]

24. Tiryakioğlu, M. On the size distribution of fracture-initiating defects in Al- and Mg-alloy castings. *Mater. Sci. Eng. A* **2008**, *476*, 174–177. [CrossRef]

25. Gumbel, E.J. *Statistics of Extremes*; Columbia University Press: New York, NY, USA, 1958.

26. Tiryakioğlu, M. On the relationship between statistical distributions of defect size and fatigue life in 7050-T7451 thick plate and A356-T6 castings. *Mater. Sci. Eng. A* **2009**, *520*, 114–120. [CrossRef]

27. Tiryakioğlu, M. Relationship between Defect Size and Fatigue Life Distributions in Al-7 Pct Si-Mg Alloy Castings. *Metall. Mater. Trans. A* **2009**, *40*, 1623–1630. [CrossRef]

28. Leitner, M.; Garb, C.; Remes, H.; Stoschka, M. Microporosity and statistical size effect on the fatigue strength of cast aluminium alloys EN AC-45500 and 46200. *Mater. Sci. Eng. A* **2017**, *707*, 567–575. [CrossRef]

29. Aigner, R.; Leitner, M.; Stoschka, M.; Hannesschläger, C.; Wabro, T.; Ehart, R. Modification of a Defect-Based Fatigue Assessment Model for Al-Si-Cu Cast Alloys. *Materials* **2018**, *11*, 2546. [CrossRef]

30. Paris, P.; Erdogan, F. A Critical Analysis of Crack Propagation Laws. *J. Basic Eng.* **1963**, *85*, 528–533. [CrossRef]

31. Richard, H.A.; Sander, M. *Ermüdungsrisse: Erkennen, Sicher Beurteilen, Vermeiden*, 2nd ed.; Vieweg+Teubner Verlag: Wiesbaden, Germany, 2012. [CrossRef]

32. Lin, J.J.; Lv, Y.H.; Liu, Y.X.; Xu, B.S.; Sun, Z.; Li, Z.G.; Wu, Y.X. Microstructural evolution and mechanical properties of Ti-6Al-4V wall deposited by pulsed plasma arc additive manufacturing. *Mater. Des.* **2016**, *102*, 30–40. [CrossRef]

33. Wu, B.; Pan, Z.; Ding, D.; Cuiuri, D.; Li, H. Effects of heat accumulation on microstructure and mechanical properties of Ti6Al4V alloy deposited by wire arc additive manufacturing. *Addit. Manuf.* **2018**, *23*, 151–160. [CrossRef]

34. Wang, F.; Williams, S.; Colegrove, P.; Antonysamy, A.A. Microstructure and Mechanical Properties of Wire and Arc Additive Manufactured Ti-6Al-4V. *Metall. Mater. Trans. A* **2013**, *44*, 968–977. [CrossRef]

35. Xie, Y.; Gao, M.; Wang, F.; Zhang, C.; Hao, K.; Wang, H.; Zeng, X. Anisotropy of fatigue crack growth in wire arc additive manufactured Ti-6Al-4V. *Mater. Sci. Eng. A* **2018**, *709*, 265–269. [CrossRef]

36. *ASTM International E 739*; Standard Practice for Statistical Analysis of Linear or Linearized Stress-Life (S-N) and Strain Life (E-N) Fatigue Data. ASTM: West Conshohocken, PA, USA, 2015.

37. Razavi, S.M.J.; Bordonaro, G.G.; Ferro, P.; Torgersen, J.; Berto, F. Fatigue Behavior of Porous Ti-6Al-4V Made by Laser-Engineered Net Shaping. *Materials* **2018**, *11*, 284. [CrossRef] [PubMed]

38. Haibach, E. *Structural Durability: Procedures and Data for Calculation of Components (Original Title in German: Betriebsfestigkeit: Verfahren und Daten zur Bauteilberechnung)*, 3rd ed.; VDI-Buch, Springer: Berlin, Germany, 2006.

39. Murakami, Y.; Beretta, S. Small Defects and Inhomogeneities in Fatigue Strength: Experiments, Models and Statistical Implications. *Extremes* **1999**, *2*, 123–147. [CrossRef]

40. Irwin, G.R. Analysis of Stresses and Strains Near the End of a Crack Traversing a Plate. *J. Appl. Mech.* **1957**, *24*, 361–364. [CrossRef]

41. Aigner, R.; Pusterhofer, S.; Pomberger, S.; Leitner, M.; Stoschka, M. A probabilistic Kitagawa-Takahashi diagram for fatigue strength assessment of cast aluminium alloys. *Mater. Sci. Eng. A* **2019**, *745*, 326–334. [CrossRef]

42. Mahdi.; Cenac, S. Estimating Parameters of Gumbel Distribution using the Methods of Moments, probability weighted moments and maximum likelihood. *Rev. Mat. Teor. Apl.* **2005**, *12*, 151–156. [CrossRef]

43. Anderson, T.W.; Darling, D.A. A Test of Goodness of Fit. *J. Am. Stat. Assoc.* **1954**, *49*, 765–769. [CrossRef]

44. Massey, F.J. The Kolmogorov-Smirnov Test for Goodness of Fit. *J. Am. Stat. Assoc.* **1951**, *46*, 68–78. [CrossRef]

45. Abidin, N.; Adam, M.; Midi, H. The goodness-of-fit test for Gumbel Distrubution: A comparative study. *Matematika* **2012**, *28*, 35–48.

46. Syed, A.K.; Zhang, X.; Davis, A.E.; Kennedy, J.R.; Martina, F.; Ding, J.; Williams, S.; Prangnell, P.B. Effect of deposition strategies on fatigue crack growth behaviour of wire + arc additive manufactured titanium alloy Ti–6Al–4V. *Mater. Sci. Eng. A* **2021**, *814*, 141194. [CrossRef]

47. Coro, A.; Abasolo, M.; Aguirrebeitia, J.; López de Lacalle, L.N. Inspection scheduling based on reliability updating of gas turbine welded structures. *Adv. Mech. Eng.* **2019**, *11*. [CrossRef]

48. Coro, A.; Macareno, L.M.; Aguirrebeitia, J.; López de Lacalle, L.N. A Methodology to Evaluate the Reliability Impact of the Replacement of Welded Components by Additive Manufacturing Spare Parts. *Metals* **2019**, *9*, 932. [CrossRef]
49. Pérez-Ruiz, J.D.; de Lacalle, L.N.L.; Urbikain, G.; Pereira, O.; Martínez, S.; Bris, J. On the relationship between cutting forces and anisotropy features in the milling of LPBF Inconel 718 for near net shape parts. *Int. J. Mach. Tools Manuf.* **2021**, *170*, 103801. [CrossRef]
50. Tolosa, I.; Garciandía, F.; Zubiri, F.; Zapirain, F.; Esnaola, A. Study of mechanical properties of AISI 316 stainless steel processed by "selective laser melting", following different manufacturing strategies. *Int. J. Adv. Manuf. Technol.* **2010**, *51*, 639–647. [CrossRef]

Article

Microstructural Impact on Fatigue Crack Growth Behavior of Alloy 718

Christian Gruber [1,2,*], Peter Raninger [1], Jürgen Maierhofer [1], Hans-Peter Gänser [1], Aleksandar Stanojevic [2], Anton Hohenwarter [3] and Reinhard Pippan [4]

[1] Materials Center Leoben Forschung GmbH, 8700 Leoben, Austria; peter.raninger@mcl.at (P.R.); juergen.maierhofer@mcl.at (J.M.); hans-peter.gaenser@mcl.at (H.-P.G.)
[2] Voestalpine BÖHLER Aerospace GmbH & Co KG, 8605 Kapfenberg, Austria; aleksandar.stanojevic@voestalpine.com
[3] Montanuniversität Leoben—Chair of Materials Physics, 8700 Leoben, Austria; anton.hohenwarter@unileoben.ac.at
[4] Erich Schmid Institute of Materials Science, 8700 Leoben, Austria; reinhard.pippan@oeaw.ac.at
[*] Correspondence: christian.gruber@mcl.at; Tel.: +43-384245922558

Abstract: Alloy 718 for forged parts can form a wide range of microstructures through a variety of thermo-mechanical processes, depending on the number of remelting processes, temperature and holding time of homogenization annealing, cogging and the number of forging steps depending on the forming characteristics. In industrial practice, these processing steps are tailored to achieve specific mechanical and microstructural properties in the final product. In the present work, we investigate the dependence of the threshold of stress intensity factor range ΔK_{th} on associated microstructural elements, namely grain size and distribution. For this purpose, a series of tests with different starting microstructures were performed at the falling stress intensity factor range, ΔK, and a load ratio of R = 0.1 to evaluate the different threshold values. Fracture initiation and crack propagation were analyzed afterward using scanning electron microscopy of the resulting fracture surfaces. In order to obtain comparable initial conditions, all specimens were brought to the same strength level by means of a two-stage aging heat treatment. In the future, this knowledge shall be used in the context of simulation-aided product development for estimating local fatigue crack propagation properties of simulated microstructures obtained from forging and heat treatment modeling.

Keywords: alloy 718; threshold value; threshold of stress intensity factor range; fracture surface; microstructure

Citation: Gruber, C.; Raninger, P.; Maierhofer, J.; Gänser, H.-P.; Stanojevic, A.; Hohenwarter, A.; Pippan, R. Microstructural Impact on Fatigue Crack Growth Behavior of Alloy 718. *Metals* **2022**, *12*, 710. https://doi.org/10.3390/met12050710

Academic Editors: Martin Leitner and Maciej Motyka

Received: 28 February 2022
Accepted: 19 April 2022
Published: 21 April 2022

Publisher's Note: MDPI stays neutral with regard to jurisdictional claims in published maps and institutional affiliations.

1. Introduction

For structural components in the aircraft industry made of alloy 718, which are manufactured via cast and wrought routes, there are high demands on mechanical properties, especially on fatigue crack resistance and fracture toughness. In addition to the characterization of mechanical properties, the microstructural influence on fatigue crack growth and on fracture mechanical parameters ought to be well known. Although components from production routes such as selective laser melting [1–4] or electron beam melting [5] exhibit significantly lower fatigue crack growth threshold and lower fracture toughness, a production route with polycrystalline microstructures using a forging or thermomechanical forming process [6–8] show improved parameters and are more suitable for components subjected to fatigue loading. Through thermomechanical processing, the microstructure of alloy 718 with low stacking fault energy and pronounced recrystallization behavior [9–12], which is typical for nickel-based superalloys, can be specifically adjusted. In this work, the influence of the microstructural constituents such as the grain size [13–15], as large as (ALA) grains, yield and ultimate tensile strength of the heat-treated state on the threshold of stress intensity factor range (ΔK_{th}) have been characterized. For this purpose, high-resolution

SEM images of the fracture surfaces of different initial states with specific microstructural parameters have been performed and considered in detail in the results. From these findings, a tentative phenomenological model for the dependence of the fatigue crack propagation properties on the microstructural and strength parameters is derived. In addition, the exposed fracture surfaces in the transition area from the long crack fatigue regime are compared to the surfaces of the residual ligament and the fracture toughness samples with the same microstructure.

2. Materials and Methods

For the series of experiments and differentiation of the individual microstructural elements, the used material was chosen from the same manufacturer and adjusted to representative extreme cases in forging production by means of variation in the process route and grain sizes. In order to maintain comparability within the test series, all samples were adjusted to a similar strength level by means of two-stage aging heat treatments; at least three experiments with the same microstructure were conducted.

2.1. Materials

The initially cast and wrought polycrystalline material made of alloy 718 was processed into four different microstructural states in terms of grain size, ALA grains, precipitates and strength by means of targeted thermomechanical processing. A difference in adjustable grain size is shown in Figure 1. The difference in equivalent circle diameter (ECD) is already well apparent from the figure and is quantified in Table 1. Further key parameters such as yield strength, tensile strength and fracture toughness in the same direction of crack propagation are also represented in the table and are the average of three measurements of the same material (Tensile test: ASTM E8 [16]; Fracture toughness: ASTM E399 [17]). Material A, B and C are double melted (Vacuum Induction Melted + Vacuum Arc Remelted) initial materials, whereas Material D was triple melted (Vacuum Induction Melted + Electro Slag Remelted + Vacuum Arc Remelted). This difference should help to adjust the size variation of the precipitates in terms of the niobium and titanium precipitates and affecting in this way the ΔK_{th}. The difference between Material A and B is used to study the grain size influence at the same strength level, whereas Material B further has a broader range of grain size distribution (heterogeneities) and a varying unknown amount of plastification due to the processing and position of the sample from a forged billet. Differences between Material B and C shall show the effect on ΔK_{th} at the same grain size with slight variation in strength; Material D represents the smallest grain size with the highest obtainable strength during the forging process for aircraft parts.

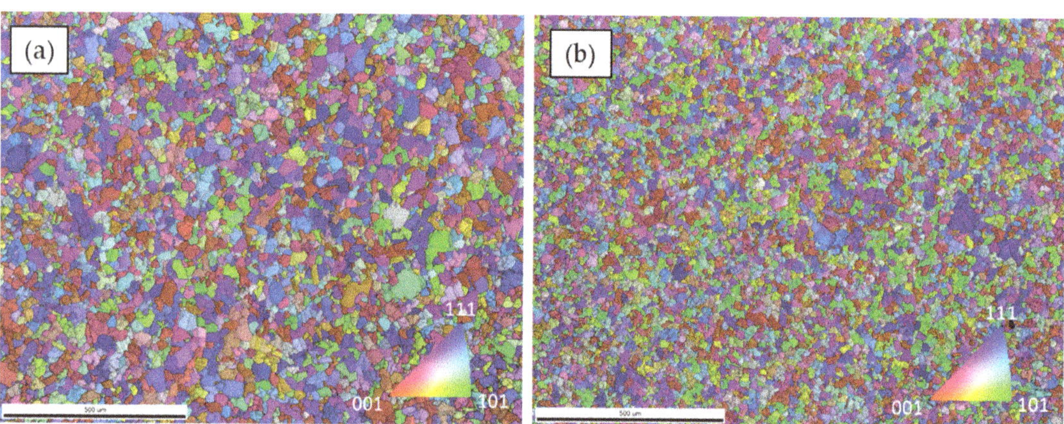

Figure 1. Electron backscatter diffraction (EBSD) grain maps with inverse pole figure (IPF) coloring and removed twins: (**a**) Material A: coarse microstructure; (**b**) Material B: fine microstructure.

Table 1. Microstructural constituents and mechanical properties from all tested materials.

Material	ECD [μm]	ALA [μm]	$R_{p0.2}$ [MPa]	UTS [MPa]	K_{IC} [MPa$\sqrt{m}$]	Validity
Material A	11.83	36.20	1237	1432	128	not valid
Material B	9.29	28.66	1274	1466	128	not valid
Material C	9.99	29.48	1174	1469	116	not valid
Material D	7.67	21.97	1390	1530	107	valid

2.2. Experiments

The experiments for determining the fatigue crack growth threshold ΔK_{th} by using the continuous load shedding (K-decreasing) procedure [18] were conducted in an air-conditioned testing room with a room temperature of $22 \pm 2\,°C$ and relative humidity of $49 \pm 3\%$. The used sample size was 100 mm × 20 mm × 6 mm with a radial crack propagation direction; this is the same direction as for the specimens from which the fracture toughness values in Table 1 had been obtained. By controlling temperature and humidity, it is ensured that any observed variation in the experimental results is not due to a variation in the environmental conditions. The experiments were performed at a resonant testing rig (Rumul Testronic 150 kN, Russenberger Prüfmaschinen AG, Neuhausen am Rheinfall, Switzerland) at a frequency of ~90 Hz by using an eight-point bending mount. The crack growth was measured using the direct current potential drop (DCPD: Matelect Crack Growth Monitor DCM-2, Matelect LTD, Newdigate road, UK) technique.

To exclude effects from manufacturing and to avoid short crack effects, the load shedding tests were started after a total crack length of a = 6.5 mm was reached by keeping ΔK constant and somewhat lower than the ΔK value used for starting the subsequent load shedding experiment. The load shedding tests were conducted by using an automatized ΔK-decreasing procedure with a normalized stress intensity factor gradient of $-0.04\ \mathrm{mm}^{-1}$ and a crack extension increment of $\Delta a = 50\ \mu m$. The stress intensity factor K was calculated from force and crack length according to ISO 12108 [19].

3. Results

In the following subsections, the results of the load shedding experiments are considered separately for each material, and the result plots are presented. Afterward, the factors influencing the threshold value and the crack propagation are compared with each other and analyzed on the basis of the corresponding fracture surfaces. The specimen geometry and the driving mode of the experiment also allowed conclusions to be drawn about the Paris region of the long crack.

3.1. Threshold of Stress Intensity Factor Range for Fatigue Growth

The threshold values were determined by the falling mode of the tests (K-decreasing) and the respective resulting parameter of each material is represented in Table 2. In addition, the beginning of the Paris regime (transition from I to II, [6]) was recorded during the experimental procedure; the constants of the Paris law (Equation (1)) were also determined for each material and their results are also given in the table.

$$\frac{da/dN}{[mm/cyc]} = C \left(\frac{\Delta K}{\left[MPa\ \sqrt{m} \right]} \right)^{m} \tag{1}$$

Note that this is a quantity equation, with ΔK to be inserted in units of [MPa$\sqrt{m}$] and da/dN in units of [mm/cyc]. Accordingly, the parameters of this equation, viz. C and m, are dimensionless.

Table 2. Summarized results for the fatigue crack growth threshold stress intensity factor range ΔK_{th} and the parameters C and m describing fatigue crack growth in the early Paris regime.

Material	ΔK_{th} [MPa$\sqrt{m}$]	C [-]	m [-]
Material A #1	8.13	7.41×10^{-14}	6.91
Material A #2	7.90	1.20×10^{-13}	6.77
Material A #3	8.23	8.91×10^{-14}	6.94
Material B #1	7.96	6.92×10^{-14}	6.96
Material B #2	7.75	9.55×10^{-14}	6.90
Material B #3	7.51	1.29×10^{-12}	6.13
Material B #4	7.00	2.69×10^{-12}	5.92
Material C #1	7.71	3.89×10^{-16}	9.59
Material C #2	7.47	8.51×10^{-15}	8.25
Material C #3	7.64	5.13×10^{-15}	8.37
Material D #1	7.00	2.75×10^{-13}	7.11
Material D #2	7.50	4.07×10^{-15}	8.81
Material D #3	6.99	2.24×10^{-14}	8.13
Material D #4	6.84	2.75×10^{-16}	9.96
Material D #5	7.33	4.90×10^{-16}	9.87

3.1.1. Material A

The first displayed and evaluated Material A represents the reference for the present investigations. This material had the coarsest microstructure (highest ECD) and also the largest ALA grains. The curves (Figure 2) nearly coincide in the initial part of the linear (Paris) regime and all have a threshold value ΔK_{th} between 7.90 and 8.23 MPa$\sqrt{m}$.

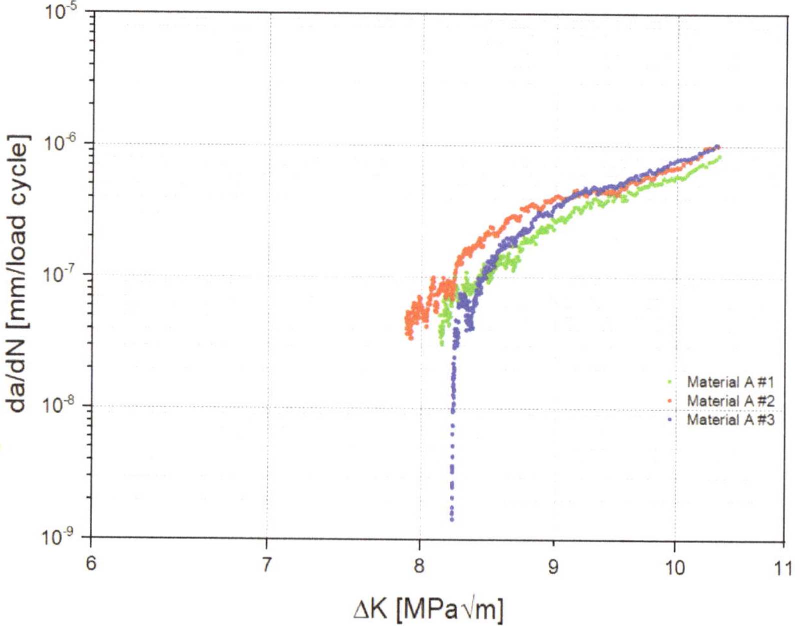

Figure 2. Fatigue crack growth curves for Material A; ECD: 11.83 mm; ALA: 36.20 mm; UTS: 1432 MPa; K_{IC}: 128 MPa$\sqrt{m}$—not valid.

3.1.2. Material B

In comparison, Material B shows a significantly larger scatter of the measured curves. This scatter is likely to be linked to plastification due to the processing of the specimens during the adjustment of the microstructure and possible local fluctuations. Material B had a significantly lower ECD at the same strength as Material A, but the grain size distribution was significantly more inhomogeneous. The scatter of the curves (Figure 3) already indicates the influence of these elements and is discussed in detail in the chapter fractography. The ΔK_{th} varies here between 7.0 and 8.0 MPa$\sqrt{m}$.

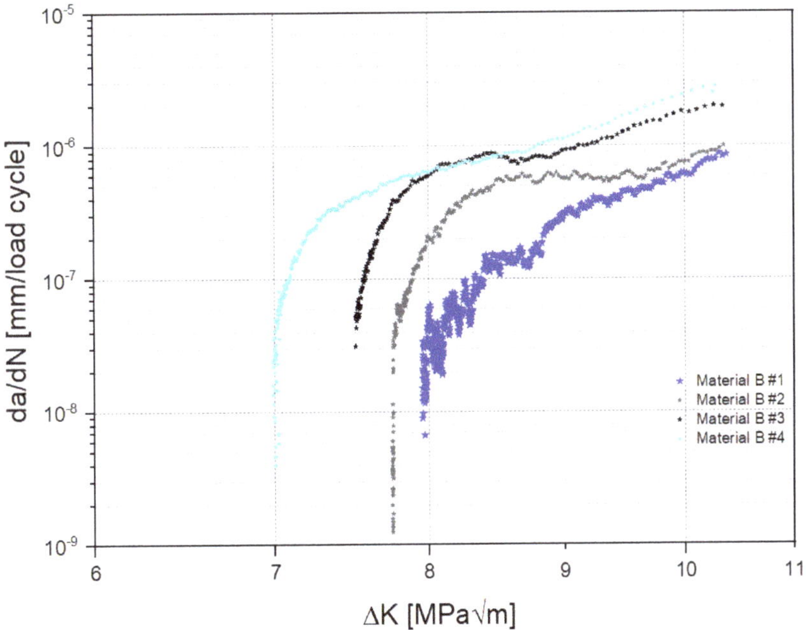

Figure 3. Fatigue crack growth curves for Material B; ECD: 9.29 mm; ALA: 28.66 mm; UTS: 1466 MPa; K_{IC}: 128 MPa$\sqrt{m}$—not valid.

3.1.3. Material C

Another state is represented by Material C, in which the microstructure is similarly fine to that of Material B, but much more homogeneous and, due to different distribution and quantity of precipitates, somewhat lower in absolute strength; therefore, the curves of these samples in Figure 4 are again more similar to each other. The abrupt end in the threshold regime has experimental reasons due to the sample geometry and a small remaining ligament with a high particle density. This higher density, combined with the size of the precipitates in Material C, led to an abrupt final rupture so that no measurement points below 4×10^{-7} mm/cyc could be recorded.

3.1.4. Material D

The lowest threshold was consistently provided by Material D, which was defined by the lowest ECD and the highest strength. After comparison with the literature [14], this behavior is explainable by roughness-induced crack closure. The curves shown in Figure 5 nearly coincide with the Paris regime; however, there is a noticeable scatter range for ΔK_{th} between 6.8 and 7.5 MPa$\sqrt{m}$. For a more precise description of the influence of the constituents from the microstructure, an evaluation of the results based on the fracture surfaces is also necessary.

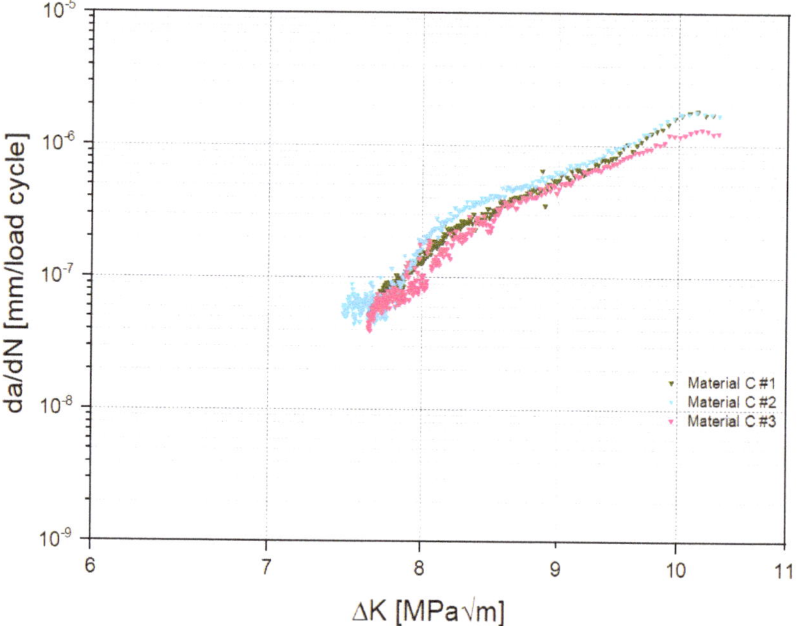

Figure 4. Fatigue crack growth curves for Material C; ECD: 9.99 mm; ALA: 29.48 mm; UTS: 1469 MPa; K_{IC}: 116 MPa$\sqrt{m}$—not valid.

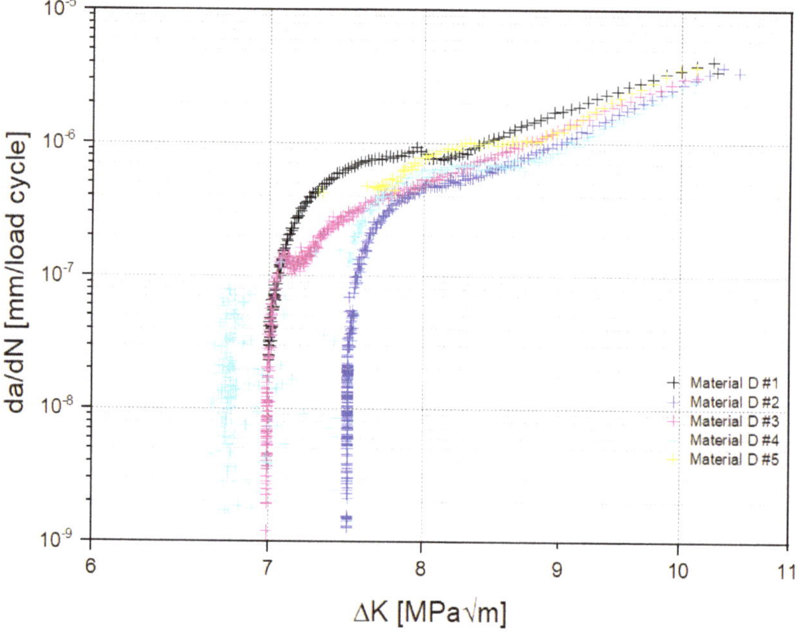

Figure 5. Fatigue crack growth curves for Material D; ECD: 7.67 mm; ALA: 21.97 mm; UTS: 1530 MPa; K_{IC}: 107 MPa$\sqrt{m}$—valid.

3.2. Fractography

In order to link the differences in the previous results to microstructural elements, the fracture surfaces were investigated with respect to the propagation of the fatigue crack. The fractographic investigations were performed with a scanning electron microscope from TESCAN (MAGNA) using an accelerating voltage of 15 kV and a secondary electron detector. First, Figure 6 shows the overview of the entire samples of Material A compared to Material B with the lowest measured threshold value of this series. The length of the crack surface of the long crack and the shortness of the overload fracture surface, determined by the position of the transition marked in the figures, indicate a microstructure effect on the threshold value since both materials have the same strength and fracture toughness, but with different grain size (ECD and ALA).

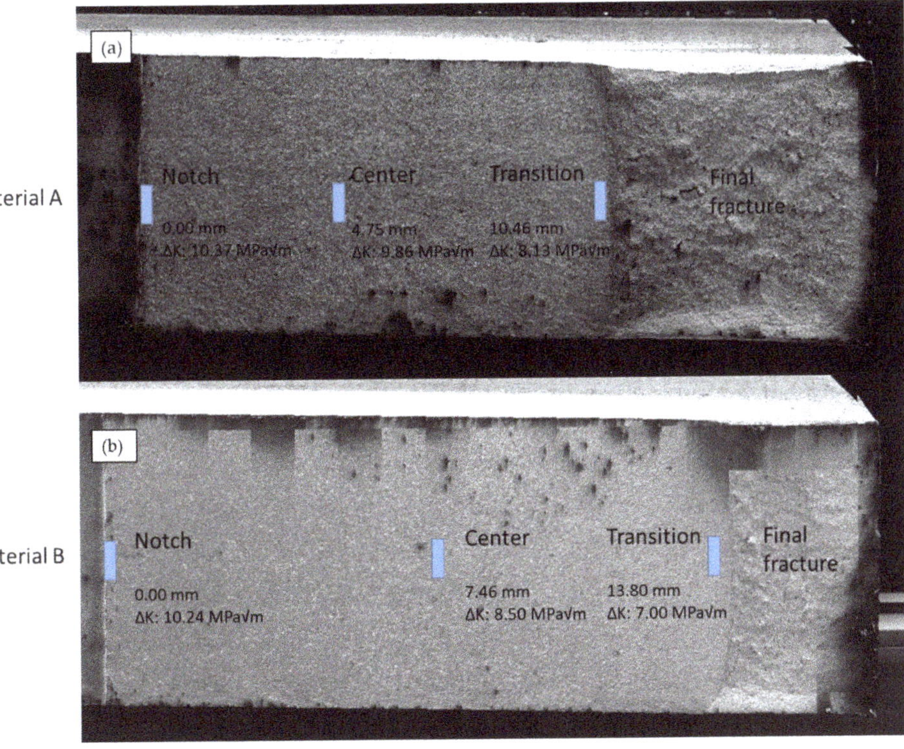

Figure 6. Overview images of the fatigue crack surfaces (20 mm × 6 mm) from the notch to the transition into final fracture: (**a**) Material A: coarse microstructure; (**b**) Material B: fine microstructure.

The marked areas in Figure 6 are pointed out at higher magnification in Figure 7. Based on the declared position of the crack propagation of the sample (notch—0%, center—50% and transition—100%), this progress can be assigned to the decreasing progress of ΔK (Figure 3 for Material A and Figure 4 for Material B). In the complete regime of crack growth with generally low ΔK throughout the test, roughness-induced crack closure is dominant, which can be deduced from the appearance of the fracture surface. In addition, no other components of the microstructure (carbides, nitrides or delta phase) were exposed or detectable in higher amounts in the region from the specimen's notch, center and the beginning of the transition. In the area of the transition to the final fracture (rupture), the first particles are exposed in Material B and indicate a change of the fracture mode from fatigue to rupture.

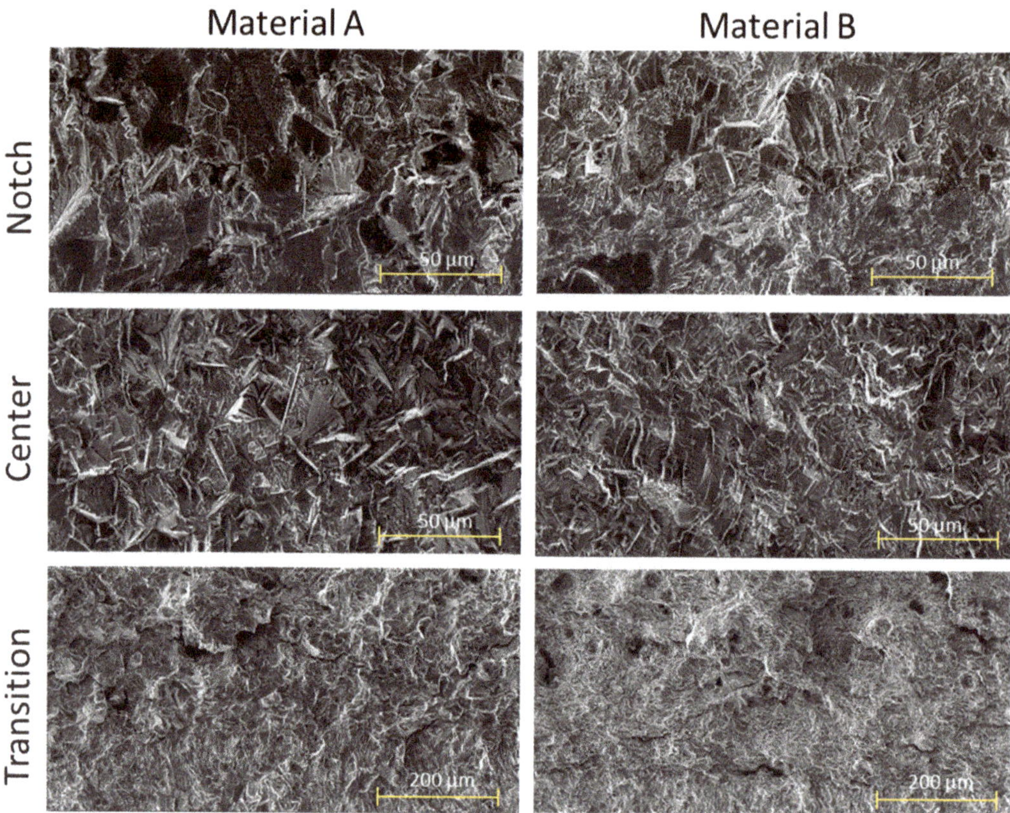

Figure 7. Comparison of fracture surfaces with comparable ΔK at the notch, center and transition area of Material A and Material B.

Due to the large variation within the series of Material B, additional experiments were carried out with load-increasing crack growth with the same R-value to analyze the differences in the threshold regime as well as in the early Paris regime. Since the ECD only reflects the mean grain diameter of a measured representative area but not its distribution, individual ALA grains are not reflected by this quantity. Figure 8 shows the impact of ALA grains on the roughness of the fracture surface and thus on the crack propagation and explains the differences in the curves in Figure 3. Furthermore, due to the process route of the corresponding pre-material billet involving cogging, a varying degree of plastic deformation, i.e., cold work, is present in the samples, which is an additional reason for the scatter of the measured curves. This cannot be investigated in detail by the applied methods but is considered in the threshold determination.

3.3. Fracture Toughness

In order to compare the threshold values and crack propagation behavior with fracture toughness K_{IC}, a comparative test series was carried out for Material B (same initial condition and same crack propagation direction). Results show a completely different picture in terms of the fracture surfaces, where mainly clusters of carbides are exposed, which are clearly the decisive microstructural feature for K_{IC}. Figure 9a provides an overview of a characteristic fracture surface, while (b) shows in detail the band structure of the particles, which is determined by the material flow during the melting process and subsequent thermomechanical treatment where the microstructure is finally set.

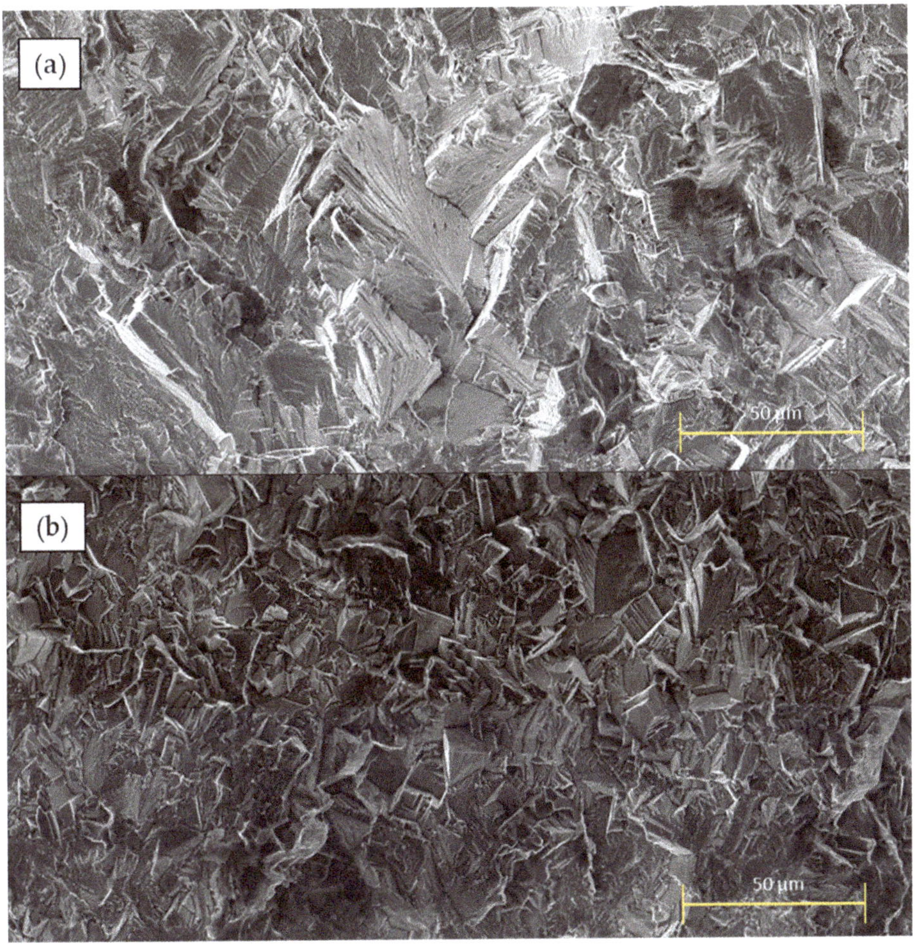

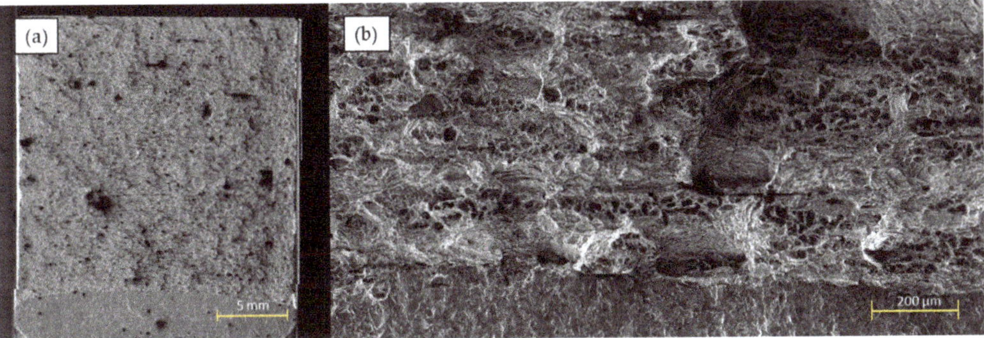

Figure 8. Grain size inhomogeneity of Material B from crack propagation samples with increasing ΔK: R = 0.1; (**a**) coarse area—ALA grains, (**b**) homogeneous fine microstructure.

Figure 9. Fractography of a CT specimen for fracture toughness testing of Material B: (**a**) overview of complete fracture sample; (**b**) detailed view showing the orientation of niobium carbides on the fracture surface in the transition area of the pre-crack to the overload fracture surface.

3.4. Modelling Approach

To derive the microstructure–property relations, the data from Tables 1 and 2 are merged and subjected to an exploratory data analysis. This analysis will serve for knowledge discovery, i.e., for identifying relevant correlations between the various microstructural, strength and crack growth parameters. Given the fact that, except for the dependence of the intrinsic threshold contribution on Young's modulus, so far, no physically based quantitative relations between the fatigue crack growth properties and microstructural and strength properties exist (cf. [18]), this pragmatic approach seems justifiable. Although it is phenomenological in nature, it might well prove useful for mechanical design purposes. A similar approach has been followed by Ashby in his classical work on materials selection in mechanical design [20]; in contrast to Ashby, who gives comparisons across various material classes, we limit ourselves here to different treatments of a single alloy, viz. alloy 718.

Regarding the kinetics of fatigue crack propagation, the well-known correlation between log C and m [20] is observed also in the present data set, see Figure 10. Note, however, that the values for m and C are exceptionally high and low, respectively. This is due to the fact that the load shedding experiments have concentrated on the near-threshold region and have started only in the very early linear (Paris) regime; keeping in mind the S-shape of the da/dN vs. ΔK curve, it is easily reasoned that the slope in the very early Paris regime is higher than in the center of that regime.

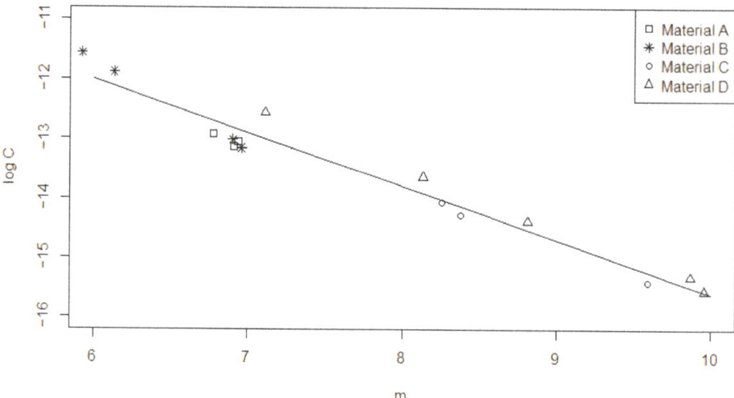

Figure 10. Linear correlation between the parameters log C and m describing the kinetics of fatigue crack growth; the samples from the plastically deformed Material B are marked by asterisks.

Proceeding to the microstructure–property relations, it turns out that ECD, ALA and UTS are nearly linearly correlated. This means that each of the two microstructural properties ECD and ALA, as well as the mechanical property UTS, will give similar performance as an explanatory variable for the fracture mechanical properties ΔK_{th}, log C, m, and K_{IC}. So, $R_{p0.2}$ is the only remaining mechanical property possibly adding explanatory value; however, no strong correlation between $R_{p0.2}$ and any of the other properties is observed; therefore, we choose, in what follows, the variables from ECD, ALA and UTS that correlate best with the respective fracture mechanical property of interest.

As can be seen from Figure 11a,b, ΔK_{th} exhibits a marked negative correlation with UTS and a positive correlation with ALA. As the threshold increases, its scatter decreases somewhat. A notable exception is Material B (indicated by triangles in the diagrams), which shows very high scatter; as mentioned above, this may be due to locally varying plastic deformation.

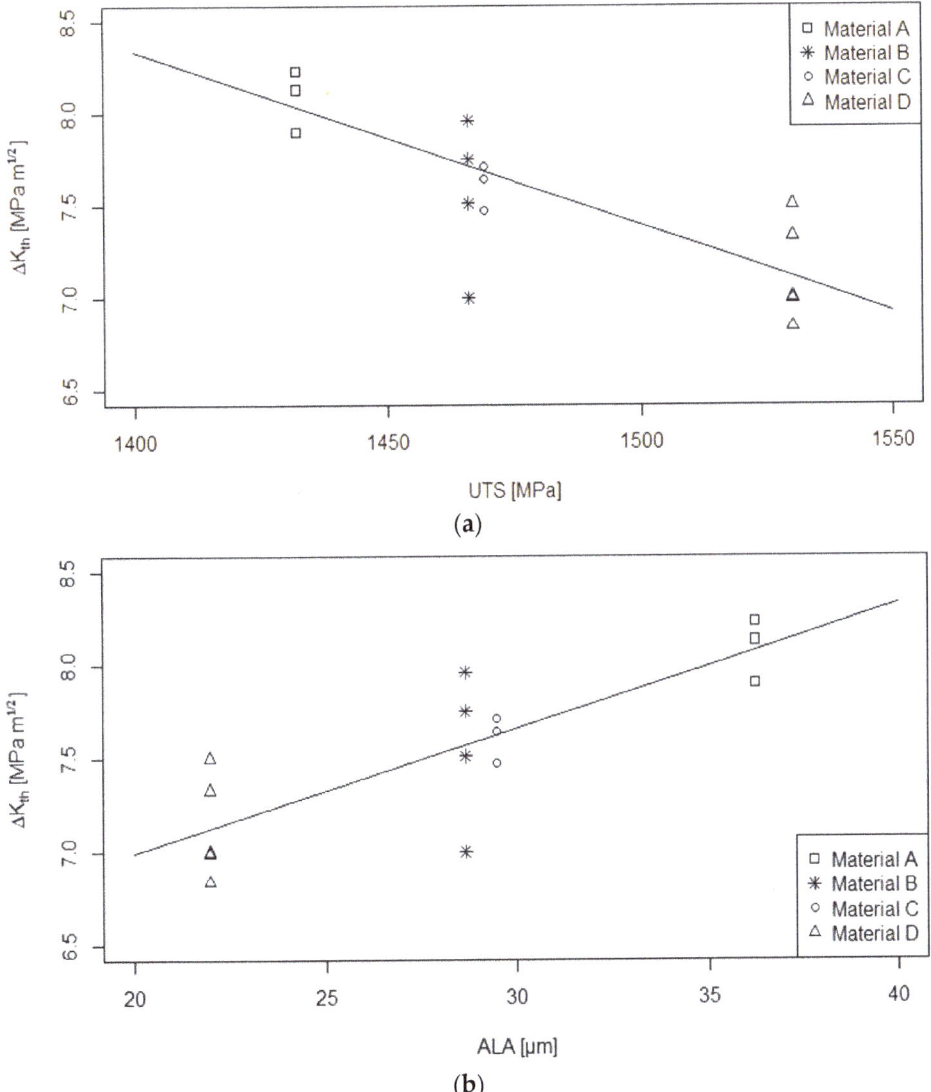

Figure 11. Linear correlations between (**a**) the mechanical parameter UTS and the fracture mechanical parameter ΔK_{th}, (**b**) the microstructural parameter ALA and the fracture mechanical parameter ΔK_{th}; the samples from the plastically deformed Material B are marked by asterisks.

In this context, it is of interest to note that there is a marked positive linear correlation between ALA and K_{IC} for Materials A, C and D, which are produced via the standard route, see Figure 12; however, K_{IC} is much higher for the plastically deformed Material B compared to Material C, which has nearly the same ALA.

Finally, the Paris exponent m describing fatigue crack growth kinetics does not show any significant correlation with the microstructural and mechanical properties; see Figure 13 as an example plot of m vs. ALA; however, the scatter of m decreases with increasing ALA (and ECD) and decreasing UTS, respectively.

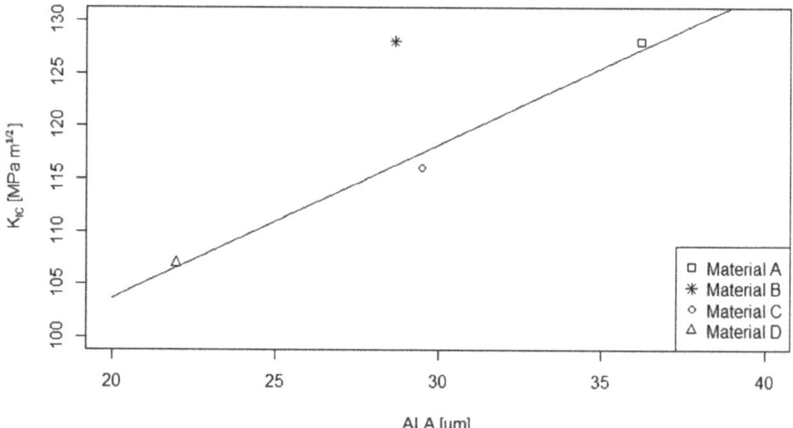

Figure 12. Linear correlation between the microstructural parameter ALA and the fracture mechanical parameter K_{IC} for materials produced via the standard route; the sample from the plastically deformed Material B is marked by an asterisk and does not follow this correlation.

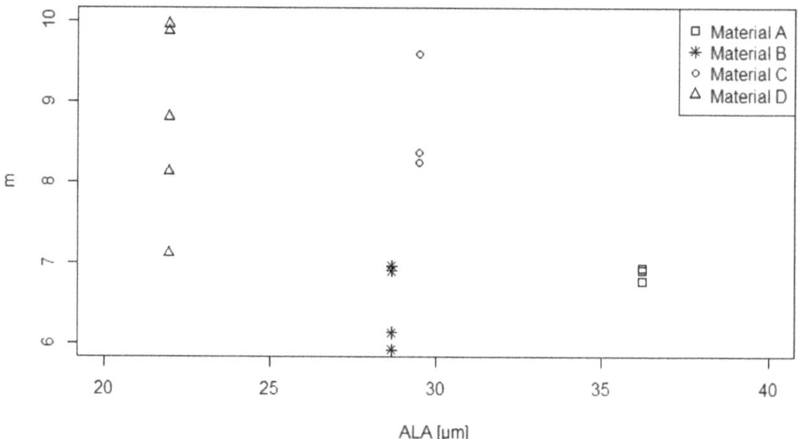

Figure 13. No clear correlation is observed between the microstructural parameter ALA and fatigue crack growth kinetics parameter m.

4. Discussion

By investigating four batches with different microstructural parameters, we sought to identify the relevant microstructural parameters influencing the fatigue crack growth parameters. Lacking physically based quantitative relations between the fatigue crack growth properties and microstructural and strength properties, a phenomenological approach has been pursued. First promising correlations between grain size, tensile strength and threshold have been observed in this first test series on wrought alloy 718. Given the still very limited amount of data and their scatter, these linear correlations may serve as a valuable first step towards modeling the influence of microstructure and strength on the fatigue crack growth properties of alloy 718.

Specifically, the difference in the mode of crack propagation between fatigue and fracture toughness testing could be shown by a thorough analysis of the fracture surfaces. The fractography confirms the hypothesis of grain size dependence of the threshold, which may be attributed to the mechanism of roughness-induced crack closure.

Clearly, further evaluations need to be supplemented in the future in order to distinguish between the influences of grain size, size distribution of γ' and γ'' precipitates and δ phase and contents of alloying elements influencing solid solution hardening on the fatigue crack growth properties, as it has been shown for the static strength in [21]; however, it will be hardly possible to derive an experimental design controlling all those parameters. Rather, it should be sought to build a database covering a large variety of different batches, and therefore different parameter combinations, and then to extract statistically relevant correlations in an attempt at knowledge discovery. In the next step, this may serve as useful phenomenological input for developing a physically based theory of microstructural influences on the threshold behavior.

In addition, a description of the mean stress influence (load ratio), as well as higher ΔK loads for a better description of the Paris regime (C and m), might be of interest; however, since the Paris regime is not of high relevance for typical rotating machinery applications of forged alloy 718, efforts should rather concentrate on the threshold behavior. The load ratio, on the other hand, is important for modeling the residual stress influence; for the specimens investigated in the present work, the residual stresses are negligible due to the small specimen size [22].

5. Conclusions

From a series of experiments with material batches of different grain sizes and production routes, we have shown that the threshold value ΔK_{th} depends only on the grain size. This means that ΔK_{th} is mainly governed by roughness-induced crack closure.

Furthermore, marked negative correlations between grain size and UTS and hence also between UTS and ΔK_{th} are observed. Interestingly, the plastically deformed material batch follows the same correlation, i.e., added plastic deformation does—other than adding notable scatter—not influence UTS and ΔK_{th}.

In contrast, plastic deformation seems to influence the fracture toughness K_{IC} advantageously. It remains to be further investigated how this finding can be related to our fractographic observations, viz., that fracture is driven by the density of second phase particles, whereas fatigue crack growth is largely independent of these particles. More specifically, not a single precipitate (carbides and nitrides) was exposed and no δ-phase was seen on the fracture surfaces of samples from threshold tests. The opposite is true for the samples from fracture toughness tests, where the particles are predominantly visible on the fracture surface and thus appear to control the fracture.

While the present work offers first interesting insights into the microstructure–property relations of alloy 718, further in-depth research is clearly needed; an outline of a possible methodological approach for knowledge discovery towards the development of a complete set of microstructure–property relations for the fatigue crack growth behavior of this superalloy has been given.

Author Contributions: Conceptualization, C.G. and P.R.; methodology experiments, J.M., A.H. and R.P.; methodology modeling, H.-P.G.; methodology fractography, A.H.; validation A.S. and H.-P.G.; writing—original draft preparation, C.G.; writing—review and editing, P.R. and A.S.; supervision, R.P. All authors have read and agreed to the published version of the manuscript.

Funding: The research leading to these results has received funding from the TakeOff program. TakeOff is a Research, Technology and Innovation Funding Program of the Austrian Federal Ministries for Climate Action, Environment, Energy, Mobility, Innovation and Technology (BMK) (867403). The Austrian Research Promotion Agency (FFG) has been authorized for the Program Management.

Institutional Review Board Statement: Not applicable.

Informed Consent Statement: Not applicable.

Data Availability Statement: The data presented in this study are available on request from the corresponding author. The data are not publicly available due to restrictions from industrial partners.

Conflicts of Interest: The authors declare no conflict of interest.

References

1. Konečná, R.; Kunz, L.; Nicoletto, G.; Bača, A. Long fatigue crack growth in Inconel 718 produced by selective laser melting. *Int. J. Fatigue* **2016**, *92*, 2. [CrossRef]
2. Wang, Z.; Guan, K.; Gao, M.; Li, X.; Chen, X.; Zeng, X. The microstructure and mechanical properties of deposited-IN718 by selective laser melting. *J. Alloy Compd.* **2012**, *513*, 518–523. [CrossRef]
3. Amato, K.N.; Gaytan, S.M.; Murr, L.E.; Martinez, E.; Shindo, P.W.; Hernandez, J.; Collins, S.; Medina, F. Microstructures and mechanical behavior of Inconel 718 fabricated by selective laser melting. *Acta Mater.* **2012**, *60*, 2229–2239. [CrossRef]
4. Kruth, J.-P.; Levy, G.; Klocke, F.; Childs, T.H.C. Consolidation phenomena in laser and powder-bed based layered manufacturing. *CIRP Ann.* **2007**, *56*, 730–759. [CrossRef]
5. Murr, L.E.; Martinez, E.; Gaytan, S.M.; Ramirez, D.A.; Machado, B.I.; Shindo, P.W.; Martinez, J.L.; Medina, F.; Wooten, J.; Ciscel, D.; et al. Microstructural Architecture, Microstructures, and Mechanical Properties for a Nickel-Base Superalloy Fabricated by Electron Beam Melting. *Met. Mater. Trans. A* **2011**, *42*, 3491–3508. [CrossRef]
6. Mercer, C.; Soboyejo, A.B.O.; Soboyejo, W.O. Micromechanisms of fatigue crack growth in a forged Inconel 718 nickel-based superalloy. *Mater. Sci. Eng. A* **1999**, *270*, 308–322. [CrossRef]
7. Clavel, M.; Pineau, A. Frequency and wave-form effects on the fatigue crack growth behavior of alloy 718 at 298 K and 823 K. *Met. Trans. A* **1978**, *9*, 471–480. [CrossRef]
8. Clavel, M.; Pineau, A. Fatigue behaviour of two nickel-base alloys I: Experimental results on low cycle fatigue, fatigue crack propagation and substructures. *Mater. Sci. Eng.* **1982**, *55*, 157–171. [CrossRef]
9. Chen, X.; Lin, Y.C.; Chen, M.; Li, H.; Wen, D.; Zhang, J.; He, M. Microstructural evolution of a nickel-based superalloy during hot deformation. *Mater. Des.* **2015**, *77*, 41–49. [CrossRef]
10. Zhang, H.; Zhang, K.; Zhou, H.; Lu, Z.; Zhao, C.; Yang, X. Effect of strain rate on microstructure evolution of a nickel-based superalloy during hot deformation. *Mater. Des.* **2015**, *80*, 51–62. [CrossRef]
11. Sommitsch, C.; Mitter, W. On modelling of dynamic recrystallisation of fcc materials with low stacking fault energy. *Acta Mater.* **2006**, *54*, 357–375. [CrossRef]
12. Gruber, C.; Raninger, P.; Stanojevic, A.; Godor, F.; Rath, M.; Kozeschnik, E.; Stockinger, M. Simulation of Dynamic and Meta-Dynamic Recrystallization Behavior of Forged Alloy 718 Parts Using a Multi-Class Grain Size Model. *Materials* **2020**, *14*, 111. [CrossRef] [PubMed]
13. Pokluda, J.; Pippan, R. Analysis of roughness-induced crack closure based on asymmetric crack-wake plasticity and size ratio effect. *Mater. Sci. Eng. A* **2007**, *462*, 355–358. [CrossRef]
14. Pippan, R.; Hohenwarter, A. Fatigue crack closure: A review of the physical phenomena. *Fatigue Fract. Eng. Mater. Struct.* **2017**, *40*, 471–495. [CrossRef] [PubMed]
15. Pippan, R.; Riemelmoser, F.O.; Weinhandl, H.; Kreuzer, H. Plasticity-induced crack closure under plane-strain conditions in the near-threshold regime. *Philos. Mag. A* **2009**, *82*, 3299–3309. [CrossRef]
16. *ASTM E8/E8M-13a*; Standard Test Methods for Tension Testing of Metallic Materials. American Society for Testing and Materials: West Conshohocken, PA, USA, 2013.
17. *ASTM E399-20a*; Standard Test Method for Linear-Elastic Plane-Strain Fracture Toughness of Metallic Materials. American Society for Testing and Materials: West Conshohocken, PA, USA, 2020.
18. Maierhofer, J.; Kolitsch, S.; Pippan, R.; Gänser, H.P.; Madia, M.; Zerbst, U. The cyclic R-curve—Determination, problems, limitations and application. *Eng. Fract. Mech.* **2018**, *198*, 45–64. [CrossRef]
19. *ISO 12108:2018*; Metallic Materials—Fatigue Testing—Fatigue Crack Growth Method. International Organization for Standardization: Geneva, Switzerland, 2018.
20. Ashby, M.F. *Materials Selection in Mechanical Design*, 5th ed.; Butterworth-Heinemann: Oxford, UK, 2016.
21. Rielli, V.V.; Godor, F.; Gruber, C.; Stanojevic, A.; Oberwinkler, B.; Primig, S. Effects of processing heterogeneities on the micro- to nanostructure strengthening mechanisms of an alloy 718 turbine disk. *Mater. Des.* **2021**, *212*, 110295. [CrossRef]
22. Maierhofer, J.; Gänser, H.P.; Pippan, R. Crack closure and retardation effects—Experiments and modelling. *Procedia Struct. Integr.* **2017**, *4*, 19–26. [CrossRef]

MDPI

St. Alban-Anlage 66

4052 Basel

Switzerland

Tel. +41 61 683 77 34

Fax +41 61 302 89 18

www.mdpi.com

Metals Editorial Office

E-mail: metals@mdpi.com

www.mdpi.com/journal/metals